3360 R / EEE

WITHDRAWN
NDSU

LOGIC, LANGUAGE, AND PROBABILITY

SYNTHESE LIBRARY

MONOGRAPHS ON EPISTEMOLOGY,

LOGIC, METHODOLOGY, PHILOSOPHY OF SCIENCE,

SOCIOLOGY OF SCIENCE AND OF KNOWLEDGE,

AND ON THE MATHEMATICAL METHODS OF

SOCIAL AND BEHAVIORAL SCIENCES

Editors:

DONALD DAVIDSON, *Rockefeller University and Princeton University*

JAAKKO HINTIKKA, *Academy of Finland and Stanford University*

GABRIËL NUCHELMANS, *University of Leyden*

WESLEY C. SALMON, *Indiana University*

LOGIC, LANGUAGE, AND PROBABILITY

A SELECTION OF PAPERS CONTRIBUTED TO
SECTIONS IV, VI, AND XI OF THE FOURTH INTERNATIONAL
CONGRESS FOR LOGIC, METHODOLOGY, AND
PHILOSOPHY OF SCIENCE, BUCHAREST, SEPTEMBER 1971

Edited by

RADU J. BOGDAN *and* ILKKA NIINILUOTO

D. REIDEL PUBLISHING COMPANY

DORDRECHT-HOLLAND / BOSTON-U.S.A.

Library of Congress Catalog Card Number 72-95892

ISBN 90 277 0312 4

Published by D. Reidel Publishing Company
P.O. Box 17, Dordrecht, Holland

Sold and distributed in the U.S.A., Canada, and Mexico
by D. Reidel Publishing Company, Inc.
306 Dartmouth Street, Boston,
Mass. 02116, U.S.A.

BC
6
I57
1971

All Rights Reserved
Copyright © 1973 by D. Reidel Publishing Company, Dordrecht, Holland
No part of this book may be reproduced in any form, by print, photoprint, microfilm,
or any other means, without written permission from the publisher

Printed in The Netherlands by D. Reidel, Dordrecht

PREFACE

The Fourth International Congress for Logic, Methodology, and Philosophy of Science was held in Bucharest, Romania, on August 29–September 4, 1971. The Congress was organized, under the auspices of the International Union for History and Philosophy of Science, Division of Logic, Methodology and Philosophy of Science, by the Academy of the Socialist Republic of Romania, the Academy of Social and Political Sciences of the Socialist Republic of Romania, and the Ministry of Education of Romania.

With more than eight hundred participating scholars from thirty-four countries, the Congress was one of the major scientific events of the year 1971. The dedicated efforts of the organizers, the rich and carefully planned program, and the warm and friendly atmosphere contributed to making the Congress a successful and fruitful forum of exchange of scientific ideas.

The work of the Congress consisted of invited one hour and half-hour addresses, symposia, and contributed papers. The proceedings were organized into twelve sections of Mathematical Logic, Foundations of Mathematical Theories, Automata and Programming Languages, Philosophy of Logic and Mathematics, General Problems of Methodology and Philosophy of Science, Foundations of Probability and Induction, Methodology and Philosophy of Physical Sciences, Methodology and Philosophy of Biological Sciences, Methodology and Philosophy of Psychological Sciences, Methodology and Philosophy of Historical and Social Sciences, Methodology and Philosophy of Linguistics, and History of Logic, Methodology and Philosophy of Science.

As in the case of its two immediate predecessors, the proceedings of the Bucharest Congress, containing the invited papers, will be published by the North-Holland Publishing Company of Amsterdam, Holland. The idea of publishing, for the first time, also a selection of contributed papers came up during the Congress. At a meeting in Bucharest, an agreement was made that a volume of papers contributed to the Congress would be published by D. Reidel Publishing Company in collaboration with the Center of Information and Documentation in Social and Political Sciences of the Academy of Social and Political Sciences of the Socialist Republic of Romania. From the twelve sections of the Congress, Section IV on the Philosophy of Logic and Mathematics, Section VI on the Foundations of Probability and Induction, and Section XI on the Methodology and Philosophy of Linguistics were chosen, since they represent important fields in which Reidel's publication activity is specialized.

All the papers presented in Sections IV, VI, and XI could not be included in one volume. Nor were they all available to the editors. In selecting papers for this volume, we have focused mainly on technical contributions containing new ideas and results in their fields, as well as competent surveys. We have also tried to provide room for different schools and research traditions in these fields. It goes without saying that the task of selecting the papers was a difficult one, and we owe an apology to many authors whose interesting papers we have not been able to include in this volume. We have included twenty-nine papers whose authors represent twelve different countries. Without claiming these to be the best contributions, we hope that our selection is at least representative of the 1971 Congress.

We wish to record our thanks to the respective publishers and editors for permission to reprint four articles which have already appeared in print elsewhere as follows: the article by Professor A. R. Anderson has appeared in *Mind*, n.s., vol. **81** (1972), pp. 348–71; the article by Professor K. Lehrer has appeared in *Philosophia*, vol. **2**, no. 4 (Oct. 1972), Israel University Press; the article by Dr. T. A. van Dijk in *Linguistics*, Mouton, Holland; the article by Dr. J. Kotarbińska in *Logique et Analyse*, Nauwelaerts, Belgium. In addition, Dr. T. Potts has retained the copyright of his own article, as it is intended to form part of a forthcoming book.

In the process of editing this volume, we have had valuable and indispensable help from many sources. We are grateful to Professor Jaakko Hintikka for his patient guidance of the lengthy editorial project; to Professor Mircea Ioanid, the Editor-in-Chief of the above-mentioned Center of Information and Documentation in Social and Political Sciences, for his sustained support of the project, and to the members of his staff for their technical assistance in preparing the volume; to Dr. Sanda

Golopenția-Eretescu and Dr. Sorin Vieru, the secretaries of Sections XI and IV, respectively, for their help in actually assembling the proceedings; and to Professors Solomon Marcus, Risto Hilpinen, and Lauri Karttunen for their helpful advice and for their suggestions. Last but not least, we thank the authors of the contributions themselves.

RADU J. BOGDAN

Central Institute for
In-Service Teachers Training,
Bucharest, Romania

ILKKA NIINILUOTO

National Research Council for the Humanities,
Helsinki, Finland

TABLE OF CONTENTS

PREFACE   v

PART I: LOGIC (Section IV)

A. R. ANDERSON / An Intensional Interpretation of Truth-Values	3
A. BRESSAN / Intensional Descriptions and Relative Completeness in the General Interpreted Modal Calculus MC^v	29
D. STOIANOVICI / Singular Terms and Statements of Identity	41
R. SUSZKO / Adequate Models for the Non-Fregean Sentential Calculus (SCI)	49
W. J. THOMAS / Doubts about Some Standard Arguments for Church's Thesis	55

PART II: PROBABILITY (Section VI)

M. BELIŞ / On the Causal Structure of Random Processes	65
L. J. COHEN / The Paradox of Anomaly	78
N. K. KOSSOVSKY / Some Problems in the Constructive Probability Theory	83
K. LEHRER / Evidence and Conceptual Change	100
I. NIINILUOTO / Empirically Trivial Theories and Inductive Systematization	108
T. SETTLE / Are Some Propensities Probabilities?	115
K. SZANIAWSKI / Questions and Their Pragmatic Value	121
G. TRAUTTEUR / Prediction, Complexity, and Randomness	124
J. M. VICKERS / Rules for Reasonable Belief Change	129

PART III: LANGUAGE (Section XI)

T. A. VAN DIJK / Models for Text Grammars	145
W. L. FISCHER / Tolerance Spaces and Linguistics	181

TABLE OF CONTENTS

S. ISARD and CH. LONGUET-HIGGINS / Modal Tic-Tac-Toe	189
V. IVANOV / On Binary Relations in Linguistic and Other Semiotic and Social Systems	196
A. KASHER / Worlds, Games and Pragmemes: A Unified Theory of Speech Acts	201
J. KOTARBIŃSKA / On Occasional Expressions	208
H. A. LEWIS / Combinators and Deep Structure	213
I. A. MEL'ČUK / Linguistic Theory and 'Meaning ⇔ Text' Type Models	223
G. ORMAN / Properties of the Derivations According to a Context-Free Grammar	226
B. PALEK / The Treatment of Reference in Linguistic Description	237
T. C. POTTS / Fregean Categorial Grammar	245
M. PRZEŁECKI / A Model-Theoretic Approach to Some Problems in the Semantics of Empirical Languages	285
I. I. REVZIN / Methodological Relevance of Language Models with Expanding Sets of Sentences	291
L. SCHWARTZ / A New Type of Syntactic Projectivity: SD Projectivity	296
J. T. WANG / On the Representation of Generative Grammars as First-Order Theories	302
INDEX OF NAMES	317
INDEX OF SUBJECTS	321

PART I

LOGIC
(Section IV)

ALAN ROSS ANDERSON

AN INTENSIONAL INTERPRETATION OF TRUTH-VALUES*

1. Introduction

In a profound and seminal paper of 1956 'Begründung einer strengen Implikation', *JSL*), Wilhelm Ackermann laid the foundation for what may reasonably be called *intensional logics*, characterized by the presence of an intensional 'if... then —' connective which demands (for provability) *relevance* of antecedent to consequent, in a sense amenable to treatment by the tools of mathematical logic. The growing body of literature devoted to the philosophical and mathematical analysis of these systems jeopardizes a prejudice which has been an important article of faith among extensionally-minded logicians, ever since Goodman and Quine got it into the mob that no decently reared Christian gentleman of respectable ancestry would consort with propositions, or others of intensional ilk. Among the banned notions is that of relevance, and the *credo* reads as follows (if memory serves): "this conception, like others in the theory of meaning, is too vague, uncertain, unsettled, indefinite, visionary, undetermined, unsure, casual, doubtful, dubious, indeterminate, undefined, confused, obscure, enigmatic, problematic, questionable, unreliable, provisional, dim, muddy, nebulous, indistinct, loose, ambiguous, and mysterious to be the subject of careful logical scrutiny."

I hold the contrary position.

This paper accordingly presents some new information concerning intensional logics and their interpretation, and is designed to bolster the minority opinion. The principal constructive result is that of Section 5, where it is shown that for certain values of propositional variables (namely, intensional, *inhaltliche* truth-values), relevant implication behaves just like material 'implication'. Mathematical details are trivial, but the fact helps explain the failure (in general) of $P \wedge (\sim P \vee Q) \rightarrow Q$ in intensional logics, and also of the principle *ex impossibilitate quodlibet*, to which Sections 6–7 are devoted.

We first clear away a few preliminaries, and then get down to business.

Notation. To facilitate comparison with classical results, we use (mainly) the notation and terminology of Church (*Introduction to Mathematical Logic*, 1956) especially Sections 10, 27, 28, 30, and 50. We assume that *well-formed formula* ('wff') is defined, and that we know what is meant by *free* and *bound* occurrences of propositional variables. We use lower-case italics for propositional variables, bold-face lower-case letters ranging over these, and bold-face capitals to range over wffs; italic capitals are used in an appropriately haphazard way for sentences or propositions (depending on context) in informal philosophical passages. Truth-functional notation and quantifiers are as usual: $\sim$, $\vee$, $\wedge$, $\supset$, $\equiv$, (p), and $(\exists p)$, and for the intensional 'if... then —' connective we use the arrow $\rightarrow$, with $\rightleftarrows$ for the intensional 'if and only if' ('iff,' an abbreviation also used metalinguistically). Where dots are used for punctuation in wffs, they are to be understood according to Church's conventions.

We differ from Church (a) in using parentheses where Church uses square brackets (reserving the latter for a different use later on), and (b) in our metalogical notation for the results of substitutions. Here we adopt a device of Curry and Feys (*Combinatory logic*, 1958) to avoid complicated statements of conditions on variables and wffs for substitutions. We want $S_\mathbf{B}^\mathbf{p}\mathbf{A} \mid$ to denote the result of putting $\mathbf{B}$ for free occurrences of $\mathbf{p}$ in $\mathbf{A}$, but in such a way that no unwanted collision of free and bound variables results. To secure this, we define the substitution-notation as follows: if $\mathbf{A}$ contains any wf part of the form $(\mathbf{q})\mathbf{C}$ in which $\mathbf{p}$ is free, and if $\mathbf{q}$ is free in $\mathbf{B}$, then *first* replace every occurrence of $\mathbf{q}$ in $(\mathbf{q})\mathbf{C}$ by $\mathbf{r}$, where $\mathbf{r}$ is the first variable in alphabetical order which occurs in neither $\mathbf{A}$ nor $\mathbf{B}$; and *secondly*, after all such replacements are made, replace free occurrences of $\mathbf{p}$ in the result by $\mathbf{B}$. The final result of these substitutions (made in the order given) is $S_\mathbf{B}^\mathbf{p}\mathbf{A} \mid$. (Notice that under these conventions $(\mathbf{p})\mathbf{A} \rightarrow S_\mathbf{B}^\mathbf{p}\mathbf{A} \mid$ will always hold.)

'$\vdash \mathbf{A}$,' '$\vdash_R \mathbf{A}$,' etc. means '$\mathbf{A}$ is a theorem', '$\mathbf{A}$ is a theorem of R', etc. Further notation will be introduced as the occasion requires.

2. Classical Extended 'Propositional' Calculus

To fix ideas we choose the axiom schemata of Church 1956, Section 27 (or Section 30).

A ⊃ . B ⊃ A
A ⊃ (B ⊃ C) ⊃ . A ⊃ B ⊃ . A ⊃ C
∼ A ⊃ ∼ B ⊃ . B ⊃ A

together with the rule: from **A ⊃ B** and **A** to infer **B**; we call this system *P*. The extended 'propositional' calculus $P^\dagger$ may be obtained from this basis by adding notation for quantification, and the axiom schemata

(p) A ⊃ S_B^pA |
(p)(A ⊃ B) ⊃ . A ⊃ (p) B (**p** not free in **A**)

together with the rule: from A to infer (p)A.

It will be important in what follows to notice that the variables p, q, r,..., of this system range over truth-values t (*das Wahre*) and f (*das Falsche*), and *not* over *propositions* or *Sinne*. The name 'propositional calculus', though of course well-entrenched historically, seems consequently to be a misnomer, and we hope the reader will forgive us in this paper for using the (as it seems to us more accurate) name: 'truth-value calculus'. For present purposes, at any rate, we shall take the position that the classical two-valued calculus serves for Fregean *Bedeutungen* alone, a position we share with, among others, Church, whose 1956 exposition of what Frege *must* have meant seems to us the clearest in the literature.

As is well known, addition of quantifiers to the truth-value calculus allows for a certain flexibility and economy in the choice of primitives, and in particular Łukasiewicz and Tarski (*Logic, semantics, metamathematics*, 1956) proposed taking material 'implication' and universal quantification as primitive for the extended truth-value calculus, defining

$$f =_{df} (p)p, \text{ and}$$
$$\sim A =_{df} A \supset f.$$

Suitable axioms for the primitives will guarantee that negation has the right classical properties, and we also define t as $\sim f$, or equivalently as $(\exists p)p$. We can then think of t {f} as having *das Wahre*, {*das Falsche*} as *Bedeutung*, and the disjunction of all values of p {the conjunction of all values of p} as *Sinn*. But these *Sinne* are for obvious reasons uninteresting in *P*, since that of t is simply t ∨ f, and that of f is t ∧ f (if the abuse of language may be forgiven), and these are the same as t and f respectively, which are already the *Bedeutungen* of *t* and *f*. The point (just stated sloppily, but clearly enough for our purpose) is that introduction of quantifiers does not help in the direction of getting from truth-values to propositions, an observation which is reinforced by the following formal facts.

In the extended 'propositional' calculus we have as a theorem

$$S_t^p A \mid \supset . S_f^p A \mid \supset (p) A,$$

which says roughly (given the amusing fiction that the horseshoe can be read 'if... then —'), that if both *truth* and *falsity* satisfy the condition A, then everything does – which says in turn that *truth* and *falsity* are all there is. And in fact this theorem and its consequences show that adding quantifiers accomplishes virtually nothing, since $(p)A$ can be shown to be *concurrent* (Church, *op. cit.*, p. 152) with

$$S_t^p A \mid \wedge S_f^p A \mid .$$

So we are still confined in effect to the *Bedeutungen* t and f, which fact might account for the general lack of interest in the extended truth-value calculus, as witnessed by the paucity of literature on the topic.

This situation is in marked contrast to that of the system R of relevant implication, and its quantificational extension $R^\dagger$ in both of which propositional variables may be construed as ranging over *propositions*, or *Sinne*. As we shall see, definitions of the wffs *t* and *f* in $R^\dagger$ lead to results which accord closely with Frege's motivation, and also help us to understand persistent and pervasive misunderstandings (so we think) of the classical truth-value calculus and its quantificational extension.

3. Relevant Implication and Entailment

In this section we give formulations of four systems of logic, and summarize facts relevant to the philosophical claims to follow.

Axiom schemata:
- A0. $A \to NA$ (where $NA =_{df} A \to A \to A$)
- A1. $A \to A \to B \to B$
- A2. $A \to B \to . B \to C \to . A \to C$
- A3. $(A \to . A \to B) \to . A \to B$

A4. $A \wedge B \to A$
A5. $A \wedge B \to B$
A6. $(A \to B) \wedge (A \to C) \to . A \to B \wedge C$
A7. $NA \wedge NB \to N(A \wedge B)$
A8. $A \to A \vee B$
A9. $B \to A \vee B$
A10. $(A \to C) \wedge (B \to C) \to . A \vee B \to C$
A11. $A \wedge (B \vee C) \to (A \wedge B) \vee C$
A12. $A \to \sim A \to \sim A$
A13. $A \to \sim B \to . B \to \sim A$
A14. $\sim \sim A \to A$
A15. $(p) A \to S_B^p A \mid$
A16. $(p)(A \to B) \to . (p) A \to (p) B$
A17. $(p) A \wedge (p) B \to (p)(A \wedge B)$
A18. $(p)(A \to B) \to . A \to (p) B$ (p not free in A)
A19. $(p)(A \vee B) \to . A \vee (p) B$ (p not free in A)
A20. $(p)(p \to p) \to A \to A$

Axiom clause: If **A** is an axiom schema, so is **(p) A**.

Rules:

R1.   From **A** and **A → B** to infer **B** (*Modus ponens*).
R2.   From **A** and **B** to infer **A ∧ B** (*Adjunction*).

We will consider the systems:
E = A1–A14, R1, R2,
$E^\dagger$ = E together with A15–A20 (and infinitely many additional axioms as in the final clause above),
R = E + A0, and
$R^\dagger$ = $E^\dagger$ + A0.

Attention in this paper will be directed primarily toward the last of these systems, but we will make remarks about the others *passim*. (For N [necessity], in A0 and A7, see Anderson and Belnap, *JSL*, 1962; A7 and A20 are redundant in the latter two systems, and if in $E^\dagger$ we redefine NA as $(p)(p \to p) \to A$, then A7 is redundant in $E^\dagger$ also.)

Obvious proofs yield for both $E^\dagger$ and $R^\dagger$ the results:

THEOREM. If ⊢ **A** then ⊢ **(p) A** (*Derived rule of generalization*).

THEOREM. If $\vdash A \rightleftharpoons B$, and D results from replacing zero or more occurrences of A in C by B, then $\vdash C \rightleftharpoons D$ (*Derived rule for replacement of relevant equivalents*).

THEOREM. If $\vdash C$, and D results from replacing zero or more occurrences of $(p)\ A$ in C by $(q)\ S_q^p A\ |$, then $\vdash D$ (*Derived rule of alphabetical change of propositional variables*).

We state some properties of the system above which will illuminate some of the differences between the extensional truth-functional connectives, and the arrow of 'if... then —'. (Bibliographical references cite either proofs, or sufficient information to construct proofs.)

1. $\rightarrow$ is not extensionally definable; *i.e.* for no purely truth-functional formula $(\ldots A \ldots B \ldots)$ is $(\ldots A \ldots B \ldots) \rightleftharpoons (A \rightarrow B)$ provable in any of the systems (Ackermann 1956, cited above).

2. All systems contain all truth-functional tautologies as theorems (Ackermann 1956, Anderson and Belnap, *JSL*, 1959).

3. In all four systems: for purely truth-functional A and B, $\vdash A \rightarrow B$ iff $\vdash A_1 \vee \ldots \vee A_m \rightarrow B_1 \wedge \ldots \wedge B_n$, where $A_1 \vee \ldots \vee A_m$ is a disjunctive normal form of A, and $B_1 \wedge \ldots \wedge B_n$ is a conjunctive normal form of B. Moreover, the latter is provable iff $\vdash A_i \rightarrow B_j$ for each i and j, iff A_i and B_j share an atom (*i.e.* a propositional variable or its denial) (Anderson and Belnap, *Philos. Studies* 1962).

4. For every A and B, $\vdash A \rightarrow B$ in R and E only if A and B share a variable (Belnap, *JSL*, 1960).

5. In R and E if $\vdash A$ and $\vdash \sim A \vee B$, then $\vdash B$ (Meyer and Dunn, *JSL*, 1969).

6. The class of theorems of E coincides with that of the system Π' of Ackermann 1956 (Meyer and Dunn, just cited).

7. There are intuitively plausible formulations of $R^\dagger$ and $E^\dagger$ (as yet unpublished) in the natural deduction style of Fitch (*Symbolic Logic*, 1952), incorporating the subscripting device of Anderson (*Zeit. math. Logik*, 1960).

4. THE DISJUNCTIVE SYLLOGISM

Thus far we have been on firm ground; the content of Sections 2 and 3 was factual, but now we venture into territory thought to be debatable. It is not, really – but that is what we must now explain. Historical background for the controversy we shall consider centres around the following question: is it appropriate, in the sense of being true to our pre-analytic intuitions, to say that a relation of necessary, relevant, logical consequence (such as we intend 'entailment' to capture) holds from the conjunction of A and $\sim A \vee B$ to B? Most students of classical truth-value and modal logic seem to feel obliged to say that it does, perhaps because of the common practice of taking this rule as primitive in formulations of the truth-value calculus. Whatever the source of the delusion, the enlightened hold with us that $A \wedge (\sim A \vee B)$ (or $A \wedge (A \supset B)$) does not have B even as a logically relevant consequence, much less a necessary one, and we note with satisfaction that $\mathbf{A} \wedge (\mathbf{A} \supset \mathbf{B}) \rightarrow \mathbf{B}$ is not a theorem of any of the systems of Section 3, nor is the corresponding rule primitive for any. The unreconstructed hold the contrary, and the dialogue between Them and Us (which we hope to take one step further in this paper) runs roughly as follows:

Us: $\mathbf{A} \wedge (\sim \mathbf{A} \vee \mathbf{B}) \rightarrow \mathbf{B}$ *must* be wrong, for if it were true, property 3 of Section 3 would give us $(\mathbf{A} \wedge \sim \mathbf{A}) \vee (\mathbf{A} \wedge \mathbf{B}) \rightarrow \mathbf{B}$, whence $(\mathbf{A} \wedge \sim \mathbf{A}) \rightarrow \mathbf{B}$, which commits a fallacy of relevance.

Them: But look; surely if $\mathbf{A} \wedge (\sim \mathbf{A} \vee \mathbf{B})$ is a tautology, then $\mathbf{B}$ must be one too, and this is what we ought to mean by entailment; it can not be that the antecedent is true and the consequent false.

Us: You're confused. We've already granted you all tautologies (see property 2), so if $\mathbf{A} \wedge (\sim \mathbf{A} \vee \mathbf{B})$, and hence $\mathbf{B}$, are both tautologies, they are both provable. But this does not show that entailment is strict implication in Lewis's sense, as you seem to say.

Them: But we mean more than that. *Modus ponens* (or 'the disjunctive syllogism', as you seem to want to call it – inaccurately, we might point out) should hold even if the A's and B's contain arrows.

Us: Well, in a sense it does, as Meyer and Dunn (*J. sym. log.*, 1969) show with the help of an unexpectedly strenuous proof: whenever ⊢ **A** and ⊢ $\sim$ **A** ∨ **B**, then ⊢ **B** (in both *E* and *R*). But that is far from showing that the conjunction of the first two entails the third. In modal logics like *E* and S4, for example, we know that whenever we have ⊢ **A**, we also have ⊢ *N***A** (necessarily **A**), but that fact does not tell us that every proposition entails its own necessity.

Them: Well…, we *still* say that *Modus ponens* is a valid entailment.

Us (reverting now to a less tendentious tone of voice, and turning our attention from *Them* to the reader): Whence comes this dogmatic tenacity on Their part? We will try to explain it by considering first some features of the classical position already noted, and then, in the following section, what happens analogously in R^t.

For the first point, we revert to the confusing terminology 'propositional calculus', for the truth-value calculus, and suppose that the variables in this calculus range over propositions, as the term suggests. The conclusion seems forced on us by the most obvious and natural quantificational axioms that there are just two 'propositions' recognized by the extended 'propositional' calculus, in view of the theorem

$$(\mathbf{p})\, \mathbf{A} \equiv S_t^p \mathbf{A} \,|\, \wedge \, S_f^p \mathbf{A} \,|\,.$$

(*We all know*, however, that there are propositions other than those expressed by t and f).

We also note that we have as a theorem

$$f \supset \mathbf{A},$$

which forces an interpretation on f: it must be thought of as the conjunction of all propositions. Even if f is taken as primitive (as, *e.g.* in Church *op. cit.*, § 10), it still has the logical force of the conjunction of all propositions, and t (primitive or defined) must have the dual force. If we adopt Gentzen's view (roughly) that the meaning of a term in a formal system is dictated by the role it plays in inference, then there is no other meaning to give to t and f, and the distinction between these two is the only one recognized, among propositions, by the theory.

So classically, we have one contradictory proposition and one necessarily true proposition, and all contradictions are identified, as are all tautologies. Why should this be?

One source of the feeling that all contradictions are the same may lie in the fact that there is a large and important range of cases where one contradiction will do just exactly as well as another, namely, in *reductio ad absurdum* proofs. Here it makes no difference *what* contradiction A leads to; if A leads to *any* contradiction, A must be false (as it is said in China: "One jot of rat's dung spoils the soup.") And just as we have classically

$$(\mathbf{A} \supset (\mathbf{B} \wedge \sim \mathbf{B})) \supset \sim \mathbf{A},$$

so we have in the systems of Section 3,

$$(A \to (B \wedge \sim B)) \to \sim A.$$

But the fact that contradictions are interchangeable in such contexts does not entail that they are interchangeable in every context. Even in the *reductio* cases *A* must be *relevant* to the contradiction, as is evidenced by the fact that in carrying out a *reductio* proof, we are sometimes puzzled initially as to *which* contradiction to try to prove. *Any* contradiction will do, but only certain contradictions follow from, or are relevantly implied by, the hypothesis we want to show false. And those are the contradictions (the ones to which *A* is relevant) which interest us.

Similarly, where *A* may be any true proposition whatever, if *B* follows relevantly from *A*, then we may conclude *B*:

$$A \to . A \to B \to B.$$

And in this context, one true *A* will do as well as another, but this fact does not excuse our ignoring differences between different true propositions.

All of these puzzles evaporate, of course, if we simply pay attention to what we all said initially, namely, that the variables range not over propositions, but truth-values, and that the interpretation forced on us by such theorems as

$$f \to B$$

is that the truth-value f must have the conjunction of all values of the variables as its *Sinn*.

The reader will no doubt have noted that the arguments we put in Their mouths above perhaps fail to do justice to Their honest intent, or Their logical subtlety or acumen. For this reason we will try to take issue later with what is no doubt the lengthiest and most exhaustive recent statement of Their position, namely, that in the article 'Entailment' by Jonathan Bennett (*Philos. Rev.* 1969). But first we turn to a discussion of what happens in $R^{\dagger}$, in order to give an account of the truths which seduce Them into Their position (and in the hope that the contrast between that position and the enlightened one can be made more striking).

5. Intensional Truth-Values

We define f in R^t just as in P^t, to wit, as short for $(p)\,p$, to be thought of intuitively as having *das Falsche* as its *Bedeutung* and the conjunction of all propositions as its *Sinn*; and t as $(\exists p)p$ (where $(\exists p) =_{df} \sim (p) \sim$), with *das Wahre* as *Bedeutung* and the disjunction of all propositions as *Sinn*. (We will be vague about 'all propositions' or, we should say more accurately, 'all values of p'. For full-scale semantical investigations this would have to be made precise, but in this paper we touch on formal semantics only peripherally.)

The first contrast with P^t to note is that R^t allows the possibility that there are lots of propositions other than those expressed by t and f. We do of course have $f \to \mathbf{A}$ and $\mathbf{A} \to t$, but we have neither $(\mathbf{A} \wedge \sim \mathbf{A}) \to f$ nor $t \to (\mathbf{A} \vee \sim \mathbf{A})$. Nor do we have such things as

$$\mathbf{A} \wedge \mathbf{B} \to .\, \mathbf{A} \rightleftharpoons \mathbf{B},$$

or

$$(p)\, \mathbf{A} \rightleftharpoons S_t^p \mathbf{A}\,|\, \wedge\, S_f^p \mathbf{A}\,|.$$

But if we restrict attention simply to t *and* f, *as is done in the classical extended truth-value calculus, the results of the classical analysis and those of* R^t *exactly coincide*. We proceed to make this claim precise.

We first summarize a number of theorems of R^t in tabular form:

$\sim$	
t	f
f	t

(i.e. $t \rightleftharpoons \sim f$ and $f \rightleftharpoons \sim t$);

$\vee$	t	f
t	t	t
f	t	f

(i.e. $t \vee t \rightleftharpoons t$, $t \vee f \rightleftharpoons t$, $f \vee t \rightleftharpoons t$, and $f \vee f \rightleftharpoons f$); similarly for the two tables below:

$\wedge$	t	f
t	t	f
f	f	f

$\to$	t	f
t	t	f
f	t	t

We prove a sample case, leaving the others to the reader. Observe first that by A15 we have

$$f \to A$$

or contrapositively from $f \to \sim A$, etc.,

$$A \to t;$$

many of the theorems needed for equivalences in the tables are consequences of these two. Secondly, standard moves enable us to prove *Permutation* $(A \to . B \to C) \to . B \to . A \to C$ in R, which, from $f \to . t \to f$ gives us in R^t

$$t \to . f \to f,$$

which is one of the relevant implications required for the arrow table; its converse comes from $A \to t$. And so on; remaining proofs are equally trivial.

(Exactly the same effect can be got by adding to R a primitive propositional constant [let's use f again], with the axiom schema $f \to A$; we leave verification of the formal details to the reader.)

Comparison of R^t with P^t will be helped by defining *truth*-value formula ('tvf') as follows: t and f are both tvfs, and if A and B are, so are $\sim A$, $A \wedge B$, $A \vee B$, and $A \supset B$ (for P^t; $A \to B$ for R^t). Then we make the claim above precise in the form of the following (which seems too lightweight to be a theorem, or even a lemma):

OBSERVATION 1. Let A be a tvf of P^t, written in $t, f, \sim, \vee, \wedge,$ and $\supset$, and get A^* from A by replacing $\supset$ by $\to$ throughout A. Then $\vdash_P^t A$ iff $\vdash_R^t A^*$. (Proof: easy.)

What this means essentially is that R^t says exactly what P^t says *about the latter's poverty-stricken little extensional domain of discourse*, a fact which can be emphasized by giving R^t some hamstrung quantifiers designed to do the work of P^t. We might, that is, add to R^t some quantifiers [**p**], [∃**p**], etc. (where [∃**p**] = $\sim$[**p**]$\sim$), to be restricted to t and f, so that we get as axioms (say)

$$[\mathbf{p}] A \to S_t^\mathbf{p} A |,$$
$$[\mathbf{p}] A \to S_f^\mathbf{p} A |,$$
$$[\mathbf{p}] (A \to B) \to . A \to [\mathbf{p}] B \text{ (}\mathbf{p}\text{ not free in } A\text{)},$$

and the like. But in fact the appropriate machinery is already available in R^t, since we can get the proper effect simply by defining the restricted quantifier:

$$[\mathbf{p}]\,\mathbf{A} =_{df} S^p_t \mathbf{A}\,|\,\wedge\, S^p_f \mathbf{A}\,|.$$

Then in view of the following easily proved

OBSERVATION 2. If $\mathbf{A}$ is a tvf, then (uninterestingly) $\vdash_{P^t} \mathbf{A} \equiv t$ or $\vdash_{P^t} \mathbf{A} \equiv f$, and (more interestingly, but not much more) $\vdash_{R^t} \mathbf{A} \rightleftharpoons t$ or $\vdash_{R^t} \mathbf{A} \rightleftharpoons f$.

And the derived rule for replacement of equivalents, we can prove in R^t the following theorems:

$[\mathbf{p}]\,\mathbf{A} \rightarrow S^p_\mathbf{B}\mathbf{A}\,|$, where $\mathbf{B}$ is a tvf,
$[\mathbf{p}]\,(\mathbf{A} \rightarrow \mathbf{B}) \rightarrow .\, \mathbf{A} \rightarrow [\mathbf{p}]\,\mathbf{B}$, $\mathbf{p}$ not free in $\mathbf{A}$.

And the rule of generalization for tvfs is forthcoming from the rule of adjunction and the definition of restricted quantification in R^t. So $[\mathbf{p}]$ acts in R^t just like $(\mathbf{p})$ in P^t.

Before returning to the discussion of Section 4, we summarize these results by saying that P and P^t are both embeddable in R^t in a way which seems to fit exactly the explanations and intuitions behind P and P^t; and we cannot forbear adding that the analytical resources of P and P^t, powerful as they are, simply wither in the light of the sun provided by R and R^t, both of which have all the power of the former twice (once as in property 2 of Section 3, and once as a consequence of the results just stated), and *vastly* more beside. Which allows us now to return to the disjunctive syllogism.

Belnap and I have already admitted elsewhere (*Philos. Studies* 1962) that the move from A and (not-*A* or *B*) to *B* is valid for intensional senses of 'or', and confessed our inability to characterize these intensional senses generally, beyond noting circularly that they are those which sanction the logical move just mentioned. Certainly the sense of 'or' in which '*A* or *B*' holds when $\sim A$ relevantly implies B is one such sense, but equally certainly there are others, which we must simply leave to one side. However, we did say that where 'or' is taken as truth-functional in the purest, strictest, fiercest sense, ('*A* or *B*' is *true* barely ['merely' – choose your own most pejorative adverb] iff at least one of '*A*' and '*B*' is true), the disjunctive syllogism is not in general valid. But now in view of the following theorems of R^t:

[p] [q] (~ p ∨ q ⇌. p → q), and
[p] [q] (p ∧ (~ p ∨ q) → q),

it may look as if we have contradicted ourselves.

Not a bit of it. We have only to look at the cases involved in the latter theorem to see that no skullduggery is going on.

(i) $p = t$ and $q = t$. Then the left conjunct of the antecedent can do the work alone (A4):

$$t \land (\sim t \lor t) \to t.$$

(So can the right conjunct, for that matter.)

(ii) $p = t$ and $q = f$. Then the right conjunct alone does the work, for by previous observations $(\sim t \lor f) \to (f \lor f)$, whence $(\sim t \lor f) \to f$ by properties of disjunction, so conjunction properties give us

$$t \land (\sim t \lor f) \to f.$$

(iii)–(iv) $p = f$. Then the left conjunct does the work, since f → A:

$$f \land (\sim f \lor t) \to t,$$
$$f \land (\sim f \lor f) \to f.$$

(And for (iii), the right conjunct could also shoulder the burden.)

So we get $A \land (\sim A \lor B) \to B$ as a theorem of R^t *for tvfs* A and B, just as we do in P^t. But note the joker: in each of the cases (i)–(iv) *one and only one* of the conjoined premisses can do all the work. Proponents of the disjunctive syllogism are, as it seems to us, bemused by the fact that it is not always the same one, which lends credence to the idea that the premisses are somehow 'functioning together' to give us the conclusion, as they clearly are for the general case of

$$A \land (A \to B) \to B.$$

Here neither A nor A → B relevantly implies B by itself, though acting in concert they do; but for the special cases (tvfs) envisaged in P^t, either A implies B by itself, or else ~ A ∨ B does, and no co-operation on the part of the conjuncts in the antecedent takes place at all.

Such at any rate is what we believe to be the correct picture concerning the issue raised in Section 4, and we now come to consider Their arguments *for* the classical position, as set forth by Bennett in 1969.

6. BENNETT ON ENTAILMENT

As we mentioned earlier, Bennett's article is the most comprehensive recent defence of the classical Lewis position on entailment we have been able to find; we therefore centre attention on that article, which summarizes the arguments from all significant sources, and some others as well. We begin by listing areas of agreement, and other matters designed to clear the ring, so that we can square off properly.

6.1. *The Enterprise*

At the outset of Section 4, we said that our central controversial problem was whether a certain move in formal logic was "appropriate in the sense of being true to our preanalytic intuitions". This language is of course vague, and it is hard to make the issue more precise other than by trying to restate it in other terms which, while equally vague, have at least the merit of being different. We might say for example that just as we could count correctly and effectively, long before we had an acceptable mathematical analysis of the sequence of natural numbers, so we could use the 'if ... then —' of entailment long before we had even the beginnings of a satisfactory mathematical treatment of the topic. And in both cases the object of the enterprise was to find a formalization of pre-analytic ideas which are already present somewhat amorphously in our common linguistic practice, in our scientific and mathematical Thought, in our Rational Intuitions, and wherever else philosophers and plain (Oxbridge-educated) men like to find them.

Or we might say with Bennett that "An analysis of the concept of entailment is answerable to careful, educated use of expressions such as ['Q can be deduced logically from P', 'From P, Q follows logically', 'There is a logically valid argument with P as sole premiss and Q as conclusion', and the like]" (p. 197 – page numbers refer to Bennett's article). He goes on to say "I shall argue that Lewis was right, and also – by implication – that his thesis is helpful and clarifying – that is, that it is a genuine analysis".

We are also after a 'genuine analysis', in a sense we think we share with Bennett, so (we feel) we are not arguing at cross purposes, though of course as regards the solution we are at loggerheads.

6.2. *Formal Arguments Against 'The Paradox'*

Bennett sets the problem as follows:

As is well known, Lewis' analysis implies that each impossible proposition entails every proposition. Accepting the analysis, I accept this result. For one thing, Lewis has an argument for it (I use '→' to abbreviate 'entails')

$$(1)\ P\ \&\sim P$$
$$(1)\rightarrow(2)\ P$$
$$(1)\rightarrow(3)\sim P$$
$$(2)\rightarrow(4)\ P\vee Q$$
$$(3),(4)\rightarrow(5)\ Q$$

If each step is valid and entailment is transitive, then each impossible proposition entails every proposition. Or, if some impossible propositions entail nothing of the form ($P\ \&\sim P$), then we get the more modest result – which is still unacceptable to all but two of Lewis' opponents – that there are millions of impossible propositions which entail every proposition. For brevity, I shall refer to the thesis that each impossible proposition entails every proposition as 'the paradox' (pp. 197-198).

He goes on to consider various points at which the argument can be blocked: (II) [his Roman numerals] the moves from (1) to (2) and (1) to (3), (III) the move (2)–(4), (IV) the move (3), (4), to (5), (V) the claim that (1) can't serve as a premiss ('contradictions entail nothing'), and (VI) the rejection of transitivity of entailment. These seem to us, as apparently they do to Bennett, exhaustive of the places at which the argument can be attacked *formally* (we defer consideration of informal attacks for the moment). We are completely in accord with Bennett's conclusions concerning (II), (III), (V), and (VI), so formal haggling in the sequel will be confined to (IV).

6.3. *Informal Arguments*

We also agree with Bennett (VIII),

that ... if it is clear that we ought to accept the paradox, it does not matter much whether this acceptance is described as our rectifying a previously inconsistent concept or as our handling more competently the consistent concept we have had all along (p. 212).

In either case we seem to have *some* pre-analytic concept, and if clarifying it demands acceptance of the Lewis argument, we must simply bite the bullet. (As we shall repeat later, however, this alternative is not forced on us). And of course Bennett is right in holding in (VIII) that simply being surprised at the Lewis paradox is no argument against it. But we must enter a demurrer at the remarks:

> At most, a judgment of counterintuitiveness may lead to a search for arguments against the thesis in question – for instance, the thesis that $(P \mathbin{\&} \sim P) \to Q$. This particular search has been underway for decades, and has so far most miserably failed (p. 213).

We cannot help observing that in the light of some sixty-odd constructive articles amplifying Ackermann's insights of 1956 (an interested reader can get the beginnings of a reasonably complete bibliography by consulting Belnap, *JSL*, 1967, and Meyer and Dunn, *JSL*, 1969), it is hard to see *what* might satisfy Bennett's exacting standards for even partial success. Each of these items is, or pretends to be, a constructive contribution to the philosophical and/or mathematical analysis of the systems of Section 3, or extensions thereof. This growing corpus is not error-free (for examples see the end of this section), but no one has claimed that the two bloopers known to the present writer are profound, far-reaching, irreparable – or even significant for the total undertaking, namely, to find a 'genuine analysis'. The claim that the literature cited above is a 'miserable failure' can only be made good by a careful criticism of the mathematical results set forth and the philosophical interpretations placed on them; there is nothing even remotely approaching such a critique in the literature.

For similar reasons we must object to Bennett's (IX):

> It is often said that the paradox infringes the principle that if $P \to Q$ then P must be 'connected in meaning' with Q. This complaint has never yet been accompanied by an elucidation of 'meaning-connection' (as distinct from a suggested representation of it in an extremely limited formal language), let alone by an attempt to show that in the given sense of 'meaning-connection' it is true both that (a) where there is an entailment there is a meaning-connection, and that (b) for some Q there is no meaning-connection between $(P \mathbin{\&} \sim P)$ and Q (p. 214).

Perhaps the results in publications listed above are to be thought of as represented in 'an extremely limited formal language', but it should be added that R, $R^\dagger$, E, $E^\dagger$, EQ, and RQ (Belnap, *op. cit.*) are *vastly* more powerful than the whole classical apparatus, which is simply the extensional fragment of these systems. They contain truth-functions, modality, quantification over individuals and propositions, the whole works – and generalization to quantification of higher order, while not yet worked out in detail, promises to be of no more than routine difficulty. Much of the literature listed above is addressed *precisely* to Bennett's points (a) and (b) just quoted. We should add that $R^\dagger$ and $E^\dagger$ can even accommodate intuitionistic 'implication' and negation in a way consonant with informal considerations adduced by Heyting and others, as pointed out in Anderson and Belnap (*J. Phil.* 1961) – and that paying attention to formulations of classical extensional logic in which the rules are entailments in the sense of *E* leads to trivial proofs of important classical results (Post completeness, Gödel completeness, Löwenheim-Skolem theorems, etc.).

We agree with Bennett that we have offered no *general* explanation of 'meaning-connection', but of course this was not our intent; we wanted rather to carry out a programme which (as we thought) was also close to Bennett's heart: to find a 'genuine analysis'. The difference is that *we* want one which agrees with that of Lewis where Lewis makes sense (almost always), but excises the unsightly warts on his theory.

This motivation is, as I believe, of some importance, and an understanding of it demands that we look at an argument *contra* which Bennett seems to feel of sufficient importance to justify stating twice in print. He writes:

It is sometimes said that $(P \& \sim P) \to Q$ is 'counterintuitive' – or 'unacceptable', 'totally implausible', 'outrageous' or the like. To call the paradox 'counterintuitive' is, apparently, to say that it seems to be logically false. Perhaps it does, but then so does 'There are as many odd prime numbers as odd numbers', yet this is true by the only viable criterion of equal-numberedness we have. Our resistance to it can be explained: most of our thinking about numbers involves only finite classes, and it never is the case that there are as many *F*'s as *G*'s if the *F*'s are a proper subclass of the *finite* class of *G*'s. If someone said. "Yes, I see all that, and I have no alternative criterion of equal-numberedness; but I still don't accept that there are as many odd prime numbers as odd numbers", we should dismiss this as mere autobiography (p. 212).

And in reviewing Anderson and Belnap ('The Pure Calculus of Entailment', *JSL*, 1962), he makes a similar point in discussing the heuristic remarks which lead to the subscript-notation for keeping track of relevance. (There we affect to prove **B** →. **A** → **A** (p. 34), and remark "in this example we indeed proved **A** → **A**, but it is crashingly obvious that we did not prove it *from* the hypothesis **B**: the defect lies in the definition, which fails to take seriously the word 'from' in 'proof *from* hypotheses'.") He writes (*JSL*, 1965):

This is in the anti-scientific spirit of: "The view that the set of primes has the same cardinality as the set of rationals fails to take 'same' seriously."

I suppose that there is some point in the parallel: in both cases the definitions (of *equicardinality* and *strictly implies*) turned out to have consequences which were not altogether expected by those who framed them (though this is a common hazard; when we get *very* clear about a topic, we sometimes catch things in our definitional nets that we had not anticipated).

But as long as we are going in for this sort of thing, we might point out some disanalogies as well. No one would deny, I suppose, that Cantor's definition of equicardinality led to surprising and 'counter-intuitive' results; indeed those concerning denumerable and non-denumerable sets turned out to be fundamental to the ensuing theory of sets, which we can all agree is beautiful, profound, entertaining, and instructive. Hilbert himself thought highly of the subject. He said that it was "the most admirable blossom of the mathematical mind and on the whole one of the foremost achievements of mankind's purely intellectual activity," and called it "a paradise created by Cantor from which nobody will ever expel us."

I do not recall hearing anyone wax equally eloquent in praise of theories developed as a consequence of the alleged fact that a contradiction entails any old thing. The *rest* of the strict implication theory is interesting enough, but results like $(P \wedge \sim P) \rightarrow Q$ are simply excrescences – harmless, perhaps, but silly.

In this connection Nuel Belnap suggests an equally apt parallel. Newton swallowed infinitesimally small magnitudes without blinking. Berkeley pointed out that they were indigestible. Bolzano, Cauchy, Weierstrass, and Dedekind agreed that infinitesimal quantities made no sense, but rather than simply grousing about the situation (*i.e.* for the case at hand, "saying undiagnostically that the... argument must be wrong in *some* way" (p. 199), a topic to which Bennett devotes nearly one-half of his article), they undertook to provide an alternative analysis which saved the many important parts of the calculus and threw away the nonsense. So Bennett's remark that "at most, a judgment of counterintuitiveness may lead to a search for arguments against the thesis in question – for instance, the thesis that $(P \wedge \sim P) \rightarrow Q$" (p. 213) is simply not true; one can also attempt to *reconstruct* a theory, so as to save the good bits and discard the bad. This is what we have tried to do.

The parallel just drawn is of course intended to be more flattering to our project than is Bennett's and we attempt to disarm those who might want to accuse us of vainlory by adding hastily that neither parallel seems to cut much ice. By focusing attention on one or another point of analogy or disanalogy, things can be made to look one way or another, depending on which axe is to be ground.

But even this observation should not be allowed to obscure the fact that R, E, and their extensions, like S4 in particular among the Lewis systems, have interesting philosophical interpretations, and the kind of stability which makes them amenable to philosophically motivated mathematical analysis. When a scientific theory engulfs its predecessors, and is constructively of interest in its own right, one may, I hope, be forgiven for believing that it may be on the right track; we return to this theme briefly in Section 8. We should add here, finally, that we agree strongly with Bennett that invective concerning implicational paradoxes is futile *in the absence of a precise, viable alternative* – but this is what we claim to have provided. To revert to Bennett's earlier metaphor, we claim not only a diagnosis, but a cure.

Bennett's sections (X)–(XIV) are devoted to informal considerations at a greater remove from the systems of formal logic with which we are concerned here, and we shall accordingly skip on to his treatment of a formal argument closely related to that with which we began.

6.4. *The Second Lewis Argument*

This is an argument, of a slightly different form, for the contrapositive of the original 'paradox' (XV).

Lewis' analysis of the concept of entailment also implies that each necessary proposition is entailed by every proposition. For this, too, Lewis has an independent argument:

$$
\begin{aligned}
&(1)\ Q \\
&(1) \to (2)\ (Q\ \&\ P) \lor (Q\ \&\sim P) \\
&(2) \to (3)\ Q\ \&\ (P \lor \sim P) \\
&(3) \to (4)\ P \lor \sim P
\end{aligned}
$$
(p. 229)

We think Bennett correctly identifies the nub of the matter:

It has been claimed that (2) is not entailed by (1) but is entailed by

$A: Q\ \&\ (P \lor \sim P);$

and if A were the argument's premiss then there would be nothing 'counterintuitive' about its having $(P \lor \sim P)$ as its conclusion. This position denies the widely accepted view that if $(Q \& R) \to S$, and R is necessary, then $Q \to S$; and anyone who takes it must be careful (p. 229).

He adds:

I suggest that any thesis which would have us reject the move from Q to $(Q \& P) \lor (Q \& \sim P)$ as invalid would *ipso facto* stand convicted of misrepresenting the common concept of entailment (p. 230).

This is a point on which we must content ourselves in simply registering disagreement. *Arguments* to the effect that necessarily true premisses don't simply go away, even if they are not explicitly mentioned, were put forward at length by Anderson and Belnap in 'Enthymenes' (*J. Phil.* 1961). Whether or not Bennett would recognize an enthymeme as distinct from a full statement of an argument we do not know. Neither do we know whether he recognizes the considerations put forward in that paper as 'arguments'; his tone suggests he would not. But we have nothing to add here to our previous discussion of enthymemes; so far as we know, no one has yet tried to show in print that our arguments won't wash.

6.5. *Speculation About a Point of Agreement, and a caveat*

Bennett apparently feels that, within limits, the choice of a particular Lewis system is irrelevant to the issue he is discussing. He does not consider the question explicitly, but his exposition suggests that the only essential point is that some reasonable sense of 'strict implication', 'possibility', 'necessity', etc., be adopted, as an adjunct to P. With this we agree entirely, and we go on to point out that necessity can be tacked onto R, just as it can to P. (Roughly, R bears the same relation to its (as it seems to us) most natural modal cousin E, as P does to S4.) But the actual moves in the 'Lewis paradox' argument are all P moves – the relevance of modality lies only in the demand that if $\vdash_P A$, then NA be a theorem of the modal system under consideration – a *very* minimal requirement. This minimal condition is also satisfied by R and E: the steps accepted in R are also accepted as necessary implications in R's modalized counterpart E, and the bad step (the disjunctive syllogism) is rejected by both.

So we agree (we think) concerning the role played by necessity in the argument – at any rate we hope so.

The *caveat* is this: some of the writings above were not available at the time of writing of Bennett's 1969 article. Whether he has been persuaded by later publications we do not know; we have no reason to think one way or the other at the moment of writing this.

But we would urge that everything of philosophical interest for the issue at hand was present, either implicitly or by explicit conjecture, in the earlier writings. The most striking recent formal results are those of Belnap (*JSL*, 1967) and Meyer and Dunn (*JSL*, 1969), both of which papers were greeted by the present writer with open arms, as confirming the fondest hopes of the several of us working on these problems. (The principal results of both papers, it should be noted, were discussed in detail in the conjectural stage, long before proofs were obtained; see Anderson, *Acta Philos. Fennica*, and Anderson and Belnap, *Math. Annalen*, 1963.) In both cases novel and nontrivial proof procedures had to be developed in order to prove what we all were sure *must* be the case, and those of us closest to the enterprise, at least, take the recent formal developments as confirming the philosophical motivations which were behind the entire undertaking from the beginning.

On one point alone we have been obliged to retreat, and that arose from an inadequate understanding of our own theory. Misled by enthusiasm, we made in Anderson and Belnap (*op. cit.* 1962) some philosophical remarks (*not* germane to the controversy under discussion here) which were simply and flatly false, as was pointed out cogently and convincingly by R. and V. Routley (*Nous*, 1969). Luckily, we were able to interest J. Alberto Coffa in trying to repair the situation, and he succeeded in explaining to us what we had meant, or *should* have meant if we had had any sense. At any rate his paper 'Fallacies of Modality' (forthcoming) straightens out that problem to our satisfaction. The other error was a small mathematical conjecture in Anderson and Belnap 1963 (*Math. Annalen*, 1963), which turned out to be trivially false – though the principal theorem there under discussion in Section 8 was finally proved by Belnap (*JSL*, 1967).

With these exceptions we still stand by our earlier guns, and we have tried in this section to prepare for the next skirmish, which concerns the principal topic left dangling above: the disjunctive syllogism (end of 6.2, above).

7. The Disjunctive Syllogism Again

This section is devoted to answering the following remark of Bennett's (where he uses "$\sim P \Rightarrow Q$ to symbolize the statement that P and Q are so related as to justify subjunctive (sometimes counterfactual conditionals" (p. 204)):

> To complete their case, Anderson and Belnap have also to maintain that if (4) is read truth-functionally, so that (2)→(3), then it is false that ((3) & (4))→(5). This is prima facie extremely implausible, *yet I cannot find that they offer any arguments for it.* Of course I am going along with the denial of
>
> $$(P \lor Q) \to (\sim P \Rightarrow Q):$$
>
> but what is now in question is the denial of
>
> $$((P \lor Q) \,\&\, \sim P) \to Q,$$
>
> which is quite different and, one would have thought, *not to be rejected except upon the basis of strenuous arguments.* (p. 205; italics supplied here and above.)

Whether or not the arguments to follow are strenuous enough for Bennett's taste we shall leave for him to decide, but we claim that they *are* arguments. Variants of the first have been stated thrice before in print, though not quite so explicitly. It runs as follows:

(a) Logicians have taught for at least two thousand years that in a valid argument from A to B, the premiss A must be relevant to the conclusion B, and that in the corresponding true proposition that if A then B, the antecedent must be relevant to the consequent; *i.e. relevance is a necessary condition for validity.*

(b) Logicians have also taught that the argument from $A \land (\sim A \lor B)$ to B is valid, and that the corresponding proposition that if $A \land (\sim A \lor B)$ then B is true, which, under plausible and almost universally accepted further assumptions, entails that the argument from $A \land \sim A$ to B is valid, and the corresponding proposition true. But $A \land \sim A$ is not in general relevant to B; ergo, *relevance is not a necessary condition for validity.*

(I take it that no one wants to dispute the penultimate claim above. I do not know of any logicians rushing to maintain, for example, that the proposition that Naples both is and is not the principal city of New South Wales is relevant to the proposition that only God can make a tree. It is in fact the *obvious* lack of relevance that makes the thing 'paradoxical'.)

Looking candidly at (a) and (b), I do not see for the life of me how *anyone* can expect to have it both ways. And if it looks as if I am saying of virtually the entire contemporary community of philosophically-minded logicians that they share in advocating a flat logical inconsistency, I can only reply that if the shoe fits.... The situation appears to be (as Wilfrid Sellars once characterized it in conversation) 'a scandal in philosophy', and it is undeniably clear from traditional and current practice that (b) is the horn that must go; relevance simply *is* required of us as sane arguers.

And if *this argument* is not enough, one can go on to point out that the Lewis position can not even be stated coherently, at least in the context of one plausible understanding of how logical connectives get their meanings, namely, what I take to be Gentzen's. The difficulty can be put in this way.

Suppose a proposition C entails A, and that it also entails B; then it seems plausible to say that C is at least as strong as, or has at least as much logical content as, the conjunction $A \wedge B$ of A and B, since that conjunction does precisely that much of C's work (whatever else C may be up to): just as, under the supposition above, $C \to A$ and $C \to B$, so $A \wedge B \to A$ and $A \wedge B \to B$; indeed $C \to A \wedge B$. This much makes sense independently of the point at issue; and so does the following:

If A and B are sufficient for C, in the sense that C follows logically from those two premises, then it would also appear that C is no stronger than the premisses A and B taken conjointly; C may well be weaker than the conjunction of A and B, but under the supposition of *this* paragraph, we are not going to get anything out of C that we would not get out of the two premisses.

Now if *both* of these conditions hold, it is hard to see how C could *be* anything other than the conjunction of A and B (or else the same person by a different name). Whether or not this characterization of C appeals to the reader, we submit that the considerations adduced in favour of it do not *prima facie* involve what Bennett calls "the [Lewis] paradox": $P \wedge \sim P \to Q$.

But when applied to 'the paradox', these views seem to lead to conclusions which are substantially more in agreement with our position than that of Lewis, Bennett *et al*. For let us suppose They are right, and that $P \wedge \sim P$ 'entails every proposition'. Then $P \wedge \sim P$ entails every proposition Q, and also $\sim Q$; and every contradictory pair of premisses Q, $\sim Q$ has $P \wedge \sim P$ as a consequence. So that '$P \wedge \sim P$' is simply odd and perhaps unfamiliar notation for the conjunction of Q and $\sim Q$.

But in *this* case there is nothing paradoxical about 'the paradox' at all, since with this understanding Q now follows from $P \wedge \sim P$ by axiom A4 of R: $\mathbf{A} \wedge \mathbf{B} \rightarrow \mathbf{A}$. Admittedly '$P \wedge \sim P$' is *outré* notation for the conjunction of Q and $\sim Q$, but if that is all that is going on, Lewis's argument has had its teeth pulled. The moral then would seem to be that the whole ruckus was foisted on us by notational confusion – and be *damned* to symbolic logic anyway!

One is left, however, with the suspicion that They will be reluctant to accept such a gutless dissolution of 'the paradox'; to fabricate dentures for the Lewis argument, They seem forced to say, "No, it is this *particular* contradiction $P \wedge \sim P$ that (irrelevantly) entails Q". Now making a distinction between one contradiction and another is perfectly all right with Us, since of course $A \wedge \sim A$ is not in general relevantly equivalent to $B \wedge \sim B$ in R (or any of its relatives), and indeed there are things true of one but not the other: *e.g.* the first relevantly implies A, whereas the second in general does not. But a distinction between different contradictions is explicitly forbidden to Them by Their own theory; so it would appear that They are in a box: if $P \wedge \sim P$ *is* the conjunction of Q and $\sim Q$, then it is hardly surprising that Q follows from it; and if it is not, it is hard to see why Their theory makes *no distinction whatever* between $P \wedge \sim P$ and $Q \wedge \sim Q$. Anything Their theory says about it will say about the other (including the remark that it *is* the other: $P \wedge \sim P = Q \wedge \sim Q$). This makes it difficult to understand Bennett's remarks (quoted in subsection 6.2 above) that "each impossible proposition entails every proposition," and "there are millions of impossible propositions which entail every proposition"; his *theory* says there is only one.

What Their theory *really* says of course is that any contradiction is the same proposition as f, which admittedly always *has the same truth-value* as any contradiction; and in $R^\dagger$ we have $f \rightarrow \mathbf{A}$. But premisses from which f may be inferred *must* amount to more than one measly little contradiction. We can only get f from what f is prepared to produce, namely *all* propositions: f has far too much muscle to be relevantly implied by one puny $P \wedge \sim P$.

What seems surprising about all this discussion of the correctness of $(P \wedge \sim P) \rightarrow Q$ is that anyone should go to such desperate lengths in order to try to justify a principle that no one needs, wants, or can use.

8. Conclusions, or What You Will

At this point we may have difficulty in getting the reader to believe that we did not, in this paper, set out to deliver a polemical diatribe, but rather to present and discuss a constructive result of sorts: the fact that the theory of relevant implication coincides with that of material 'implication' where their areas of competence overlap, namely for tvfs. This area is precisely that of $P^\dagger$'s expertise, and $P^\dagger$ is accordingly the feebler of the two, since anything $P^\dagger$ can do, $R^\dagger$ can do better – a fact which commends $R^\dagger$ to our attention.

In addition to swallowing up all of P and $P^\dagger$, the cluster of systems mentioned in Section 3 can claim other merits: they allow us to make the *Sinn-Bedeutung* distinctions which have been hovering in the backs of all our minds since we began to read Frege, and they have the stability required to prompt the investigations in the burgeoning literature mentioned above. The fact that a fair portion of the papers cited are written, in part anyway, in the vernacular of mathematics should occasion no dismay. The same is true of classical extensional and modal logic, and I am not alone in believing that the philosophical and mathematical motivations and results for intensional (and other) logics confer dignity on each other.

The extreme attitudes toward such developments are I suppose represented by Quine and Wittgenstein. Quine has on occasion felt that intensions are scurvy customers, of interest only to the intellectually depraved:

The theory of meaning even with the elimination of the mysterious meant entities, strikes me as in a comparable state to theology – but with the difference that its notions are blithely used in the supposedly most scientific and hard-headed brands of philosophy (*Proc. Amer. Acad. of Arts and Sci.*, 1951).

That is, no one in his right mind would attempt to accommodate Fregean *Sinne* in a formal system. And Wittgenstein seems to feel that mathematical techniques as applied to logic should be shunned as the work of the devil:

The curse of the invasion of mathematics by mathematical logic is that now any proposition can be represented in a mathematical symbolism, and this makes us feel obliged to understand it (*Remarks on math.*, 1956).

I think on the contrary that we should neither turn our backs on recalcitrant problems, nor throw away precision tools when they are available. The sane attitude seems to me that expressed by Meyer (*Logique et Analyse*, 1968):

From the point of view of scientific progress, what counts is not one's personal preference for one version of 'entails' over another; what counts is that significant distinctions, once introduced, shall not be conflated, confounded, or dismissed on the simple ground that they could not be made at a previous state in the development of a science.

Meyer's attitude is perhaps a *trifle* too tolerant for my tastes, but his heart is in the right place.

University of Pittsburgh

NOTE

* Reprinted from *Mind*, n.s. vol. **81** (1972), pp. 348–371, by kind permission of the publisher and editor.

ALDO BRESSAN

INTENSIONAL DESCRIPTIONS AND RELATIVE COMPLETENESS IN THE GENERAL INTERPRETED MODAL CALCULUS MC$^\text{v}$

1. Introduction

Bressan's communication at the 1971 International Congress for Logic, Methodology, and Philosophy of Science (Bucharest) to be spoken of here, concerns the main results of Memoir 3 in Bressan's book (1972). The same communication was essentially based on some results that were presented at the analogue of the above congress in 1964 (Jerusalem)[1] and constitutes the main achievements of Memoir 1 (and some of Memoir 2) in Bressan (1972). Therefore here I shall first hint at Memoir 1 and 2 in Bressan (1972) briefly – cf. Sections 2–6 – and then at Memoir 3 in more detail. Of course, for a thorough treatment of the subject mentioned in the title and, more generally, for an extended theory on **MC$^\text{v}$** and the language **ML$^\text{v}$** on which **MC$^\text{v}$** is based, the reader is advised to take Bressan (1972) into account.

Incidentally, while **ML$^\text{v}$** has attributes and function types of all finite levels, the work (Bressan, 1973) being printed was done to construct the analogue of Bressan (1972) for a type-free language, **ML$^\infty$**, and a calculus **MC$^\infty$** based on **ML$^\infty$**.

Let us now say why and how the book (Bressan, 1972), to which the present communication is intimately related, was written – cf. the source footnote (*) in Bressan (1972, p. 103). The author, who professionally is a mathematical physicist, worked on axiomatization of classical mechanics according to E. Mach and P. Painlevé – cf. Painlevé (1922). The use of extensional logic in this connection – cf. Hermes (1938) and (1959a) – gave rise to serious difficulties – cf. Rosser (1938) or Bressan (1962, N11), or else Bressan (1972, N21). So, modalities seemed indispensable to overcome them. Hermes (1959b) sketched a solution of the problem using some modal operators. However the language on which this solution is based has considerable limitations (e.g. excludes the use of contingent identity and certain kinds of functions).

Bressan (1962) wrote to cast Painlevé's axiomatization (Painlevé, 1922), from a general point of view, into a form which is rigorous in connection with both the language used, and the proper axioms (of classical particle mechanics) stated. Incidentally, Painlevé himself seems to admit that his (1922) book is lacking in these two aspects – cf. footnotes 53 and 68 in Bressan (1972, pp. 110, 111).

In 1959 the author was rightly advised by some American logician not to use modal logic to solve the aforementioned language problem since this logic was not yet developed enough. Therefore he decided to use possibility (and necessity) concepts, not directly (i.e. by means of a modal language), but by means of an extensional language, through a device related to the extensional semantical rules given by Carnap (1956) for certain modal languages. More precisely, in Bressan (1962) the set **CMP** of the mechanically possible cases, briefly the **CMP**-cases, is introduced.[2] On the one hand, concepts such as *mass point* are considered to be classes. On the other hand, ordinary contingent concepts are changed. As an example, the concept $pos_K^*(M, t)$, i.e. the concept of the position of the mass point M at the instant t (in the inertial spatial frame K), is changed into the concept $pos_K(M, t, \gamma)$, which is the position of M (in K) at the instant t in the **CMP**-case γ. Hence some common sentence p^* such as $P = pos_K^*(M, t)$ is translated into a sentence p_γ – which in the above example is $P = pos_K(M, t, \gamma)$ – open with respect to the parameter $\gamma (\in \mathbf{CMP})$. Then $\Diamond p^*$ and Np^* can obviously be translated into $(\exists \gamma) p_\gamma$ and $(\gamma) p_\gamma$ respectively.

The language used in Bressan (1962) is adequate but not usual. The author felt that a general and rather usual modal language such as **ML^v** and an efficient logical calculus based on **ML^v**, such as **MC^v**, should be constructed. This job was accomplished, in a first version, in 1964 and the main results of it were presented at the aforementioned Jerusalem Congress.[3] The second version is Bressan (1972).

Let us mention that Bressan (1972) constitutes a (modal) logical basis for many physical theories; among them are classical mechanics according to Painlevé and also (a general non-quantistic) physical theory in general relativity whose axiomatic foundations are laid down in Bressan (1964).

Nowadays the interest of the field to which Bressan (1972) refers, is witnessed by the works of various authors. Among them let us mention Gallin in connection with the essentially modal axiom AS18.19(N3)[4] and Scott who at the Irvine meeting in 1968, gave advices published in Scott (1969) concerning modal semantics, that are in agreement with $\mathbf{ML}^v$.[5] As the author substantially says, he limits himself to advices[6] (which however are clear), and from his postscript he seems not to have very much faith in the possibility of application of the kind of modal languages he is advising.[7] Perhaps this is connected with the fact that Scott (1969) refers to nothing similar to the notion of absolute concepts or quasi absolute concepts – cf. Bressan (1972, N24), that are basilar to apply $\mathbf{ML}^v$ to science and everyday life. Some examples of these applications are given in Bressan (1972, NN 19–22 and 24).[8]

2. General considerations on $\mathbf{ML}^v$

The v-sorted interpreted modal language $\mathbf{ML}^v$ is based on the type system $\tau^v \cup \{0\}$ where 0 is the sentence type and, briefly speaking, the set τ^v is defined inductively by the following conditions: (a) The v individual types 1 to v are in τ^v, and (b) for $t_0, \ldots, t_n \in \tau^v$ the function type $(t_1, \ldots, t_n : t_0)$ and the attribute type $(t_1, \ldots, t_n)$ are in τ^v. $\mathbf{ML}^v$ has the variables v_{tn} and constants c_{tn} ($n = 1, 2, \ldots$) for every $t \in \tau^v$, it contains the all sign $\forall$, the necessity sign N, (contingent) identity $=$, and the description operator $\imath$. Furthermore, $(\imath v_{tn}) p$ is well formed and has the type t for every matrix p. Let $\mathbf{EL}^v$ be the extensional part of $\mathbf{ML}^v$.

The semantic analysis of modalities for $\mathbf{ML}^v$ is based on the concept of a set Γ of 'possible cases' γ. They are called 'possible worlds' by the tradition and 'Γ-cases' by Bressan who had used this generalization over state descriptions (Carnap, 1956) in Bressan's (1962) monograph on foundations of classical particle mechanics – cf. Note 2. The semantics for $\mathbf{ML}^v$ also comply with Carnap (1956) in that every designator is conceived of as having both an extension and an intension (the latter determines the former in every $\gamma \in \Gamma$); furthermore, these semantics are directly based on $\mathbf{S}_5$ as those in Carnap (1956) and Carnap (1963). (However any finite number of mutually compatible possibility – and necessity-operators of the kinds used in S3, S4, Feys' system T, and Brouwer's system B can be defined within $\mathbf{MC}^v$ – cf. Bressan (1972, Chap. 13) – following in part Kripke (1963). Thus one realizes, from a general point of view, a suggestion by Lemmon – cf. Lemmon (1959).

Let us add that Carnap (1956) says that "the problem of whether or not it is possible to combine modalities and variables in such a way that the customary inferences of the logic of quantification – in particular, specification and existential generalization – remain valid is, of course, of greatest importance" (pp. 195–196). The calculus **MC**^{v} solves this problem in a general way. This makes **MC**^{v} efficient in connection with both physical axiomatization problems – cf. Part II of Memoir 1 in Bressan (1972) – and other problems of traditional modal logic – cf. Bressan (1972, NN 53, 55).

The semantics for **ML**^{v} are uniform in that the following facts hold:

(1) In connection with every logical type t, only one kind of variables is used; in particular, no distinction between class variables and property variables is made (in conformity with Carnap's method of extension and intension).

(2) The quasi intensions (QI-s) of type $(t_1, ..., t_n)$ $[(t_1, ..., t_n : t_0)]$ are constructed starting out from those of the types $t_1, ..., t_n$ $[t_0, ..., t_n]$ in a way independent of $t_0, ..., t_n$.

(3) For $t \in \tau^v$, all designators of type t are treated semantically in the same way (disregarding whether they are variables, constants, function expressions, or definite descriptions), and more precisely the intensional designatum $des_{\mathscr{M}\mathscr{V}}(\Delta)$ at a model $\mathscr{M}$ and value assignment $\mathscr{V}$ of any designator Δ of type t is a QI_t, i.e. a QI of type t.[9]

(4) All designators are meaningful in all Γ-cases and these cases are considered all on a par, having in mind axiomatizations of sciences such as classical mechanics, physics, or chemistry.

The situation with e.g. geology, geography, and astronomy is different in that for them the 'real case' is relevant. For it is useful to introduce a (privileged) constant ρ representing the 'real Γ-case' (Section 7).

Let us add that the QIs for matrices are the subsets of Γ. Of course, the matrix p is said to hold in $\gamma(\in \Gamma)$ at (the model) $\mathscr{M}$ and (value assignment) $\mathscr{V}$ if $\gamma \in des_{\mathscr{M}\mathscr{V}}(p)$. Furthermore, a semantic system for **ML**^{v} has the form

(1) $\qquad \langle \Gamma, D_1, ..., D_v, \alpha^v \rangle$ with $\alpha_t^v = \alpha^v(t) \in QI_t$ for $t \in \tau^v$,

where the sets D_1 to D_v are thought of as individual domains; D_1 to D_v and Γ determine the class QI_t for $t \in \tau^v$, and α^v is a function of domain τ^v. For $t \in \tau^v$ α_t^v is used (following Frege) to express the extensions of descriptions in case they do not fulfill the condition of exact uniqueness.

3. Some Axioms of MC^v

The designation rules for ML^v are such that axiom schemes AS12.1–23 in Bressan (1972) hold. They are those of a lower predicate calculus based on S5, in particular we have

AS12.8 $(N)(\forall x)\, \Phi(x) \supset \Phi(\Delta)$

and

AS12.13 $(N)\, Nx = y \supset [\Phi(x) \equiv \Phi(y)]$,

where (as below) x and y are distinct variables, where (N) is any string of universal quantifiers and N's, whose scope is the whole part of the formula at its right, where (as below) $\Delta\,[y]$ is free for x in $\Phi(x)$, and where $\Phi(\Delta)\,[\Phi(y)]$ results from $\Phi(x)$ by substituting $\Delta\,[y]$ for x at the free occurrences of x. AS's12.10–12 substantially say that contingent identity is an equivalence relation. Let us write explicitly

AS12.14 $(N)\, F = G \equiv (\forall x_1, ..., x_n)\,[F(x_1, ..., x_n) \equiv G(x_1, ..., x_n)]$

and

AS12.15 $(N)\, f = g \equiv (\forall x_1, ..., x_n)\, f(x_1, ..., x_n) = g(x_1, ..., x_n)$,

where (as below) F and $G\,[f$ and $g]$ are attributes [functors] distinct from one another and (of course) from the distinct variables x_1 to x_n. Furthermore, let $F\,[f]$ not occur free in the matrix p [the term Δ]. Then we have

AS12.16 $(N)(\exists F)(\forall x_1, ..., x_n)\,[F(x_1, ..., x_n) \equiv p]$,
AS12.17 $(N)(\exists f)(\forall x_1, ..., x_n)\, f(x_1, ..., x_n) = \Delta$,
AS12.18 (i) $(N)\, p(\exists_1 x)\, p \supset x = (\imath x)\, p$,
         (ii) $(N) \sim (\exists_1 x)\, p \supset (\imath x)\, p = (\imath x)\, x \neq x$,

and

AS12.19 $(N)(\exists F)(\forall x_1, ..., x_n)\,\{[\Diamond\, F(x_1, ..., x_n) \equiv NF(x_1, ..., x_n)] \wedge$
$\wedge\,[F(x_1, ..., x_n) \equiv p]\}$.

The last axiom – cf. Note 3 – has no direct or indirect extensional analogue, and is basilar to define some analogues of Γ-cases within ML^v itself (Section 6).

The syntactical theory for the calculus MC^v based on the language ML^v and ASs12.1-23, is developed in a thorough way in Bressan (1972, Memoir 2). In particular metatheorems such as the deduction theorem, and theorems for rules G and C (i.e. universal generalization and the formal analogue of an act of choice) are considered.

Incidentally if $\vdash p$, i.e. p is provable in MC^v, then $\vdash (N)p$. Furthermore, ASs12.16,18 yield $\vdash (\exists F)(\forall x_1, ..., x_n) N[F(x_1, ..., x_n) \equiv p]$ – cf. Bressan (1972, N 40, (46)).

4. On the extensional translation of ML^v into EL^{v+1}

In Bressan (1972, N 15) the extensional translation Δ^η (into EL^{v+1}) of any designator Δ in ML^v is defined.

The semantical system for EL^{v+1} has the form

(2) $\quad \langle D_1, ..., D_v, a^{v+1} \rangle \quad$ with $\quad a_t^{v+1} = a^{v+1}(t) \in E_t \quad$ for $\quad t \in \tau^v$,

where E_t is the class of the extensions of type t and D_i and a^{v+1} have in (2) obvious meanings similar to those of D_i and a^v respectively in (1). The designatum $des_{MV}(\Delta)$ of any designator Δ in EL^{v+1} at M and V is defined in a common way – cf. Bressan (1972, N16).

Now use 'κ' for $v_{v+1,1}$ (*the case variable*) and identify D_{v+1} with Γ. Then the extensional translation $M = \mathcal{M}^\eta[V = \mathcal{V}^\eta]$ of $\mathcal{M}[\mathcal{V}]$ can be defined in such a way that – cf. Theor. 16.1 in Bressan (1972)

(3) $\quad des_{\mathcal{M}\mathcal{V}}(\Delta) = des_{MV}(\Delta^\eta) \quad$ *for every designator* Δ *in* ML^v

and that, *in case Δ is a matrix, Δ holds in $\gamma (\in \Gamma)$ at $\mathcal{M}$ and $\mathcal{V}$ iff Δ^η holds at M and $V^{(\gamma)}$* where $V^{(\gamma)} = (V - \{(\kappa, V(\kappa))\}) \cup \{(\kappa, \gamma)\}$.

By (3) the translation $\Delta \to \Delta^\eta$ of ML^v into EL^{v+1} is in the strong sense according to Carnap, i.e. Δ^η characterizes the intension of Δ.[10] It constitutes the proof of the so-called thesis of extensionality – see Carnap (1956, p. 141) – in connection with the general modal language ML^v. However in conformity with a statement by Carnap – see Carnap (1956, p. 142), the expressions in ML^v are simpler than their translations into the extensional language EL^{v+1} as are the corresponding deductive manipulations. Thus ML^v [MC^v] is technically more efficient than EL^{v+1}

[the extensional part EC^{v+1} of MC^{v+1}, which is an extensional calculus based on EL^{v+1}].

5. On the concept of absolute attributes

The concept $Abs_{(t_1, \ldots, t_n)}$ – briefly *Abs* – of absolute attributes of type $(t_1, \ldots, t_n)$ is very important to apply ML^v (or MC^v) to physics and to several situations of everyday life – cf. Bressan (1972, Chapter 4). First, we define *modally constant attributes* ($MConst_{(t_1,\ldots,t_n)}$) and *modally separated attributes* ($MSep_{(t_1,\ldots,t_n)}$) of type $(t_1, \ldots, t_n)$:

DEF. 5.1 $\quad F \in MConst_{(t_1, \ldots, t_n)} \equiv_D (\forall x_1, \ldots, x_n) \times$
$\times [\Diamond F(x_1, \ldots, x_n) \equiv NF(x_1, \ldots, x_n)]$,

DEF. 5.2 $\quad F \in MSep_{(t_1, \ldots, t_n)} \equiv_D (\forall x_1, y_1, \ldots, x_n, y_n) \times$
$$\times \left[F(x_1, \ldots, x_n) \wedge F(y_1, \ldots, y_n) \wedge \Diamond \bigwedge_{i=1}^{n} x_i = y_i \supset N \bigwedge_{i=n}^{n} x_i = y_i \right].$$

Now we can write

DEF. 5.3 $\quad F \in Abs_{(t_1, \ldots, t_n)} \equiv_D F \in MConst_{(t_1, \ldots, t_n)} \wedge F \in MSep_{(t_1, \ldots, t_n)}$.

The (natural) absolute concept *Nn* of natural numbers can be defined on purely logical grounds – cf. Bressan (1972, N 27) while others such as the one of *mass point* are naturally grasped by experience. Some of them are assumed as primitive concepts in axiomatic theories.

Incidentally *Abs* can be generalized into the concept *QAbs* of quasi absolute attributes and the analogue holds for *MConst* and *MSep* – cf. Bressan (1972, N 24). The concept *QAbs* is useful e.g. in connection with living beings.

Of course an absolute concept, e.g. *Nn*, determines the corresponding extensional concept by means of its extensionalization, e.g. $Nn^{(e)}$, while the converse is false.

There is a double use of common nouns – see Bressan (1972, Chapter 5) – in that in some context they are used extensionally and not absolutely while in some other contexts the converse holds. This is relevant e.g. for the axiomatization of Mechanics according to Painlevé (1922) as was pointed out in Abs_{64} – cf. Note 1.

6. On the Analogue El of Γ-cases, defined within $\mathbf{MC}^v$ itself

The analogue El of Γ mentioned in the title is defined within $\mathbf{MC}^v$ in Bressan (1972, N 48) together with the matrix l_u that contains free the only variable u and that means: *u is the Γ-case ($u \in El$) which is taking place*. To mention some properties of El and l_u, and in particular how Np and $\Diamond p$ can be characterized by means of them, we first define *there is a strictly unique x such that p* ($(\exists^\wedge_1 x) p$):

DEF. 6.1 $(\exists^\wedge_1 x) \Phi(x) \equiv_D (\exists x) \Phi(x) \wedge (\forall x, y) [\Phi(x) \Phi(y) \supset Nx = y]$,

where x and y are distinct variables, and $\Phi(y)$ [$\Phi(x)$] results from $\Phi(x)$ [$\Phi(y)$] by substituting y for x [x for y]. We have – cf. Bressan (1972, pp. 201, 202, 204)

(4) $\vdash El \in Abs$, $\vdash l_u \supset u \in El$, $\vdash (\exists^\wedge_1 u) l_u$, $\vdash u \in El \supset \Diamond l_u$,

(5) $\vdash \Diamond (l_u p) \equiv u \in El \wedge N(l_u \supset p)$,

(6) $\vdash (\forall u) [u \in El \supset \Diamond (l_u p)] \equiv Np \equiv (\forall u) N(l_u \supset p)$

(7) $\vdash (\exists u) [u \in El \wedge N(l_u \supset p)] \equiv \Diamond p \equiv (\exists u) \Diamond (l_u p)$.

Using l_u, two new description operators depending on the parameter u can be defined. The first, ι_u, is intensional, unlike ι whose extensional character appears, among other things, from A12.18 (Section 3); the second, η_u, is a combination of ι and ι_u that has better substitution properties than both ι and ι_u:

DEF. 6.2 $(\iota_u x) p \equiv_D \Diamond (l_u p (\exists^\wedge_1 x) p)$, $\wedge \Phi [\iota_u x) \Phi(x)]$

so that $(\iota_u x) p$ substantially is the *x such that p holds in the possible case u provided such an x be strictly unique*.

DEF. 6.3 $(\eta_u x) p \equiv_D (\iota x) [qx = (\iota_u x) p \vee \sim qx = (\iota x) p]$, where q is $(\exists^\wedge_1 x) p$.

Among the various properties of ι_u and η_u – cf. Bressan (1972, N 51) – let us mention the following

(8) $\vdash l_u (\exists^\wedge_1 x) \Phi(x) \supset \Phi [(\iota_u x) \Phi(x)] \wedge \Phi [(\eta_u x) \Phi(x)]$.

The operators ι_u and η_u are useful in themselves – see e.g. the solution (b') of the puzzle (b) in Bressan (1972, p. 222), which is based on ι_u –, and especially in case u is identified with the constant ρ [N 7].

7. ON THE CALCUS MC_ρ^v USEFUL TO DEAL WITH THE REAL WORLD

The constant ρ representing the real world is only subject to axiom A52.1 in Bressan (1972):

A52.1 $\rho \in El$.

Of course I_ρ, ι_ρ, and $\daleth_\rho$ have all properties of I_u, ι_u, and $\daleth_u$; in particular the theorems obtained from (4) to (8) by substituting ρ for u hold – cf. Bressan (1972, N52).

The sentence I_ρ is the analogue for MC_ρ^v of the propositional constant n introduced by Meredith and Prior (1964, p. 215) in a quite different way. So I_u appears to be a generalization of n. Let us add that in Bressan (1972, N 53) I_ρ is proved to substantially fulfill the axioms on n stated in Meredith and Prior (1964).

Among the philosophical puzzles solved by means of ι_ρ in Bressan (1972, N 55) let us mention the following assertion in ordinary language;
(a) *The (modally prefixed) number of living presidents (living at the instant t) could be larger than it is (in reality)*.

The most natural translation of (a) into MC_ρ^v is

(ā) $\Diamond\, (\daleth n)\, q > (\iota_\rho n)\, q$, where $q \equiv_D n \in Nn \wedge Liv\ Pres \in n$

and where $\vdash Nn \in Abs$ – cf. Theor. 45.1 in Bressan (1972) – and *Liv Pres* denotes (in ML^v) the extensional concept of living presidents.

It is useful to introduce $\mathbb{R}p$, i.e. *p occurs in the real case*:

DEF. 7.1 $\mathbb{R}p \equiv_D \Diamond (I_\rho p)$.

The main properties of $\mathbb{R}$ are stated in Bressan (1972, N 52). Now let us remark that I_ρ or $\mathbb{R}$ is useful in dealing with some nonmodal sentences of ordinary language that are of interest in sciences such as geology, geography, and astronomy. Moreover a double use of such sentences is put in evidence in Bressan (1972, N 53). More in particular the factual sentence of ordinary language
(b) *The artificial satellite S has the distance d from Venus at the instant t* seems, at first sight, to have a natural translation into MC^v of the form

p_b where

$$p_b \equiv_D S \in \text{ArtSat} \land \text{dist}_V(S, t) = d.$$

This occurs by the following pragmatic situation, p_b is meant to hold in the Γ-cases where it is asserted or read, and it is read in the real case. However all Γ-cases are on a par for $\mathbf{MC}^v$, so that the meaning of p_b in $\mathbf{MC}^v$ appears clearly for instance by remarking that a correct use of p_b is made when p_b is assumed in order to prove, using rule C, a counterfactual conditional such as

(c) *If ... S had the distance d ..., then the observation ω would be made.* The factual sentence (b) can be correctly translated into $\mathbf{MC}^v$ by means of $\mathbb{R}p$.

8. On the relative completeness of $\mathbf{MC}^v_\rho$

In Bressan (1972, N 31) the aforementioned extensional translation $\Delta \to \Delta^\eta$ of $\mathbf{MC}^v$ into $\mathbf{EC}^{v+1}$ is proved to leave the entailment relation invariant. Furthermore, using El it is possible – see Bressan (1972, NN 56–62) – to define within $\mathbf{MC}^v$ certain analogues of the QIs and a translation of $\mathbf{EC}^{v+1}$ into $\mathbf{MC}^v$, the star translation $\Delta \to \Delta^*$, in such a way that a certain basic conditioned equivalence between any designator Δ in $\mathbf{MC}^v$ and $(\Delta^\eta)^*$ holds – cf. Theor. 62.1 in Bressan (1972) – and that in particular $\vdash (N)\ \Delta \equiv (\Delta^\eta)^*$ in case Δ is a closed matrix. This allows us to prove quickly Theor. 63.1 in Bressan (1972) which (in particular) asserts that $p_1, ..., p_n \vdash p_0$ in $\mathbf{MC}^v$ iff $p_1^\eta, ..., p_n^\eta \vdash p_0^\eta$ in $\mathbf{EC}^{v+1}$.[11]

Now let us say that the extensional translation $\Delta \to \Delta^u$ reflects the semantical theory for $\mathbf{MC}^v$. Hence to prove $\vdash \Delta^\eta$ in $\mathbf{EC}^{v+1}$ is practically equivalent to proving the logical validity of Δ in the semantical meta-language for $\mathbf{ML}^v$. It can be substantially concluded – cf. Bressan (1972, N 64) – that *the matrix Δ is provable in $\mathbf{MC}^v$ iff its logical validity is provable in the extensional metalanguage for $\mathbf{ML}^v$*. This equivalence theorem constitutes the property of relative completeness for $\mathbf{MC}^v$ mentioned in the title.

Let us remark that the only axioms of the modal calculus $\mathbf{MC}^v$, that are quite different from the axioms of common extensional and modal calculi and that have to be added to the other axioms of $\mathbf{MC}^v$ in order to obtain the aforementioned relative completeness are AS12.19 (Section 3) and another axiom which depends on certain particular features of

ML^v – A25.1 in Bressan (1972). To avoid these features and the latter axiom is a matter of routine. Incidentally they have in fact been avoided in MC^∞ – cf. Bressan (1973, NN 8, 9, 16, 17).

Università di Padova

NOTES

[1] Cf. the abstract of Bressan's communication – briefly Abs_{64} – on p. 96 in the Program and Abstracts of the 1964 Congress for Logic, Methodology, and Philosophy of Science (Jerusalem).

[2] The main results of Bressan (1962) were presented at the 1960 International Congress for Logic, Methodology, and Philosophy of Science at Stanford.

[3] It must be said that for a long time this oral communication and what is written in Abs_{64} – cf. Note 1 – remained the only public presentation of MC^v. It is true that, among other things, some basic axioms and definitions are explicitly written in Abs_{64}. I mean axioms AS12.18 on descriptions, the new essentially modal axiom AS18.19 (Section 3), and the definition of absolute concepts – cf. Def. 5.3 for $n=1$. However the author, an Italian mathematical physicist, met with considerable expression difficulties, nearly all of the linguistic kind. These difficulties could be overcome only in the academic year 1967–68 during his stay in Pittsburgh. There N. Belnap undertook the linguistic correction of the first version of Bressan (1972) – which contained the first two among the three Memoirs in Bressan (1972). Belnap taught (the second version of) Bressan (1972) in his 1970 fall course for graduate students at the University of Pittsburgh. These students, some of whom worked out some publications based on or strictly related to Bressan (1972), also contributed, together with Belnap, to the improvement of Bressan (1972).

[4] The author has been informed that recently Gallin rediscovered AS18.19 – cf. Note 3 – in connection with a modal language different from ML^v in semantics, and that he proved the independence of AS18.19 from certain other axioms. Bressan's conjecture of the independence of AS12.19 from the other axioms of MC^v, which he gives for granted e.g. in the conclusions about the relative completeness of MC^v – cf. Bressan (1972, N 64) or this paper, Section 8.

[5] Cf. Scott (1969). It is evident that e.g. the axiom on description AS18.19 (N 3), explicitly written in Abs_{64} – cf. Note 1 – conforms with Scott (1969).

[6] Higher level types or functions are not considered in Scott (1969). Furthermore no axiom system is laid down there. In particular AS18.19 is not hinted at.

Incidentally AS12.19 has basic consequences in MC^v (NN 6–8), and its two analogues for the type-free modal calculus MC^∞ have the analogue consequences and additional important consequences – cf. Bressan (1972, N 21).

[7] This seems due to some criticism of Montague – cf. the postscript in Scott (1969). Incidentally Belnap writes in his Foreword to Bressan (1972), p. 1: "Perhaps one should credit the author's near total isolation from the logical community for allowing him to proceed with the elaboration of his fresh ideas unobstructed by premature criticism...".

[8] As it appears from Abs_{64} – cf. Note 1 – in 1964 Bressan communicated that a certain double use of nouns can be put in evidence by means of absolute properties (which are completely defined in Abs_{64}), and that "by essentially absolute uses of 'real number'

one can build a theory based on ML^ν, which contains E. Mach's definition of mass, keeps the advantages of Bressan (1962), and is quite similar to Painlevé (1922)."

[9] $\mathscr{M}[\mathscr{V}]$ is a function whose domain is the class of constants [variables], and $\mathscr{M}(c_{tn})$, $\mathscr{V}(v_{tn}) \in QI_t$ for $n = 1, 2, \ldots$.

[10] The possibility of such a result was not obvious, according to an assertion by Carnap in Carnap (1963, III, 9, IV, p. 894).

[11] An axiom asserting the existence of at least two [infinitely many] objects of type t is understood to hold in $EC^{\nu+1}$ [in MC^ν and $EC^{\nu+1}$] (only) for $t = \nu + 1$ [$t = 1, \ldots, \nu$]. This remark is missing in Bressan (1972).

BIBLIOGRAPHY

[1] Bressan, A., 'Metodo di assiomatizzazione in senso stretto della Meccanica classica. Applicazione di esso ad alcuni problemi di assiomatizzazione non ancora completamente risolti', *Rend. Sem. Mat. Univ. di Padova* **32** (1962) 55–212.

[2] Bressan, A., 'Una teoria di relatività generale includente, oltre all'elettromagnetismo e alla termodinamica, le equazioni costitutive dei materiali ereditari', *Rend. Sem. Mat. Univ. di Padova* **34** (1964) 1–73.

[3] Bressan, A., *A General Interpreted Modal Calculus* (with foreword by Nuel D. Belnap, Jr.) New Haven and London, Yale Univ. Press, 1972.

[4] Bressan, A., 'The Interpreted Type-Free Modal Calculus MC^∞. Part 1: The Type-Free Extensional Calculus EC^∞ Involving Individuals, and the Interpreted Language ML^∞ on which MC^∞ is Based; Part 2: Foundations of MC^∞', *Rend. Sem. Mat. Univ. di Padova*, 1973, in press.

[5] Carnap, R., *Meaning and Necessity*, Univ. of Chicago Press, Chicago, 1956.

[6] Carnap, R., 'Replies and Systematic Expositions', in *The Philosophy of R. Carnap* (ed. by Paul A. Schilpp) Tudor Publishing Co., New York, Library of Living Philosophers, 1963, pp. 859–999.

[7] Hermes, H., 'Eine axiomatisierung der allgemeinen Mechanik'. *Forschungen zur Logik und zur Grundlegung der exakten Wissenschaften*, Vol. 3, Verlag von Hirzel, Leipzig, 1938.

[8] Hermes, H., 'Zur Axiomatisierung der Mechanik', in *Proceedings of the International Symposium on the Axiomatic Method*, Berkeley 1957–58, North-Holland Publishing Co., Amsterdam, 1959a, p. 250.

[9] Hermes, H., 'Modal Operators in an Axiomatisation of Mechanics', *Proceedings of the Colloque International sur la méthode axiomatique classique et moderne*, Paris, 1959b, pp. 29–36.

[10] Kripke, S. A., 'Semantical Analysis of Modal Logic I: Normal, Proposition al Calculi', *Zeitschrift für mathematische Logik und Grundlagen der Mathematik* **9** (1963) 67–96.

[11] Lemmon, E. J., 'Is There Only One Correct System of Modal Logic?', *Proceedings of the Aristotelian Society* (Suppl.) **33** (1959) 23–40.

[12] Meredith, C. A. and Prior, A. N., 'Investigations into Implicational S5,' *Zeitschrift für mathematische Logik und Grundlagen der Mathematik* **10** (1964) 203–20.

[13] Painlevé, P., *Les axiomes de la méchanique*, Gauthier-Villars, Editeur, Paris, 1922.

[14] Rosser, J. B., 'Review of H. Hermes,' *Journal of Symbolic Logic* **3** (1938) 119–20.

[15] Scott, D., 'Advice on Modal Logic', in *Philosophical Problems in Logic* (ed. by K. Lambert), D. Reidel, Dordrecht, 1969.

DRAGAN STOIANOVICI

SINGULAR TERMS AND STATEMENTS OF IDENTITY

Statements containing singular terms have been a source of problems and of significant disagreements among logicians both in past and in recent times. Those constituted by two singular terms coupled by 'is' have been of particular interest as possible counterexamples to the unrestricted applicability of the traditional subject-predicate analysis. Do they belong to a special category of statements of identity? Or are we to countenance the predicability of singular terms and thus adhere to the subject-predicate pattern in this case too? In what follows I want to make some comments on this point. Specifically, I will argue against the idea of the dispensability of the concept of identity in favour of an unexceptional subject-predicate reading of statements having the form '*a* is *b*', where both '*a*' and '*b*' are singular. But I will not claim, then, that all statements made by using sentences of this form are statements of identity, but will join with authors attempting to trace some differences pertaining to the kinds of singular terms involved and to the context of utterance.[1] Finally, I shall make a few general remarks on the meaning and workability of the eliminationist's idea, trying to dissociate the technical aspect of the matter from its philosophical significance.

1. The attempt to make plausible the ordinary subject-predicate analysis for what in contemporary logic is put apart as a distinctive category of statements of identity may stem up from the feeling that one can in this way obviate the somewhat strange and bewildering problems pertaining to the concept of identity by simply realizing their futility. It may seem attractive, indeed, to treat the 'is' in 'The Evening Star is the Morning Star' or in 'Napoleon is Bonaparte' as being not a peculiar 'is' of identity, but the same 'is' as that occurring in ordinary subject-predicate statements, in which a general term is predicated; for the latter, unlike the (purported) statements of identity, are felt to have a transparent and familiar meaning. In a singular subject-predicate statement, e.g., no matter whether the 'is' is treated as a third constituent besides the subject

and the predicate, or as an inseparable part of the predicate, what is being done is ascribing a characteristic to the individual referred to, or otherwise specified by, the singular subject-term. In view of our familiar thinking in terms of things having properties and standing in relation to one another, there is no problem in understanding that an individual *a* has a characteristic *P*, while it may not be equally simple to make sense of the assertion that *a* is identical to *b*. Why, then, shall we not admit that in saying '*a* is *b*', where both '*a*' and '*b*' are singular, we are in fact attributing to *a* certain characteristic, i.e., why shall we not assimilate this kind of statement to that in which the predicate is constituted of a general term, serving to ascribe characteristics to the individual specified by the subject-term? The concept of identity seems thus to be eliminable in favor of the simple and clear idea of affirming or denying characteristics of things.

I think there is something definitely wrong with this sort of argument in so far as it suggests assimilating one kind of statement to another in order to transfer the structural clarity of the latter to the interpretation of the former. Let us take the case of statements constituted by two proper names linked by 'is', as in Russell's example 'Napoleon is Bonaparte'. The eliminationist's proposal invites us to countenance the predicability of singular terms and to construe such predicates as 'is-Bonaparte' on a par with ordinary general ones, i.e., as serving, like these, to characterize the individual specified by the subject-term.

It may be thought that this proposal would get some support from the idea that singular terms, including proper names, do have a sense, that their meaning does not reduce to reference. Both in its general intent and in its details, this idea is as yet under discussion, though one cannot refuse it some plausibility and intuitive attractiveness. We need not try to reject it, anyway. For admitting it does not provide firm ground for the contention that in statements of the form '*a* is *b*' where both '*a*' and '*b*' are proper names, '*b*' is simply (monadically) predicated of *a*, *if this is to mean that one ascribes in this way a certain characteristic to a.* Returning again to Russell's example, it is obvious that being Bonaparte is not at all a characteristic of Napoleon in the sense in which being a man, or a great strategist or having won or lost such and such a battle are characteristics attributable to him. It would be absurd to include 'is-Bonaparte' in an enumeration, however detailed, of the characteristics of Napoleon. (It would be neither absurd nor redundant, indeed, to mention in such an enumeration that his family name was 'Bonaparte', but if this is what the statement 'Napoleon is Bonaparte' means, then one has sided with a metalinguistic reading of this statement, unacceptable, of course, for whoever hopes to eliminate the copula of identity along the lines sketched above).

Thus, one is confronted with a sort of dilemma: either one clings up to the idea that there is a clear and unproblematical sense of 'attributing a characteristic to an individual' typified by the predicative use of general terms, but then one will hardly be able to claim that the same job is done in the case of the alleged predicative use of singular terms such as proper names; or, if one is willing to save at all costs the claim that the predication of general terms and the alleged predication of singular terms are essentially the same thing, one is forced to admit that their common trait is not that clear and undisputable notion of a characteristic being attributed to an individual, that was initially acknowledged in, and abstracted from, the predicative occurrences of general terms. In the first case, the contrast between the two kinds of statement is not abolished. In the second, the project of thoroughly eliminating the bewildering copula of identity in favor of the admittedly straightforward copula of predication is self-defeating; for statements of identity and ordinary subject-predicate statements are then reduced to a common denominator at the cost of abandoning that specific reading of the latter which seemed to confer upon them a privilege of clarity and to recommend them as representative instances of a universally recognizable pattern.

2. If these objections are sound, we may conclude that the concept of identity resists the attempted reduction, at least in statements where the copula is flanked by two proper names. The same remarks hold, obviously, for those statements in which one of the two singular terms is a proper name and the other a demonstrative.

The case seems, however, to be somewhat different when at least one of the two singular terms flanking the copula is a definite description. On the one hand, some statements of this kind seem to be less reluctant to a monadic reading than those whose both terms are proper names. This makes them, in a sense, less interesting for a discussion on identity. But on the other hand, the importance of the concept of identity was thought by some logicians to reveal itself primarily in statements containing definite descriptions, rather than in those where the copula occurs between two proper names. We may recall in this sense that Bertrand Russell explicitly brought about the unequal significance of the two cases for the understanding of the concept of identity (or, better, the unequal significance of the concept of identity for an adequate interpretation of the two kinds of statements). In *Principia Mathematica*, for instance, he entertained the view that 'Napoleon is Bonaparte' is a proposition about names, while emphasizing that 'Scott is the author of *Waverley*' cannot be interpreted in the same way; it requires a totally different analysis, in view of the "descriptive phrase" occurring inside it, an analysis relying essentially on the concept of identity.

For the purpose at hand we can set aside the much controverted Russellian method of paraphrasing sentences containing definite descriptions as well as his treatment of proper names. We will only retain the fact that according to his view a sentence having the form 'a is the so-and-so' (where 'a' is a proper name) expresses an identity. Now is there any disagreement between this view and the claim that such a sentence may be analysed logically into a subject and an ordinary predicate?

At this point it is perhaps important to notice that a sentence of the form 'a is the-so-and-so' *may* be used to express an identity, i.e., to operate an identification, and that it may be used in another way as well. Russell's examples and comments illustrate the former case. Thus, in a context where somebody enquires who the author of *Waverley* is, the sentence 'Scott is the author of *Waverley*' is used in order to express an identity, i.e., to say that the author of *Waverley* is the same man (or the same writer) as Walter Scott. But the other case is no less common – the case when 'a is the so-and-so' serves simply to ascribe to a a certain characteristic. Thus, the statement 'a is the first publisher in Romanian of Plato's *Protagoras*', addressed to someone who knows a but doesn't know, say, that Plato's dialogue has been published in Romanian at all, could hardly be interpreted as revealing to the listener the identity of a. Such a statement will rather be understood as serving simply to ascribe to a a characteristic, indeed, one uniquely possessed by him.

To make clearer the contrast between the two cases, we may take advantage of some distinctions delineated in recent discussions over names and descriptions. In the first case, the definite description is used referringly, its role is to single out for attention an individual on account of which something is to be said. But since the other term in the sentence is a proper name, admitting that proper names never serve to predicate characteristics of things forces us to recognize that in this case we have to do with a statement of identity. (In such statements of identity as these, the referential role of the description is better displayed when the word order is 'the-so-and-so is *a*', e.g., 'The author of *Waverley* is Scott'. But this is inessential, of course; when the order is reversed, the peculiar intent of the statement is shown by proper accentuation). In the second case, however, the referential function accrues to the proper name, while the description occurs attributively; now, the description serves for ascribing a characteristic to the individual referred to by the proper name. It is a characteristic which belongs to it (him) exclusively, and so it *might* be relied upon for identifying it (him), but this is not what is being done in *this* context.

So it is not improper to treat statements of the second kind, at least for some purposes, on a par with ordinary (singular) subject-predicate statements. Such an analysis discloses enough of their structure when, for example, we are not interested in the uniqueness of the property conveyed by the (predicatively occurring) definite description. It may be carried out properly whenever one of the two singular terms is a definite description occurring non-referringly.

Conversely, I think we are justified in saying that a statement having the form '*a* is *b*' cannot be interpreted as a subject-predicate statement if both singular terms in it occur referringly. And this is sometimes the case, for instance, when the question arises over the identity of an individual satisfying a definite description. In such cases as these, as well as in those where both terms are proper names, we have to do with genuine statements of identity. Their proper interpretation is indeed beset with difficulties, but these are not removed by adhering to a reading that does not take account of the peculiar intent of this kind of statement.

3. Attempts at dispensing with the concept of identity may originate, however, in considerations of quite another sort. While admitting that the distinction between statements of identity and ordinary subject-predicate statements may be justified and perhaps even illuminating when we are primarily interested in what is precisely done in one case

and in the other, one might still contend that it introduces an unnecessary complication in the formulation of rules of inference. In other words, one's preference for a uniform reading of 'is' in all sentences where it copulates two terms, singular or general, may be rooted in one's desire for economy in the repertory of techniques used in discovering logical relations between propositions. And this may well be taken to constitute the primary concern of the logician.

This line of putting together two kinds of logical entities, in spite of some obvious differences among them, only if they are somehow amenable to a unique set of logical rules, is not without precedent in traditional logic. Historians of logic are faced with a difficult task when trying to extract a coherent view of the status of singular terms and sentences in ancient and medieval logic. While the contrast between singulars and universals is sharply traced in the Aristotelian thesis of the impredicability of the individual, in later formulations of the theory of syllogism it has been customary to assimilate singular statements to universal or particular ones.

We have seen that in attempting to dispense with the copula of identity, one has to countenance the predicability of singular terms of all kinds. Another counterintuitive idea put to the same end is that of the quantifiability of singulars. A contemporary author advanced recently a way of deriving the properties of identity by making use of the so-called 'wild' quantity of singular terms. On this approach, inferences involving singular statements, even statements of identity, are brought under the well-known syllogistic rules for universal and particular statements.[2]

My aim here is not to comment on the technical aspect of the matter. There may well be more than one serviceable analysis of a statement or of a set of statements in view of eliciting their logical relations. The several possible ways of laying down the principles of syllogistic reasoning offer such an example. In some of them singular terms and statements are subjected to such analyses as allow the application to them of the same rules as obtain for a category or another of general ones. But one should not disregard the fact that such an accommodation of singulars is always achieved by way of analogy or convention. Thus, it is not equally proper to say with respect to a singular subject-term that it is distributed as is with respect to a general subject-term preceded by the word 'any' or 'every'. By the same token, to say that singular terms do have quantity, but that this quantity is somehow indifferent or 'wild' amounts in fact to adopting in their regard a convention enabling us to extend over them the applicability of a logical category whose meaning is properly understood only in relation to sortal universals. Disregarding on purpose the peculiarity of statements of identity in view of achieving a diminution on the side of rules of inference is another instance of the same policy.

But, if this is what is really being done in such reductions and simplifications, then while appreciating the inventiveness of manoeuvres by which these are effected, one should at the same time be aware of their possible philosophical drawbacks. It is by no means an accident that just those problems pertaining to the dichotomy of singulars and universals and to the meaning of identity have been of great concern to philosophically minded logicians, such as Leibniz, Frege, Russell or Wittgenstein. All of them were preoccupied in a way or another in working out a system of logical categories and a symbolism apt to aid and discipline our insights in the way things are or are dealt with in thinking and language. In one of his letters to Russell, in 1913, Wittgenstein wrote: "... Identity is the very Devil and *immensely important*.... It hangs... directly together with most fundamental questions...". This dramatical emphasis put on the significance of the concept of identity did not arise, apparently, from definite difficulties in applying logical rules to arguments involving statements of identity; it points rather to the difficulty of explaining the peculiar place and function of this concept in our thinking and speaking of things and of their properties and relations.

When what is sought for is not simply a minimal and manageable set of rules for deciding on the validity of inferences, but an instrument for conceptual clarification, i.e., when logical research is pursued in connection with philosophical analysis, some seemingly futile distinctions may achieve great prominence and become essential marks of an adequate conceptual apparatus. Problems concerning singular terms and identity are among those which confronted logicians when they were attempting to work out 'ontologically perspicuous' or epistemologically perspicuous logical symbolisms, such as not to disguise or distort important features of our conceptual equipment, or such as to distort them in (what one hopes will be) an illuminating way.

University of Bucharest

NOTES

[1] While formulating the following summary remarks, I will not explicitly mention specific contributions of one writer or another to the discussion of the points at issue, although part of what I say is in close agreement with some of them.

[2] The proponent of this peculiar eliminationist view is Fred Sommers, in a short paper 'Do We Need Identity?', *Journal of Philosophy* **66** (1969) 499–504.

ROMAN SUSZKO

ADEQUATE MODELS FOR THE NON-FREGEAN SENTENTIAL CALCULUS (SCI)

This note contains the proof of the following theorem: every model, adequate for SCI, is uncountable.

The SCI-language L, considered here, is constructed from sentential variables in $\{p_i\}$ by means of sentential connectives of negation $\neg$, conjunction $\wedge$, and identity $\equiv$. SCI-consequence C is defined by truth-functional axioms, logical axioms for identity and the rule of modus ponens. A set of formulas $X \subseteq L$ is a theory if $X = C(X)$. A theory X is consistent or complete if, correspondingly, $X \neq L$ or X is maximal consistent.

An SCI-model is a structure $\langle A, -, \cap, \circ, P \rangle$ where $\langle A, -, \cap, \circ \rangle$ is an arbitrary algebra under $-, \cap, \circ$, of the same type as the SCI-language and P is a subset of A such that for all a, b in A:

$$-a \in P \quad \text{iff} \quad a \notin P$$
$$a \cap b \in P \quad \text{iff} \quad a, b \in P$$
$$a \circ b \in P \quad \text{iff} \quad a = b.$$

For each model $M = \langle A, P \rangle$, we define the consequence C_M on L as follows: $\alpha \in C_M(X)$ iff for every h in $Hom(L, A)$, if $h(X) \subseteq P$ then $h(\alpha) \in P$. (For general information on SCI and its models, see references.)

COMPLETENESS THEOREM with respect to the class of models K:

$$C = \underset{M \in K}{\text{Inf}}\ C_M.$$

Here, $K=$ the class of all models or $K=$ the class of all countable models or $K=$ the class of all quotients L/T, i.e., L modulo $(\alpha \equiv \beta) \in T$, for complete theories T.

ADEQUACY THEOREM: There exists a model M of the power of the continuum, adequate for C, i.e., such that $C = C_M$.

AUXILIARY LEMMA: $C = C_{\langle A, P \rangle}$ iff for every complete theory T there exists h in $Hom(L, A)$ such that $\breve{h}(P) = T$ where $\breve{h}=$ the h-counter-image operation.

1. Truth-valuations

A mapping $t: L \to 2 =$ two-element Boolean algebra under $-, \cap$, is called a truth-valuation (TV), if for all $\alpha, \beta, \gamma, \delta$ in L:

(1) $\quad t(\neg \alpha) = - t(\alpha)$
(2) $\quad t(\alpha \wedge \beta) = t(\alpha) \cap t(\beta)$
(3) $\quad t(\alpha \equiv \alpha) = 1$
(4) $\quad$ if $t(\alpha \equiv \beta) = 1$, then $t(\alpha) = t(\beta)$
(5) $\quad$ if $t(\alpha \equiv \beta) = 1$, then $t(\neg \alpha \equiv \neg \beta) = 1$
(6) $\quad$ if $t(\alpha \equiv \beta) = t(\gamma \equiv \delta) = 1$, then
$\quad\quad t((\alpha \wedge \gamma) \equiv (\beta \wedge \delta)) = t((\alpha \equiv \gamma) \equiv (\beta \equiv \delta)) = 1$.

Symmetry, $t(\alpha \equiv \beta) = t(\beta \equiv \alpha)$, and transitivity, $t(\alpha \equiv \gamma) = 1$ whenever $t(\alpha \equiv \beta) = t(\beta \equiv \gamma) = 1$, follow. Define:

$\quad L_0 =$ the set of all variables,
$\quad L_{n+1} = L_n \cup \{\neg \alpha, \alpha \wedge \beta, \alpha \equiv \beta \mid \alpha, \beta \in L_n\}$
$\quad E_0 = \emptyset, E_{n+1} =$ the set of all equations $\alpha \equiv \beta$ in L_{n+1}.

A mapping $h: L_n \to 2$ is called a partial truth-valuation (PTV) of rank $n = 0, 1, 2, \ldots (PTV_n)$, if for α, β, γ in L_{n-1}:

(7) $\quad h(\neg \alpha) = - h(\alpha)$
(8) $\quad h(\alpha \wedge \beta) = h(\alpha) \cap h(\beta)$
(9) $\quad h(\alpha \equiv \alpha) = 1$
(10) $\quad h(\alpha \equiv \beta) = h(\beta \equiv \alpha)$
(11) $\quad$ if $h(\alpha \equiv \beta) = h(\beta \equiv \gamma) = 1$, then $h(\alpha \equiv \gamma) = 1$
(12) $\quad$ if $h(\alpha \equiv \beta) = 1$, then $h(\alpha) = h(\beta)$

and for $\alpha, \beta, \gamma, \delta$ in L_{n-2}:

(13) $\quad$ if $h(\alpha \equiv \beta) = 1$, then $h(\neg \alpha \equiv \neg \beta) = 1$
(14) $\quad$ if $h(\alpha \equiv \beta) = h(\gamma \equiv \delta) = 1$, then
$\quad\quad h((\alpha \wedge \gamma) \equiv (\beta \wedge \delta)) = h((\alpha \equiv \gamma) \equiv (\beta \equiv \delta)) = 1$.

(I) (1) There is a one-one correspondence between TV and the family of all complete theories. Truth-valuations are characteristic functions of complete theories.

(2) There is a one-one correspondence between TV and the family of infinite increasing sequences of PTV. If t is in TV then the restriction of

t to L_n is in PTV_n. If for all n, $h_n \in PTV_n$ and h_{n+1} is an extension of h_n then there is a unique $t \in TV$ such that $t(\alpha) = h_n(\alpha)$ for $a \in L_n$ and all n; $t = \lim_n h_n$.

2. Extreme Extensions of Partial Truth-Valuations

A binary relation R on L_n may be identified with the set of all equations $\alpha \equiv \beta$ in E_{n+1} such that α bears R to β.

Let $h \in PTV_n$. Define $F(h)$ as the set of all equations $\alpha \equiv \beta$ in E_{n+1} such that $h(\alpha) = h(\beta)$. Obviously, $F(h)$ is an equivalence relation on L_n.

An equation η in E_{n+1} is said to be generated by h in PTV_n if either

(1) η is $\neg \alpha \equiv \neg \beta$ and $h(\alpha \equiv \beta) = 1$ or
(2) η is $(\alpha \wedge \gamma) \equiv (\beta \wedge \delta)$ or $(\alpha \equiv \gamma) \equiv (\beta \equiv \delta)$ and, $h(\alpha \equiv \beta) = h(\gamma \equiv \delta) = 1$.

Let $G(h)$ be the set of all equations in E_{n+1}, generated by h. Again, $G(h)$ is an equivalence relation on L_n.

Given h in PTV_n, define $h^0, h^+ : L_n \to 2$ as follows. Both are extensions of h, i.e., for α in L_n:

$$h^0(\alpha) = h^+(\alpha) = h(\alpha).$$

Furthermore, for all $\neg \alpha$, $\alpha \wedge \beta$, $\alpha \equiv \beta$ in $L_{n+1} - L_n$, we set:

$$h^0(\neg \alpha) = h^+(\neg \alpha) = -h(\alpha)$$
$$h^0(\alpha \wedge \beta) = h^+(\alpha \wedge \beta) = h(\alpha) \cap h(\beta)$$
$$h^0(\alpha \equiv \beta) = 1 \quad \text{iff} \quad (\alpha \equiv \beta) \text{ is in } G(h)$$
$$h^+(\alpha \equiv \beta) = 1 \quad \text{iff} \quad (\alpha \equiv \beta) \text{ is in } F(h).$$

(II) h^0 and h^+ are in PTV_{n+1}. If $g \in PTV_{n+1}$ and g is an extension of h then

(1) $g(\alpha \equiv \beta) = 1$ whenever $h^0(\alpha \equiv \beta) = 1$ and,
(2) $h^+(\alpha \equiv \beta) = 1$ whenever $g(\alpha \equiv \beta) = 1$.

3. Special Truth-Valuations

We say that the formulas α, β are similar and the equation $\alpha \equiv \beta$ is special iff either α, β are variables or both α, β have the same principal connective, i.e., both α, β are negations or conjunctions or equations. Notice that all generated equations, that is, in $G(h)$ are special. If h is in PTV_{n+1} then h is called special iff

(1) $h(p_i \equiv p_j) = 1$ for all i, j;
(2) α, β are similar whenever $h(\alpha \equiv \beta) = 1$;
(3) $h(\alpha \equiv \beta) = 0$ and $h(\alpha) = h(\beta)$, for some $\alpha \equiv \beta$ in E_{n+1}.

(III)   There exists a special g in PTV_2.

Proof. Obviously, there exists an h in PTV_1 such that $h(p_i \equiv p_j) = 1$ for all i, j. Let $g = h^0$. Since g extends h it follows that $g(p_i \equiv p_j) = 1$ for all i, j. If $g(\alpha \equiv \beta) = 1$, then $\alpha \equiv \beta$ is in $G(h)$ and, hence α, β are similar. Consequently, $g(p \equiv (p \equiv p)) = g((\neg p) \equiv (p \equiv p)) = 0$. On the other hand, either $g(p) = g(p \equiv p)$ or $g(\neg p) = g(p \equiv p)$.

(IV)   If $n > 1$ and h is special in PTV_n then h^0 is special in PTV_{n+1}.

Proof. We infer, like in the proof of (III), that α, β are similar whenever $h^0(\alpha \equiv \beta) = 1$ and, $h^0(p_i \equiv p_j) = 1$ for all i, j. Obviously, there exist γ, δ in L_{n-1} such that $h(\gamma) \neq h(\delta)$. Hence, $\neg \gamma$ and $\delta \wedge \delta$ are in L_n and

$$h^0(\neg \gamma) = h(\neg \gamma) = h(\delta \wedge \delta) = h^0(\delta \wedge \delta)$$

where the middle equation holds by 'truth-tables'. But, the equation $(\neg \gamma) \equiv (\delta \wedge \delta)$ is in E_{n+1} and $\neg \gamma$, $\delta \wedge \delta$ are not similar. Therefore, $h^0((\neg \gamma) \equiv (\delta \wedge \delta)) = 0$.

(V)   If $n > 0$ and $h \in PTV_n$ then h^+ is not special.

Proof. Suppose that α is in $L_{n-1} - E_{n-1}$. Then, α and $\neg \alpha$ are in L_n. The equations $\alpha \equiv (p \equiv p)$ and $(\neg \alpha) \equiv (p \equiv p)$ in E_{n+1}, are not special. Since $h(\alpha) = 1$ or $h(\neg \alpha) = 1$, hence $h^+(\alpha \equiv (p \equiv p)) = 1$ or $h^+((\neg \alpha) \equiv (p \equiv p)) = 1$, respectively.

(VI)   If $n > 1$ and h is special in PTV_n then there exists a special h^* in PTV_{n+1} such that h^* is an extension of h and $h^0 \neq h^*$.

MODELS FOR THE NON-FREGEAN SENTENTIAL CALCULUS 53

Proof. By hypothesis, $h(\gamma \equiv \delta) = 0$ and $h(\gamma) = h(\delta)$ for some $\gamma \equiv \delta$ in E_n. Then, the equation $\neg \gamma \equiv \neg \delta$ is in $E_{n+1} - G(h)$. Let R be the least equivalence relation on E_{n+1} over $G(h) \cup \{\neg \gamma \equiv \neg \delta\}$ and define $h^*: L_{n+1} \to 2$ as follows. (1) $h^*(\alpha) = h(\alpha)$ if α is in L_n. (2) For all $\neg \alpha$, $\alpha \wedge \beta$, $\alpha \equiv \beta$ in $L_{n+1} - L_n$ we set: $h^*(\neg \alpha) = -h(\alpha)$, $h^*(\alpha \wedge \beta) = h(\alpha) \cap \cap h(\beta)$ and, $h^*(\alpha \equiv \beta) = 1$ iff $\alpha \equiv \beta$ is in R. Then h^* is an extension of h and $h^* \in PTV_{n+1}$. We have shown above that $R \neq G(h)$. Hence, $h^* \neq h^0$. We must show that h^* is special. All equations in $G(h)$ are special and $\neg \gamma \equiv \neg \delta$ also is special. Therefore, all equations in R are special. Consequently, if $h^*(\alpha = \beta) = 1$ then α, β are similar. Furthermore, like in the proof of (IV), find γ and δ in L_{n-1} such that $h(\gamma) \neq h(\delta)$. The formulas $\neg \gamma$ and $\delta \wedge \delta$ are in L_n and again $h^*(\neg \gamma) = h(\neg \gamma) = h(\delta \wedge \delta) = h^*(\delta \wedge \delta)$. So, we have an equation $(\neg \gamma) \equiv (\delta \wedge \delta)$ in E_{n+1} where $\neg \gamma$, $\delta \wedge \delta$ are not similar. Hence, $h^*((\neg \gamma) \equiv (\delta \wedge \delta)) = 0$.

4. Special Complete Theories

(VII) There is a one-one map $2^\omega \to TV^* =$ the set of all $t \in TV$ such that $t(p_i \equiv p_j) = 1$ for all i, j.

Proof. Given $c \in 2^\omega$, define an increasing sequence of partial truth-valuations $h_n \in PTV_n$ as follows. Let h_2 be special in PTV_2 and let h_0, h_1 be restrictions of h_2 to L_0 and L_1, respectively. Suppose, h_n is defined for some $n > 1$. Then, choose h^* as in (VI) and put $h_{n+1} = h_n^0$ or $h_{n+1} = h_n^*$ if, correspondingly, $c_{n-1} = 0$ or $c_{n-1} = 1$. Define t_c in TV^* as $\lim_n h_n$. Since, $h_n^0 \neq h_n^*$ it follows that the map $c \to t_c$ is one-one.

A complete theory is said to be special if it contains all equations $p_i \equiv p_j$ for all i, j. If $\langle A, P \rangle$ is a model and $h \in Hom(L, A)$ then h is called special if $h(p_i) = h(p_j)$ for all i, j.

(VIII) The collection of all special complete theories is uncountable.

THEOREM. If $C = C_{\langle A, P \rangle}$, then A is uncountable.

Proof. Suppose, $C = C_{\langle A, P \rangle}$ and A is countable. If T is a special complete theory, $h \in Hom(L, A)$ and $h(T) = P$, then $(p_i = p_j) \in T$ and hence $h(p_i \equiv p_j) \in P$, i.e., $h(p_i) = h(p_j)$. Thus, h is special. Therefore,

$\{\check{h}(P) \mid \text{special } h \in Hom(L, A)\}$ is a countable set. Use auxiliary lemma to infer that the family of all special complete theories is countable. Contradiction.

Stevens Institute of Technology,
Hoboken, New Jersey

BIBLIOGRAPHY

Bloom, S. L. and Suszko, R., 'Semantics for SCI', *Studia Logica* **28** (1971) 77–81.
Bloom, S. L. and Suszko, R., 'Investigations into the Sentential Calculus with Identity', *Notre Dame Journal of Formal Logic* **13** (1972) 289–308.

WILLIAM J. THOMAS

DOUBTS ABOUT SOME STANDARD ARGUMENTS FOR CHURCH'S THESIS*

1. Introductory Remarks

'Church's Thesis' (**CT** hereinafter) is the statement that a function of natural numbers is effectively computable if and only if that function is recursive. Like any biconditional, **CT** may be regarded as a conjunction of two conditionals. The arguments treated in this paper have been advanced in support of the so-called 'problematic conjunct' of **CT**, *viz.* that all effectively computable functions are recursive. When I use **CT** then, I shall intend the problematic conjunct. So far as I have been able to determine, no one thinks that **CT** is amenable to anything like a conclusive proof of its truth, though I think that everyone recognizes that it could, at least in principle, be refuted by a suitable counterexample. Even Kreisel, who takes **CT** to be a statement of intuitionistic mathematics, nowhere considers the possibility that **CT** might be a theorem, though he speaks of 'finding axioms inconsistent with' **CT**.[1] Perhaps we ought to be surprised, or at least puzzled by this fact, since some of the arguments which have been given for **CT** look like deductive arguments from premises that are supposed to be beyond dispute. Of course, if these arguments really were deductively valid and dependent only upon indisputable premises, the propounders of the arguments would be likely to show more confidence in them than anyone has done.

The standard arguments given for **CT** are of two general kinds: there are arguments from essentially empirical data, or 'heuristic' evidence, and there are those which are based on what may be called '*a priori* considerations'. Of the former sort, the simplest is the argument which has as its premise the fact that many effective (in the intuitive sense) functions are known to be recursive, and none are known which are not recursive. I shall argue that the other arguments from empirical considerations reduce to the no-known-counter-example argument, and that that argument ought not to convince us. In regard to the *a priori* arguments, I shall argue that the known examples are either essentially unclear or are based on question-begging presuppositions.

2. 'Heuristic evidence'

2.1. The basic piece of 'heuristic evidence', the lack of a known counter-example, provides no very convincing support for **CT**. Consider that evidence of this sort has never been thought convincing in mathematics. One could argue in the same fashion for the truth of, say, the Goldbach or the Fermat conjecture, but surely no one regards either of those conjectures as having a known truth-value. Indeed, if a mathematician even entertained mere suspicions as to the truth-value of either of these conjectures, those suspicions would be founded on intuition (perhaps arising from a study of suggestive results in number theory) but certainly not on the cornucopia of confirming cases available to him. Now, if we are to allow 'heuristic evidence' as convincing in the matter of the decision about **CT**, there ought to be some justification for treating *this* question as somehow very different from other mathematical questions, different in a way that licenses our taking serious account of 'heuristic' considerations. Such a justification does not seem to be forthcoming.

Wang[2] objects to the no-known-counter-example argument on other grounds: he observes that the 'evidence' also 'supports' (to the extent that it supports **CT**) hypotheses about effectiveness which identify that notion with classes of functions less inclusive than general recursiveness, as well as hypotheses which identify effectiveness with more inclusive classes (e.g. the class of arithmetical functions).

2.2. Kleene[3] has argued for **CT** from his 'Recursion Theorem'. The Recursion Theorem says, essentially, that recursive functionals preserve recursiveness and relative recursiveness. The Theorem is applied to the question of **CT** by arguing that the Theorem shows that one could not construct a counter-example to **CT** by using recursive methods, which methods are all that are available, it is claimed.

While I think that the Recursion Theorem does shed some light on what a counter-example to **CT** would have to be like, Kleene's claim that the Theorem and its attendant argument 'virtually excludes doubt' of the truth of **CT** must be rejected. The Theorem does not establish the non-existence of a heretofore unknown non-recursive computable function which ought to be included among the initial functions of a recursive theory of computable functions. Indeed, this possibility is specifically allowed for in the second half of the Recursion Theorem, where *relative* recursiveness is shown to be preserved.

Another possibility not excluded by the Recursion Theorem is that we might discover an effective operation (or functional) on recursive functions which is non-recursive and which leads outside the class of recursive functions. Of course if we did have such an operation, we would also have some good candidates for new initial functions.

The 'heuristic' data to which the argument from the Recursion Theorem makes appeal is precisely the lack of a counter-example to **CT**. The claim that recursive methods (functionals) are all that are available for the construction of a counter-example is supported only by the evidence to which that first argument appeals. If we possessed a counter-example, then we would, obviously, have an effectiveness preserving non-recursive functional.

2.3. The argument from (as far as is known) common closure properties is this: recursiveness and effectiveness are closed under (as far as we know) precisely the same operations.[4] The data to which appeal is made in this argument, if I understand it correctly, is the same as the data on which our first argument depends. For, as in the case of the argument from the Recursion Theorem, if there were an effectiveness preserving operation (functional) which led outside recursiveness, then we would have a counter-example to **CT**, and conversely. Thus, it is appropriate to be dissatisfied with this argument for the same reasons which lead one to reject the argument from the Recursion Theorem.

2.4. Perhaps the most famous of the 'heuristic' arguments is the argument from diverse formulations. The idea is that there are quite a number of distinct formal reconstructions of the notion of an effectively computable function. It has turned out that these alternative reconstructions are equivalent. As Kleene says,[5] the fact that the diverse formulations all characterize the same class of functions is 'a strong indication that this class is fundamental.' I doubt whether anyone would be inclined to quarrel with Kleene on that point, but the connection between this admitted fundamentality and **CT** is not clearly made out in any published version of the argument I have been able to discover.

One can, with Kreisel,[6] agree that recursiveness is a fundamental notion while doubting that that is reason to think it the *right* fundamental notion. Kreisel suggests that a systematic error may underly the diverse formulations which turn out to be equivalent.

The argument has this in common with the three arguments we have just been discussing: the underlying idea in all four of these arguments is that no maneuver of a certain class suffices to lead outside the class of general recursive functions while at the same time preserving effectiveness. The same objection offered against the earlier arguments would seem equally appropriate here: one can ask whether the class of maneuvers considered contains all the effective maneuvers. The evidence that it does seems to be just the absence of a counter-example.

3. 'A PRIORI' CONSIDERATIONS

Church and Turing have each offered arguments for **CT** which are based on *a priori* considerations. I shall treat Turing's argument in some detail and then I will indicate how Church's arguments succumb to the objection (offered earlier) to the 'heuristic' arguments.

3.1. The strategy of Turing's argument[7] is to analyze the respects in which his formal notion might be too restrictive, into several categories and then to argue that no essentially new functions would be computable by Turing machines modified so as to be unrestricted in the respects of each category. The categories are: the complexity of observed symbols, the complexity of the atomic acts which constitute a computation, the topological properties of the symbol space, and the character of the symbolic representation of the arguments and values of the functions computed.

While the analysis of the intuitive notion of effective computation into categories seems innocent enough, the use of these categories as distinct from one another in the argument for **CT** is objectionable. The argument proceeds by treating possible improvements in respect of each category *ceteris paribus* and giving reasons why no such improvements could be made. This procedure, in which the categories are treated as completely separable from each other, begs some important questions. It should be clear that if a machine were endowed with a structurally more complex symbol space than the Turing machine tape, and if we entertain seriously the possibility that the machine so endowed may be more powerful than the Turing machine, we cannot assume that the atomic acts of which the machine is capable will be no more complex than Turing machine steps. To make use of its more complex symbol space, our hypothetical machine will have to be able to move about in it. Similarly, more complex symbol-observation might entail a more sophisticated repertoire of atomic movements. While I think this objection is fatal to Turing's argument, even if the separability of the categories be granted, there are other defects in the treatment of them provided by Turing's argument.

In regard to the complexity of observed symbols, it is argued that increasing the complexity will not broaden the class of computations, since if any finite number of distinguishable atomic symbols are allowed as observable, what we have is still a Turing machine. Moreover, if we allow compound arrays of symbols as observable (in one atomic observation) there is still no gain, since compound arrays can be observed by a Turing machine by successively observing the component symbols, and storing the information acquired thereby.

The argument presupposes that compound symbols are formed according to some uniform manner from atomic symbols, and that the uniform manner is Turing performable. The claim that no compound symbol (where that term is understood to denote an expression which can be effectively observed) fails to be observable (by a succession of observations and storing operations) by a Turing machine, is equivalent to **CT**.

As to the complexity of machine operations, it is argued that because there must be some finite bound to the complexity of admissible acts, any act more complex than a Turing machine act must be reducible to a succession of Turing machine operations. This argument presupposes that the symbols are Turing observable, and that the symbol space can be negotiated by a Turing machine. Neither presupposition seems warranted except by appeal to **CT**.

The argument concerning the character of the symbol space is the most involved of the discussions of the four categories. The errors which occur in this argument are more numerous and more interesting than those already considered. Unfortunately, in this brief address I shall only be able to offer a rather general objection and to indicate cursorily some specific errors.

The argument is intended to show that the essential 1-dimensionality of the symbol space of the Turing machine is no more a limitation than that imposed by any symbol space suitable for effective computation. The 'suitability' requirement is stated picturesquely, if a little vaguely, by Kleene: "The symbol space must be sufficiently regular in structure so that the computer will not become lost in it during the computation."[8] The strategy of the argument is to argue that any suitable symbol space S can be mapped into the Turing machine tape in such a way that the Turing machine can mimic a machine M based on S.

It is asserted that the suitability requirement entails there being, for each 'cell' of the symbol space, some finite number of ways of moving from that cell. It is then claimed that one can, without loss of generality, assume that the limit is the same for all cells![9] One has only to reflect on a tree-like symbol space, where at each node (cell) at the n-th rank there are $n+2$ branches, to see that generality is indeed lost by this assumption. Nevertheless, let us grant the assumption, calling the limit n.

We are then told, plausibly enough, if one grants the assumption, that a space S can be enumerated without repetitions so that: for every way of moving M_i, $(i=0,...,n)$ from a given cell, there is a computable function μ_i such that, if x is the index (in the enumeration) of some cell, then $\mu_i(x)$ is the index of the cell reached by the motion M_i. I think that this assumption may also be granted, though it is important to keep in mind that all we grant is that the μ_i are *computable* (say by M), not that they are recursive. S is then mapped into the Turing machine tape by picking a square of the tape as a distinguished square, which we may call the 0-th square; then the j-th cell in the enumeration of S corresponds to the j-th square (on the tape) to the right of the 0-th square.

It is then claimed, without further argument, that μ_i must be Turing computable, and that hence M can be mimicked by a Turing machine. Certainly if μ_i were Turing computable, then we would have to agree, so must be whatever function M computes. But what reason, other than **CT**, is there to believe μ_i to be Turing computable? As I pointed out earlier, it has been tacitly assumed that M can (somehow) differ from a Turing machine only in respect of its symbol space. But of course if

M can really use its structurally more complex symbol space, it will have to differ from a Turing machine also at least in respect of the kinds of atomic acts it can perform.

I think the argument in regard to the notational system used to represent arguments and function values is straightforward and correct.

3.2. Church has two arguments for his Thesis.[10] The first argument has to do with formal theories. We may think of effectively computable functions as being representable in a formal theory. But, Church tells us, formal theories are recursively enumerable. Functions representable in r.e. theories are recursive.

The second argument has to do with "symbolic algorithms". A symbolic algorithm for the computation of $\varphi(x)$ is a sequence of expressions, the length of which is a function l of x, such that the first expression can be found effectively from x, and the $n + 1$st expression, if there is one, can be found effectively from x and the first n expressions. Further, it must be possible to effectively determine when the $l(x)$-th expression has been reached, and when it has, to read from it the value of $\varphi(x)$. Now, if it be granted that the functions and predicate called 'effective' in this description of an algorithm turn out to be recursive (under a Gödel-numbering), then φ is recursive.

These two arguments may be straightforwardly shown to be equivalent.[11] The same objection applies to both forms of the argument: What reason, other than **CT**, is there for supposing the 'effective' functions involved in the arguments to be recursive? That is, after all, the point at issue.

4. Concluding Remarks

The arguments considered in this paper are the principle arguments offered in support of **CT**. I think that I have shown that the arguments are not, under close examination, very convincing.

One might wonder how **CT** came to be so widely accepted, if the evidence which supports it is truly so scarce or of such low quality. Perhaps the widespread acceptance of **CT** is due to no more than its initial plausibility leading to its virtually unanimous early approval by those few members of the mathematical community who were conversant with recursive function theory at the time of its inception. Universal acceptance of an unfounded (or poorly founded) doctrine is no stranger to the community of logicians; consider, e.g. the universal acceptance, for over two millennia, of Aristotle's syllogistic logic for representing all possible arguments.

University of North Carolina at Charlotte

NOTES

* This paper is adapted from Chapter IV of my Ph.D. Dissertation *Church's Thesis and Philosophy*. Special thanks are due to my advisors, Professor Raymond Nelson and Professor Howard Stein.

[1] G. Kreisel, 'Mathematical Logic' in *Lectures on Modern Mathematics* (ed. by T. L. Saaty), Wiley, New York, 1965, vol. III, p. 147.

[2] Hao Wang, *Survey of Mathematical Logic*, Science Press, Peking 1964, p. 88.

[3] S. C. Kleene, *Introduction to Metamathematics*, Van Nostrand, Princeton, 1952, p. 348. Cf. S. C. Kleene, *Mathematical Logic*, Wiley, New York, 1967, pp. 240–1 n. 171 and Kleene (1952), p. 352.

[4] Kleene (1952), pp. 319–20.

[5] Kleene (1952), p. 320.

[6] Kreisel *op. cit.*, p. 144.

[7] A. M. Turing, 'On Computable Numbers, with an Application to the Entscheidungsproblem', reprinted in *The Undecidable* (ed. by M. Davis), Raven Press, Hewlett, New York, 1965, pp. 135ff. Cf. Kleene (1952), pp. 377ff.

[8] Kleene (1952), pp. 379–80.

[9] Kleene (1952), p. 380.

[10] Alonzo Church, 'An Unsolvable Problem of Elementary Number Theory', reprinted in *The Undecidable* (ed. by M. Davis), pp. 101–102. Cf. Kleene (1952), pp. 322–3.

[11] Kleene (1952), p. 323, does essentially this when he says that we may regard the ordered pair (x, the first expression in the sequence) as an axiom of a first order theory.

PART II

PROBABILITY
(Section VI)

MARIANA BELIŞ

ON THE CAUSAL STRUCTURE OF RANDOM PROCESSES

1. Introduction

In order to cope with an unknown environment, living systems had to discover the properties of the different objects and phenomena and the relations between them. Among the different types of relations already known, the causal is one of the most important, as it makes possible to find the origin of a phenomenon, to predict a future one or even to reproduce it.

The main features of the causal connection are the following.

In a causal connection, the cause generates the effect; accordingly, the cause precedes the effect in time.

The transformation of the cause into the effect implies quantitative as well as qualitative changes of the phenomenon.

A causal process takes place in certain *conditions* which can be favorable or not to the process. The conditions differ from the causes as they do not generate the effect but only influence its occurrence. Some conditions influence the process at its beginning (initial conditions) and some others during its development.

The causal connections form a chain or take place in series; that is, the effect which results in a given causal connection can be a cause in another causal connection, and so on.

The causal connections take place also in parallel, various causal connections occurring simultaneously.

In general, a causal connection represents a multiple connection between a group of causes and a group of effects (the one-to-one connection is a particular case). The various causes which work together in a given process form a 'causal structure'.

Several types of causal connections are known.

1.1. *Determinate Causal Connections*

In this case, for a given causal structure, the effect is always the same.

Knowing the causal structure, the effect is predictable with certainty and the connection cause-effect is encoded by a dynamical law.

1.2. *Random Causal Connections*

In this case, for a given causal structure, the effect is variable, depending on some phenomena which cannot be noticed by a human observer. Among these random processes, a certain class shows a regularity of the issues in a long run of experiments. They were called *random processes with statistical regularities* to distinguish them from the rest of the random processes.

A tremendous amount of experimental and theoretical work was devoted to clarify the nature of the random processes, and to find a formalism able to encode this type of process.

The subjective and the objective aspect of random phenomena have been analysed: is randomness an objective property of some natural phenomena or is it only our lack of knowledge which makes them imprevisible? "Chance is only the measure of our ignorance", wrote Poincaré (1896) in his *Calculus of Probabilities*. "Fortuitous phenomena are, by definition, those of the laws of which we are ignorant". But he added: "Is this definition very satisfactory? When the first Chaldean shepherds followed with their eyes the movement of the stars, they didn't yet know the laws of astronomy, but would they have dreamed of saying that the stars move by chance? If a modern physicist is studying a new phenomenon and if he discovers its law on Tuesday, would he have said on Monday that the phenomenon was fortuitous?"

Another difficulty is the 'a priori' estimation of the numerical value of the chance of occurrence of random events, based on the available information upon the experiment.

The probability of 'isolated' events is another tricky problem. The method of 'the bet' (E. Borel, 1939; O. Onicescu, 1940) implies also an information-processing mechanism for the evaluation of the chances. What is this mechanism?

Even in the case of repeatable events the connection between the mathematical theory of probability and the phenomena of the real world is not well established. "What are the features of the real world and the objective properties of the phenomena which lead to the values of the probability given by the mathematical theory"? asked A. I. Khintchine (1952). What is the true connection between the classical definition of probability and the frequencies of the event in the long run? "Bernoulli's theorem exhibits algebraic rather than logical insight" wrote J. M. Keynes in his *Treatise on Probability* (1921).

The development of decision theory in our days and the necessity of building technical decision systems give a new impulse to look through the problem of the nature of random processes.

2. Models of causal processes

A causal process can be represented by a communication system S, which maps a set of input signals (causes) into a set of output signals (effects).

An input signal $c(t)$ is a k-tuple vector whose components $c_1(t)$, $c_2(t), ..., c_k(t)$ are time variable. The output signal is a m-tuple vector $e(t) = e_1(t), ..., e_m(t)$. The range of $c(t)$ is the input space $R[c(t)]$. By varying $c(t)$ over the input space, the output vector $e(t)$ varies over the output space, according to the relations:

$$S^1(c_1, ... c_k, e_1, ... e_m) = 0$$
$$\cdots\cdots\cdots$$
$$\cdots\cdots\cdots \quad (1)$$
$$S^m(c_1, ... c_k, e_1, ... e_m) = 0$$

or, more compactly,

$$S(c, e) = 0 \quad (2)$$

which represents the input-output relation of the system, that is a set of input-output pairs.

From the point of view of our model this represents the set of cause-effect pairs which are connected by means of the process. It follows that to a given cause correspond many effects depending on the initial conditions of the process. These initial conditions represent the state of the system at the moment (t_0), that is a n-tuple vector $s(t_0)$ ranging over the state space $\sum$ of the system. Given the input vector c and the state of the system $s(t_0)$ the output is uniquely determined according to the function:

$$e = S(c, s(t_0)) \quad (3)$$

which represents the input-output-state relation of the system or the dynamical law of the process.

The input signals as well as the state of the system are time variable. From the point of view of an external observer these variations can be measurable or not; this leads to two distinct models of the transmission system.

If the variations are measurable they are encoded by a dynamic law. In this case the transmission system is noiseless (Figure 1).

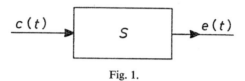

Fig. 1.

If the variations are not measurable because of their order of magnitude or their time constant, they are included in a source of noise $n(t)$. In this case the transmission system is a noisy one (Figure 2).

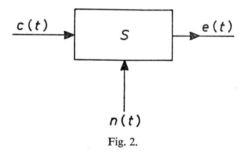

Fig. 2.

The source of noise also includes irrelevant signals belonging to other transmission channels which work simultaneously (in parallel). These signals, measurable or not, superimpose on the useful signal to be transmitted (the cause) or on the actual state of the system (the conditions).

In fact the noisy transmission system is the most general one because of the inability of the observer to measure all the variable and irrelevant signals which occur during a transmission. But, as we shall see later, some transmissions are practically unaffected by the noise which can not alter the input-output connection of the system in a given state.

3. The Stability of Various Causal Connections

As we said before, the presence of the noise affects differently the various causal connections.

Some of them are unaffected by noise; that is, the input-output-state relation maintains itself, in spite of the variations mentioned above.

Some others are more or less affected by the noise which can change to a large extent the state of the system and therefore the input-output dependency.

The various causal connections are then more or less *stable* with respect to the noise.

According to the general principles of system's stability, a system is stable if:

$$(\forall t)\,(\forall \varepsilon)\,(\exists \delta(\varepsilon)\,(\|s_0\| < \delta(\varepsilon) \Rightarrow (\|s(s_0, t_0; t)\| < \varepsilon))$$
$$t > 0, \varepsilon > 0 \qquad (4)$$

where s_0 is the initial state and $s(s_0, t_0; t)$ is the final state.

On the contrary, a system is unstable if there are no variations of the initial state (even very small) for which the final state could be maintained inside a sphere of radius ε.

Causal connections fall into one of these categories as the dependency cause-effect is affected or not by small variations of the initial conditions or by irrelevant phenomena.

A causal process is *stable* if, for a given causal structure, the effect is the same or has little variations, even if irrelevant phenomena occur during the process or if the initial conditions have variations. From the point of view of our model, this causal connection is represented by a noiseless system (the noise can be present but it doesn't affect the transmission).

A causal process is *unstable* if infinitesimal variations of the initial conditions or the occurrence of irrelevant phenomena can change the state of the system and therefore the effect to a large extent. As these infinitesimal variations are generally unmeasurable by an external observer, the change of the effect in the presence of the same causal structure appears to be due to 'chance' and the process is called a 'random' one.

The inability of the observer to measure noise phenomena led to the opinion that the imprevisible character of the issues of a random process is only a result of our ignorance. In fact our ignorance, or better our difficulty to measure noise phenomena with the device we have, appears just because the process is unstable. The noise is also present in the case of some stable processes, but it does not matter.

The instability of some causal connections represents the objective feature of random processes. As Poincaré has intuitively noticed, the movement of the stars is not a random phenomenon because it is stable and unaffected by noise whereas the tossing of a coin is a random one because the effect strongly depends upon infinitesimal variations of the initial conditions.

Unstable causal connections, having a group of constant (permanent) causes and conditions, show in the long run a stability of the issues. They form the class of 'random processes with statistical regularities'.

The stability of the different systems was studied especially in the technical area and more recently, due to cybernetics, this concept was extended to biological systems. From the above considerations the concepts of stability and instability appear to be much more general, extending to all phenomena of the real world. It is possible that a deeper analysis of this property shows that causal connections are not only of two kinds (stable and unstable) but they have various degrees of instability getting to the limit case of full stability.

4. Encoding the Regularities of Random Processes

According to the relation (3), the output of a causal transmission system (that is, the effect) is uniquely determined by the input (the cause) and the state of the system (the conditions). Any changes of the input or state of the system involve variations of the output. These variations are more or less important, according to the degree of stability of the system.

In general, an unstable process involves many issues, mutually exclusive, whose occurrence depends on small, unmeasurable phenomena which mix up with stable and permanent causes. If the process involves a unique issue, its instability leads to two alternatives: the existence or the absence of this issue.

The particular issue of an unstable process cannot be predicted with certainty by an external observer because: (1) of his inability to measure noise phenomena, and (2) of the great influence of noise phenomena on the issue.

However, it is to be expected that "... in a long run of experiments, the action of regular and constant causes prevails over those of irregular causes...", as Laplace wrote three centuries ago. This means that in a long series of experiments the 'noise components' will cancel each other and every issue will occur proportionally to the stable causes which determine it in the whole causal structure of the process. By repeating the experiment and counting the number of times each issue has occurred, we have then a measure of its 'chance of occurrence', that is, its probability. This experimental or 'frequency' method has some shortcomings. First, it is not always possible to repeat the experiment in exactly the same conditions; besides, it is an 'a posteriori' method, giving information about the probabilities of the issues only after having performed the experiment.

Nowadays, decision problems raise the necessity of 'a priori' methods for estimating the chances of occurrence of random events, without doing the experiment. This naturally leads to the analysis of the causal structure of a random process in order to detect the stable causes which determine each issue, giving its 'chance of occurrence'.

Both methods will be presented next from the point of view of our communication causal model.

4.1. *The 'A Posteriori' Method: Summing up the Issues*

Let us consider the noisy communication system (Figure 2) provided with input signals. These signals originate from the causal structure of the

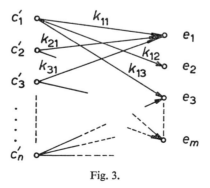

Fig. 3.

experiment to be performed, and they have to be transmitted through the channel in order to produce the output signals (the effects). More exactly, if the stable structure of an experiment comprises $c'_1, c'_2, \ldots c'_n$ causes and conditions (Figure 3), then each effect from the set $e_1, e_2, \ldots e_m$ of mutually exclusive effects is connected to a subset of the input elements. We introduce the coefficients k_{ij} to indicate the weight of these connections, that is, the measure in which the cause or the condition c'_i influences the effect e_j.

The stable causal determination of an effect can be estimated by the ratio of the sum of the weights of favourable causes and conditions to the sum of all the weights of the given causal structure. For the effect e_j, for example, this ratio is:

$$P_j = \frac{\sum_i k_{ij}}{\sum_{i,j} k_{ij}} \qquad (5)$$

$$i = 1, \ldots, n$$
$$j = 1, \ldots, m.$$

The ratio P_j represents *the possibility of occurrence* of the effect e_j. It represents in our model *the amplitude* of the input signal to be transmitted.

A random experiment is then modelled by a noisy transmission channel whose input signals have the amplitudes $u_1 = P_1, u_2 = P_2, \ldots, u_m = P_m$, and whose output signals must be proportional to these amplitudes if the transmission is correct.

A method widely used in technical transmission systems to extract the signal from noise is the 'accumulation' method. It consists in repeating many times the transmission of the same signal and summing up the output. In the long run, the noise components cancel each other and the useful signal reinforces progressively. It is proved that if the transmission is repeated N times, the signal to noise ratio at the output is $\sqrt{N}$ times better than at the input.

The 'frequency method' is actually based on this principle. A group of stable causes and conditions are assembled, the experiment is performed, and the result is recorded. Each issue must occur proportionally to its causal possibility, which represents the input signal. But this signal is

ON THE CAUSAL STRUCTURE OF RANDOM PROCESSES 73

embedded in noise, and many experiments have to be performed in order to extract the information from noise at the output.

By considering the input signals given by (5), after N transmissions (N very great), the amplitudes of the output signals will be proportional to those of the input:

$$u'_1 = NP_1, u'_2 = NP_2, \ldots u'_j = NP_j, \ldots u'_m = NP_m$$

(to simplify, we have considered a transfer factor equal to one).

But, in the case of our causal transmission, the output signals (the issues of the experiment) can only take the values 1 or 0 (an event does exist or not). The different 'amplitudes' of the output signals are then represented by the different number of times each issue has occurred, that is for the effect e_j:

$$u'_j = \sum_{k=1}^{N} \alpha_{jk} \tag{7}$$

$\alpha_{jk} = 1$ (e_j occurs)
$\alpha_{jk} = 0$ (e_j does not occur).

For N very great, if the event e_j has occurred n_j times, the amplitude of the respective output signal is, taking into account (6) and (7):

$$u'_j = NP_j = n_j \tag{8}$$

that is, the number of times an issue has occurred (n_j), is equal to the accumulated output signal of the causal transmission channel. From (8) we get:

$$\frac{n_j}{N} = P_j \tag{9}$$

that is, the relative frequency of occurrence of an issue is equal to its causal possibility P defined in (5). This number represents the chance of occurence of the issue, that is, its probability.

4.2. *The 'A Priori' Method: The Analysis of the Causal Structure*

Instead of placing the observer at the output end of the transmission channel in order to watch the output signals (the issues of the experiment), the 'a priori' method places him at the input end in order to analyse the causal structure of the experiment. If the observer is able to evaluate the possibilities of occurrence of the different issues (5), he knows their chances of occurrence, that is, their probability. In other words, from the point of view of our model, if the amplitudes of the signals to be transmitted are known, it is no more necessary to extract them from noise by an experimental method.

In the particular case when the possibilities of occurrence of the effects are equal, $P_1 = P_2 = ... = P_m$, they occur an equal number of times, $n_1 = n_2 = ... = n_m = N/m$, where N is the total number of experiments, and m is the total number of issues. We have obviously, $n_1 + n_2 + ... n_m = N$. By using (9) we get:

$$P_j = \frac{n_j}{N} = \frac{N/m}{N} = \frac{1}{m} = P_1 = P_2 = ... P_m. \tag{10}$$

If a subset of q issues represents the same event, then its possibility of occurrence is q times greater, that is q/m.

We see that in these particular cases, the possibilities of occurrence of the different issues depend only of the number of issues, m (or m and q), of the experiment.

This represents the 'classical' or 'direct' method of evaluating the probabilities. It is also an 'a priori' method but only for the particular case of equal possible events. Concerning the general case, Laplace has written in his second "general principle". "... But this supposes equally possible events. If they are not, we have to determine first their possibilities of occurrence whose evaluation is one of the most tricky problems of the theory of probabilities."

The analysis of the causal structure of an experiment enables one to evaluate the possibilities of occurrence of the different issues, according to formula (5). According to this formula, the observer has to know the stable causes and conditions which determine a certain issue and their degree of influence, represented by the coefficients k. The correctness of this evaluation depends upon his knowledge about the phenomenon, his past experience or the analogies he can establish with other similar phenomena. In this sense it is a subjective measure because it depends upon the personal information of the observer. Obviously it does not depend upon his personal wishes or feelings. As E. T. Jaynes (1968) wrote: "... the most elementary requirement of consistency demands that two persons with the same relevant prior information should assign the same prior probabilities".

In practice, the information-processing mechanism used by men to evaluate probabilities is influenced by their feelings; moreover, the information they have about the real world is seldom complete. That is why subjective probabilities generally differs from the objective ones. A pure information-processing system using a correct algorithm must be able to generate a subjective probability which approaches the objective one in the measure in which the information it gets about the phenomenon is greater.

The evaluation of the possibilities of occurrence according to formula (5) enables one to calculate the prior probability even in the case of experiments with a single issue. The method holds even in the case of "non-testable" information, in the sense used by E. T. Jaynes.

In order to calculate the possibilities of occurrence according to formula (5), we have to estimate the values of the coefficients k_{ij} in the

TABLE I
Evaluation of prior probabilities

Causes and conditions	Effects e_1	e_2	e_j	e_m
c'_1	k_{11}	k_{12}	....	k_{1m}
c'_2	k_{21}	k_{22}	....	k_{2m}
c'_i	....	....	k_{ij}	....
c'_n	k_{n1}	k_{n2}	....	k_{nm}

Table I. For simplicity, the numerical values of these coefficients can be normalized between 0 and 1.

Depending upon the knowledge the observer has about the process, the values of these coefficients will reflect more or less the real influence cause-effect, that is, the subjective probability will approach more or less the objective one.

In the following we shall take three examples of evaluating prior probabilities by considering the causal structure of the phenomenon: the tossing up of a coin (Table II), the weather forecast (Table III) and the issue of a tennis game (Table IV).

TABLE II

The tossing up of a coin

Causes and conditions \ Effects	e_1 (head)	e_2 (tail)
c'_1 (the coin is homogenous)	0.5	0.5
c'_2 (the coin is symmetric)	0.5	0.5
c'_3 (its surfaces are plane)	0.5	0.5
c'_4 (the tossing mechanism is invariable)	0.5	0.5

TABLE III

Weather forecast

c' \ e	e_1 (sun)	e_2 (rain)	e_3 (snow)
c'_1 (dark clouds at the east line of the horizon)	0.2	0.7	0.1
c'_2 (wind from the east)	0.1	0.8	0.1
c'_3 (high temperature)	0.5	0.5	0
c'_4 (we are in August)	0.7	0.3	0

TABLE IV

The tennis game

c \ e	e_1 (gambler A wins)	e_2 (gambler B wins)
c'_1 (technical knowledge)	0,8	0,2
c'_2 (age of the gamblers)	0,4	0,6
c'_3 (place of the game)	0,7	0,3
c'_4 (the weather)	0,5	0,5

From Table II we get: $P_1 = 2/4 = 0.5 = P_2$.
From Table III: $P_1 = 1.5/4 = 0.375$; $P_2 = 2.3/4 = 0.575$; $P_3 = 0.2/4 = 0.05$.
From Table IV: $P_1 = 2.4/4 = 0.6$; $P_2 = 1.6/4 = 0.4$.

Polytechnic Institute,
Department of Electronics,
Bucharest

BIBLIOGRAPHY

Borel, E.: *Valeur pratique et philosophique des probabilités*, Gauthier Villars, Paris 1939.
Jaynes, E. T.: 'Prior probabilities', *IEEE Trans. on SSC* **4**, nr. 3, 1968.
Keynes, J. M.: *A Treatise on Probability*, Mcmillan and Co., London, 1921.
Khintchine, A. I.: *Metod Proizvolnih functi i borba protiv idealizma v teori veroiatnostei*, Academia Nauka – SSSR – Moscow, 1952.
Laplace, P. S.: *Essai philosophique sur les probabilités*, Gauthier Villars, Paris, 1921.
Onicescu, O.: 'La probabilité d'un évènement isolé', *Bull. Sect. Sci. Acad. Roum.* **22**, nr. 6, 1940.
Poincaré, H.: *Calcul des Probabilités*, Paris, 1896.

L. JONATHAN COHEN

THE PARADOX OF ANOMALY

In the present paper I shall discuss one of the less obvious criteria of adequacy that any satisfactory analysis of confirmation must meet, viz. the requirement that if such an analysis purports to explicate the concept of confirmation or support which is in common use in natural science it must be capable of explicating or elucidating the concept of what scientists call an 'anomaly'.

By calling a fact an 'anomaly' what is meant, roughly, is that, though the fact conflicts with a certain theory, the theory's acceptability is unaffected because it is well substantiated by other evidence. Moreover, even where no such expression as 'anomaly' is actually used, the concept of anomaly is often implicitly present in the minds of scientists. For example, Newton's theory of gravitation was thought at first, even by Newton himself, to give a markedly incorrect value for the forward movement of the apse of the moon. It was not till 1752 that Clairaut showed how the theory could be made to produce results that agreed with the observed movements. Yet no serious thinker in the meantime either qualified or rejected Newton's theory because of its apparent failure to accord with known fact.[1] Similarly no serious thinker rejected Newton's theory because it failed to explain the movement of the perihelion of Mercury, although this movement was in the end explained only by Einstein's general relativity theory.

Further examples are unnecessary: accounts of them are easily found in the recent literature[2] of the philosophy of science. Nor can such anomalies be written off as mere errors of observation. In every case the relevant observations are thought to be repeatable and to deserve being taken seriously. Sometimes scientists seem just to suppose that the anomaly is only an apparent one: the anomalous fact must somehow be reconcilable with the theory, though nobody knows how to achieve this at the time. Sometimes instead they seem to suppose that the awkward fact is the result of some as yet unknown law which restricts the operation of the laws asserted by the theory. Sometimes perhaps scientists are just reluctant to give up one theory unless a better one is available. But in every case three features characterise the situation:

(i) A statement A (describing the anomaly) is thought to be observably true;

(ii) A apparently conflicts with a theory T that is thought to be otherwise satisfactory; and

(iii) The counter-evidence to T that is described by A is thought to be quite unimportant in relation to the merits of T.

So an analysis of scientific reasoning which is to explicate the concept of anomaly must apparently be capable of allowing a high level of support to some hypotheses even on the basis of observable evidence that contradicts them. Indeed, this level of support must apparently be high enough to justify accepting the hypothesis if a better one is not available.

Clearly any philosopher, who holds that a falsified hypothesis must *ipso facto* always be rejected, cannot explicate the concept of anomaly. If Popper has held this view, as seems likely,[3] his account of scientific reasoning is defective in at least that respect. But so too is any account that treats evidential support for a hypothesis and its consequences as a logical probability. I.e., if we suppose evidential support to be assessed by a dyadic function $s[H,E]$, taking names of sentences as fillers of its argument-places, then that function does not share the same logical syntax as a dyadic function $p(X, Y)$ satisfying the axioms of the mathematical calculus of probabilities. For we have the following situation. The available evidence is describable in a self-consistent conjunctive sentence $A \& B$, where A reports the anomalous fact, B reports the rest of the evidence, T states a scientific theory, and A logically implies the falsehood of T. Since A implies the falsehood of T, so does $A \& B$. Hence, if $s[H, E]$ is a logical probability, we must have $s[T, A \& B] = 0$. But we are required to have $s[T, A \& B] > 0$, since T is to have a high level of support on the available evidence.[4] Therefore $s[H, E]$ is not a logical probability.

For philosophers attached to probabilistic versions of inductive logic this result will constitute a paradox. Perhaps therefore they will object to the result in one or other of the following four ways.

(1) They might object that a probabilistic inductive logic can still represent the situation: it can point to the fact that the situation is just one where we have $s[T, A] = 0$ and $s[T, B] > 0$. I.e. they might claim

that all we need to represent is that the theory has zero probability on the evidence of the anomalous fact and a higher probability on the rest of the evidence. But this does not resolve the paradox, because it does not represent the peculiarity of the situation. For any contingent H, there will always be pairs of sentence, E_1 and E_2 such that $s[H, E_1] = 0$ and $s[H, E_2] > 0$. What makes the anomaly situation peculiar, in those terms, is that we also need to have $s[H, E_1 \& E_2] > 0$, since H is supposed to have a level of support, on the actual evidence, that justifies accepting it, while $E_1 \& E_2$ describes all the actual evidence.

(2) It might be objected instead that what needs splitting up and considering in separate parts is not the evidential sentence but rather the hypothesis. The paradox arises, on this view, because the fillers of the first argument-place in a support-function are thought to include generalisations, like Newton's laws of motion. Such a generalisation is exposed to contradiction by a description of the anomaly, and the same hypothesis-sentence seems then capable of having both zero, and also greater-than-zero, support. But there are in any case well-known difficulties about assigning realistic support-values to universal sentences. And, if instead the only fillers of the first argument-place are singular sentences, then what we have to deal with here is not a single pair of sentences, viz. an evidential sentence $A \& B$ and a generalisation T, but rather a very large and possibly infinite number of pairs of sentences, where each pair consists of the same evidential sentence $A \& B$ and one or other of the substitution-instances, $S_1, S_2, \ldots$, of T. It can thus be supposed that, while very many of these substitution-instances are well supported by $A \& B$, a few of them – conflicting with the anomalous fact or facts – may have zero support. But again the objection misses an essential feature of the anomaly situation. If there is to be an anomaly at all, even the substitution-instance that conflicts with it must have greater than zero support on the available evidence. What made lunar movements appear anomalous in relation to Newton's theory of gravity was that the theory was taken to generate *well-supported* predictions which were observably false. If Newton had merely come up with a range of logically independent predictions of which a lot were well-supported and a few were ill-supported, that would not in itself have seemed an oddity that demanded further explanation. (This is why Carnap's theory of instance-confirmation cannot explicate anomaly.)

(3) It might be objected that inductive logic should represent the ideal situation in scientific reasoning rather than the actual one, and that ideally there would be no anomalies. But, while this is certainly a self-consistent point of view, it runs counter to the fact that suitable anomalies are not only acknowledged *de facto* from time to time in the actual history of science but are also normally regarded as being *de jure* acceptable in any assessment of the current state of enquiry in a particular field. An inductive logic that resolved the paradox of anomaly in this way would be incapable of explicating the support-assessments that are commonly invoked in day-to-day decisions about which theory to accept and which to reject at a particular time. The concept of anomaly plays a vital part in the assessments that are to be explicated. It is not a head under which certain assessments may be regarded as degenerate cases that are unworthy of explication.

(4) It might be objected that the concept of a scientific anomaly is itself to be regarded as a (higher-order) anomaly for inductive logic. On this view, anomalies are too unimportant a feature of science to need taking into account if their existence conflicts with an inductive logic that is otherwise acceptable. But, first, it is a highly controversial issue whether or not a probabilistic inductive logic is indeed otherwise acceptable. And, secondly, the concept of anomaly cannot be left out of account since the history of science reveals it to be altogether too integral a feature of the ongoing dialectic of scientific enquiry.

I conclude therefore that the paradox of anomaly cannot be resolved unless one rejects the thesis that evidential support in natural science is measurable by a logical probability. Nor can resort to statistical probability resolve the paradox. When occasionally a man lives to the age of 100, we describe such an event as an improbable one, not as an anomaly: it falsifies no theory. But the concept of anomaly is accommodated perfectly well by the type of inductive logic that I have developed elsewhere (in *The Implications of Induction*, 1970) as a sophisticated version of the Bacon-Herschel-Mill tradition. Think of a relevant variable, for hypotheses of a certain type, as a set of situational characteristics, variation from one to another of which falsifies some hypotheses of this type. Think of the set of known relevant variables, for hypotheses of a certain type, as being ordered by the numbers of such hypotheses that these variables have falsified and by alphabetical order otherwise. Then you can conceive of a hierarchy of tests for hypotheses of a given type, in which more and more of the relevant variables are manipulated, in all their possible combinations with one another, as the tests get more and more thorough. In the less thorough tests only the more important variables will be manipulated, but in the most thorough tests even the less important variables will be. The evidential support that exists for a hypothesis can then be graded ordinally by reference to the thoroughness of the test it has passed or failed to pass. So, when E reports both the result of a test t that H has passed, and also of a test t_{i+1} that H has failed, you would have $s[H, E] = i$, where $i > 0$ though E contradicts H. That is, H can have higher than zero-grade support even on evidence that falsifies H. Now, the various kinds of phenomena that a theory in a certain field of scientific enquiry should explain constitute the relevant variables for it.[5] The well-known kinds of phenomena that have frequently defeated past efforts at explanation will be the most important variables, and any kinds of phenomena that are regarded as anomalous will be relatively unimportant variables. Hence a scientific theory will be able to pass even fairly thorough tests and have quite a high level of evidential support, and yet still be in conflict with those parts of the evidence that are taken to constitute anomalies for it. The paradox of anomaly disappears.

The Queen's College, Oxford

NOTES

[1] See the remarks by Florian Cajori in the appendix to his edition of Newton's *Mathematical Principles*, 1962, p. 648ff.

[2] E.g. R. G. Swinburne, 'Falsifiability of Scientific Theories', *Mind* **73** (1964) p. 434ff; and I. Lakatos, 'Falsification and the Methodology of Scientific Research Programmes', in *Criticism and the Growth of Knowledge* (ed. by I. Lakatos and A. Musgrave), Cambridge University Press, 1970, p. 138ff.

[3] Cf. I. Lakatos, 'Popper on Demarcation and Induction', Note 17, forthcoming in *The Philosophy of Sir Karl Popper* (ed. by P. Schilpp).

[4] I assume here, for the sake of simplicity, that a statement of the anomalous evidence directly contradicts some logical consequence of a single scientific theory. In practice the situation is sometimes rather more complicated. A contradiction may only be derivable if certain other well-supported laws (often called 'background information') are also invoked as premises. But the problem is just the same. T is now a conjunction of the theory with other generalizations but it still enjoys a high level of support from the available evidence.

[5] For a more detailed account of this inductive logic see my *The Implications of Induction* (1970) §§5–13; and also my 'The Inductive Logic of Progressive Problem-shifts', *Revue Internationale de Philosophie* **95/96** (1971) 62–77.

N. K. KOSSOVSKY

SOME PROBLEMS IN THE CONSTRUCTIVE PROBABILITY THEORY

1. INTRODUCTION

In this paper we shall study normed decidable Boolean algebras for the purpose of constructing the initial parts of a constructive theory of probability.

A series of authors, in particular, Glivenko (1939), Kolmogorov (1948) and Markov (1949) developed an approach to the construction of a theory of probability, in which the elements of a Boolean algebra are called events and probability is a measure on the elements of the Boolean algebra (normalized so that the whole space has measure[1]). It is difficult to indicate precisely when this idea was made clear; some authors connect its origin with the name of George Boole.

There is still another approach to the construction of the foundations of a theory of probability which is essentially different from those already mentioned and which is being fruitfully developed at the present time: we are talking about papers of Kolmogorov (1965), Martin-Löf (1966), and other authors.

In Section 2 of this paper we shall introduce the concepts of decidable Boolean algebras, and probability spaces (i.e. normed decidable Boolean algebras with the norm of the unit element equal to 1). A normed decidable Boolean algebra β is turned into a metric space, and the points of its constructive completion are called events over β (as a metric one takes the norm of the symmetric difference of elements of β). It turns out that many properties of Boolean operations established in classical mathematics are also provable in constructive mathematics. The situation is different with respect to unions and intersections of sequences of events over β.

We shall give a necessary and sufficient condition that there exists in a probability space a monotonic sequence of events not possessing an intersection or union of all its terms. This condition allows us to assert that in interesting probability spaces it is usually possible to have monotonic sequences of events not having an intersection or union of all its terms.

In Section 3 we shall introduce and study the concept of a graduated frame over a probability space and the concept of an integrable *FR*-construct over a probability space (the latter concept is a generalization to the case of probability spaces of the constructive analogue, introduced in Šanin (1962), of the concept of a Lebesgue-integrable function). We shall define and study certain operations over integrable *FR*-constructs.

Since the whole investigation is carried out for a probability space which is not assumed to have a 'pointwise structure', the definition of analogues of Lebesgue sets of integrable *FR*-constructs is the main point of the following constructions. The definition of these analogues is planned in an essentially different way from the definition of their 'prototypes'.

For any nontrivial probability space (i.e. one having more than two elements), one constructs an integrable *FR*-construct such that, first, there cannot be an algorithm which associates with an arbitrary *FR*-number a Lebesgue set, and, second, for any *FR*-number such a Lebesgue set is quasi-realizable.

One can also prove that, under certain conditions imposed on a probability space, one can construct an integrable *FR*-construct such that there does not exist a Lebesgue set corresponding to the number 0.

We shall introduce a definition of 'convergence almost everywhere' of a sequence of integrable *FR*-constructs to an integrable *FR*-construct. This definition, which has its source in a theorem of Egorov, provable in classical mathematics, is not of a pointwise character, which is essential for the construction of a constructive theory of normed Boolean algebras.

We shall prove that, for every integrable *FR*-construct f (which is, by definition, a sequence of graduated frames, together with a regulator of convergence in itself with respect to the integral metric), one can construct a sequence of graduated frames which will converge almost everywhere to f.

In Section 4 we shall formulate constructive analogues of weak and strong laws of large numbers for random variables possessing mathematical expectations (synonym: 'integrable *FR*-constructs').

Thus, in the present paper we shall propose a constructive approach to a construction of some initial parts of probability theory on the basis indicated above. As far as possible, the author tended to follow the approach of Goodstein (1951) to the construction of certain fragments of constructive mathematics, characterized by the use of concepts of an approximative nature and of little logical complexity.

Sections 2 and 3 of the present paper are based on the author's paper (1970).

2. PROBABILITY SPACE

2.1. All the notation, concepts, and relations used in Šanin (1962) will be used in what follows without explanation.

A decidable set, together with a decidable, symmetric, reflexive, transitive relation (called an equality relation) on the elements of the set, will be called a *p-set of type* $[P, P^=]$.

Let A be an *alphabet* and let B be a p-set of type $[P, P^=]$ of words in the alphabet A. Instead of the words 'b is an element of the p-set B' we shall usually write '$b \in B$'. The expression $P \overset{B}{=} Q$ will signify that P and Q are equal elements of the p-set B.

By a *decidable Boolean algebra* we shall mean any quintuple of objects of the form $(A, B, \underset{\circ}{\cup}, \underset{\circ}{\cap}, \underset{\circ}{\mathsf{C}})$ such that A is an arbitrary alphabet not containing the letter $\square$; B is a p-set of type $[P, P^=]$ of words in the alphabet A; $\underset{\circ}{\cup}, \underset{\circ}{\cap}$, and $\underset{\circ}{\mathsf{C}}$ are algorithms over the alphabet $A \cup \{\square\}$ which are equality-preserving mappings into the p-set B, and have the properties of commutativity, associativity and two-sided distributivity; and, moreover, for any elements P, Q, and R of the p-set B, the following conditions are satisfied:

$$! \underset{\circ}{\cup}(P \square Q), \quad ! \underset{\circ}{\cap}(P \square Q), \quad ! \underset{\circ}{\mathsf{C}}(P);$$

$$\Lambda \in B, \quad \underset{\circ}{\cup}(P \square Q) \in B, \quad \underset{\circ}{\cap}(P \square Q) \in B, \quad \underset{\circ}{\mathsf{C}}(P) \in B;$$

$$\underset{\circ}{\cup}(\underset{\circ}{\cap}(P \square Q) \square Q) \overset{B}{=} Q; \quad \underset{\circ}{\cap}(P \square \underset{\circ}{\cup}(P \square Q)) \overset{B}{=} P;$$

$$\underset{\circ}{\cup}(P \square \underset{\circ}{\mathsf{C}}(P)) \overset{B}{=} \underset{\circ}{\mathsf{C}}(\Lambda); \quad \underset{\circ}{\cap}(P \square \underset{\circ}{\mathsf{C}}(P)) = \Lambda;$$

$$\underset{\circ}{\cap}(P \square \underset{\circ}{\mathsf{C}}(\Lambda)) \overset{B}{=} P; \quad \underset{\circ}{\cup}(P \square \Lambda) \overset{B}{=} P.$$

Here the symbol 'Λ' denotes the empty word.

By a *normed decidable Boolean algebra* we mean any pair of the form

$\{\beta_0, \mathscr{P}_0\}$ where β_0 is a decidable Boolean algebra, and $\mathscr{P}_0$ is an algorithm which is an equality-preserving mapping of the p-set B into a set of duplexes such that, for any Q_1 and Q_2 for which $Q_1 \in B$ and $Q_2 \in B$,

$$! \mathscr{P}_0(Q_1); \quad 0 \underset{B}{\leq} \mathscr{P}_0(Q_1);$$

$$\mathscr{P}_0(\underset{\circ}{\cup}(Q_1 \square Q_2)) + \mathscr{P}_0(\underset{\circ}{\cap}(Q_1 \square Q_2)) \underset{B}{=} \mathscr{P}_0(Q_1) + \mathscr{P}_0(Q_2);$$

$$\mathscr{P}_0(\Lambda) \underset{B}{=} 0.$$

If a normed decidable Boolean algebra β is such that $\mathscr{P}_0(\underset{\circ}{C}(\Lambda)) \underset{B}{=} 1$, we shall call it a *probability space*.

Remark. An example of a probability space is the set of complexes on $0 \triangle 1$ together with appropriate operations. This probability space will be denoted by K (cf. Šanin, 1962; Section 7.4).

In what follows, if nothing is said to the contrary, we shall consider a fixed normed decidable Boolean algebra $\{(A, B, \underset{\circ}{\cup}, \underset{\circ}{\cap}, \underset{\circ}{C}), \mathscr{P}_0\}$ which will be denoted in the sequel by β.

In what follows, if t_1 and t_2 are terms, then the terms

$$\underset{\circ}{\cup}(t_1 \square t_2), \underset{\circ}{\cap}(t_1 \square t_2), \underset{\circ}{\cap}(t_1 \square \underset{\circ}{C}(t_2))$$

and

$$\underset{\circ}{\cup}(\underset{\circ}{\cap}(t_1 \square \underset{\circ}{C}(t_2)) \square \underset{\circ}{\cap}(t_2 \square \underset{\circ}{C}(t_1)))$$

will be denoted, respectively, by the expressions

$$(t_1 \underset{\circ}{\cup} t_2), (t_1 \underset{\circ}{\cap} t_2), (t_1 \backslash t_2)$$

and

$$(t_1 \underset{\circ}{\triangle} t_2).$$

2.2. We construct an algorithm ρ_0 such that for any words Q_1 and Q_2 ($Q_1 \in B, Q_2 \in B$) in the alphabet A

$$\rho_0(Q_1 \square Q_2) \simeq \mathscr{P}_0(Q_1 \underset{\circ}{\triangle} Q_2).$$

The algorithm ρ_0 transforms the normed decidable Boolean algebra into a constructive metric space Ω_0 with metric ρ_0; moreover, the algorithm $\mathscr{P}_0$ is a one-place Lipschitz operator from the space Ω_0 into the space of real duplexes, and 1 is a Lipschitz coefficient of this operator.

SOME PROBLEMS IN THE CONSTRUCTIVE PROBABILITY THEORY 87

Let us construct an algorithm Δ' such that, for any elements Q_1 and Q_2 of the p-set $\overset{\circ}{B}$,

$$\underset{\circ}{\Delta}{'}(Q_1 \square Q_2) \simeq (Q_1 \underset{\circ}{\Delta} Q_2).$$

2.3. To the constructive metric space Ω_0 we apply the method of standard FR-completion (cf. Šanin, 1962). The resulting complete constructive metric space is denoted Ω. The metric in Ω is denoted by ρ. FR-constructs in the space Ω_0 (i.e. points in the space Ω) will be called *events over β*.

As a new separation sign for the construction of events over β we shall use the symbol $\textcircled{1}$. Therefore the alphabet $A \cup \{\textcircled{1}\}$ denoted by A', is the alphabet of the space Ω.

To the algorithms $\overset{\circ}{\mathscr{P}}_0$, $\overset{\circ}{\cup}$, $\overset{\circ}{\cap}$, $\overset{\circ}{\mathsf{C}}$, Δ' and the number 1 we apply the standard algorithms for extending Lipschitz operators (cf. Šanin, 1962). The resulting Lipschitz operators, for which 1 is a Lipschitz coefficient, will be denoted by $\mathscr{P}$, $\cup$, $\cap$, C and Δ.

Let S be an event over β. The real duplex $\mathscr{P}(S)$ will be called *the probability of S*.

The algorithms $\cup$, $\cap$, C and Δ will be called the algorithms for constructing the union, intersection, complement, and symmetric difference of events over β.

In what follows, if t_1 and t_2 are terms, then the terms $\cup(t_1 \square t_2)$, $\cap(t_1 \square t_2)$, $\Delta(t_1 \square t_2)$, $\cap(t_1 \square \mathsf{C}(t_2))$ will be denoted, respectively, by the expressions $(t_1 \cup t_2)$, $(t_1 \cap t_2)$, $(t_1 \Delta t_2)$, $(t_1 \setminus t_2)$.

2.4. Now we shall introduce operations over algorithmic sequences of events over β. Let α be an algorithm which is applicable to every natural number and transforms any natural number into some event over β. We denote by $^\cup\alpha$ the algorithm such that

$$^\cup\alpha(0) \simeq \alpha(0), \quad \forall k(^\cup\alpha(k+1) \simeq (^\cup\alpha(k) \cup \alpha(k+1))).$$

We shall say that an event m over β is the *union* of all terms of the sequence α if $0 \, lim_E \, G$, where G is the algorithm such that, for every k,

$$G(k) \simeq \rho(m \square \, ^\cup\alpha(k)),$$

and E is the metric space of real duplexes.

2.5. Let us note one essential difference between constructive and classical mathematics. In classical probability theory there is a theorem about the existence of the union of any sequence of events, but in constructive mathematics the situation is different (Theorem 2.5.2).

As a preliminary step, let us prove a certain strengthening of a theorem of Specker on the existence of monotonic bounded algorithmic sequences of rational numbers for which there cannot be an algorithm which is a regulator of convergence in itself (Specker, 1949).

2.5.1. For any algorithm H satisfying the condition

$$\forall n (\exists k (H(k) <_B 2^{-n}) \& (H(n) >_B 0)),$$

one can construct an algorithm S transforming natural numbers into duplexes, an algorithm α of type with values in natural numbers and a strictly monotonic algorithm N such that:

(1) $\quad S(k) = \sum_{j=0}^{k} \alpha(j \square k) \cdot H(N(j+1)) < 1;$

(2) $\quad \forall jk (0 \leqslant \alpha(j \square k) \leqslant \alpha(j \square k+1) \leqslant 1);$

(3) $\quad$ There cannot be an algorithm which is a regulator of convergence in itself of the algorithm S.

We shall say that a probability space possesses property S^- if one can construct an algorithm α, transforming natural numbers into events over β, such that $\alpha(i) \leqslant \alpha(i+1)$ for any natural number i, and, at the same time, there cannot be an event over β which is the union of all terms of the sequence α.

2.5.2. In order that a probability space possess the property S^- it is necessary and sufficient that there be realizable a constructive sequence γ of pairwise disjoint events over β such that

$$\forall n \exists k (0 <_B \mathscr{P}(\gamma(k)) <_B 2^{-n}).$$

Remark. The theorem just proved testifies to the fact that a great many interesting probability spaces possess property S^-. In particular, this property is possessed by the probability space K (cf. Šanin (1962), assertion 14.3.2).

2.6. Let α be an algorithm such that α transforms natural numbers into events over β. We denote by $^\frown\alpha$ the algorithm such that

$$^\frown\alpha(0) \simeq \alpha(0),$$
$$\forall k\, (^\frown\alpha(k+1) \simeq (^\frown\alpha(k) \cap \alpha(k+1))).$$

We shall say that an event m over β is the *intersection* of all the events over β which are terms of the sequence α if $0\ lim_E\ G$, where G is an algorithm such that for every k

$$G(k) \simeq \rho(m\ \square\ ^\frown\alpha(k)),$$

and E is the space of all real duplexes.

We shall say that a probability space possesses property S^+ if one can construct an algorithm α, transforming natural numbers into events over β, such that $\alpha(i) \supseteq \alpha(i+1)$ for any i and, at the same time, there cannot be an event over β which is the intersection of all terms of the sequence α.

2.6.1. In order that a probability space possess property S^+ it is necessary and sufficient that it possess property S^-.

3. INTEGRABLE FR-CONSTRUCTS

3.1. By a *partition of the probability space* β we mean any list of elements $a_1, \ldots, a_q$ of the p-set B which possesses the following properties:

(1) $\mathscr{P}(\bigcup\limits_{i=1}^{q} a_i) = 1,$
$\quad\ \ _B$

(2) For any natural numbers i, j in $1 \triangle q$,

$$i \neq j \supset \mathscr{P}_B(a_i \cap a_j) = 0.$$

Let σ and τ be letters different from the letters of the alphabet $A \cup \{\square, 0, |, -, /\}$. By a *graduated frame over* β we mean any word of the form

$$a_1 \tau a_2 \tau \ldots \tau a_q \sigma R_1 \tau R_2 \tau \ldots \tau R_q,$$

where R_i is a rational number, a_i is an element of the p-set $B(1 \leq i \leq q)$, and the list $a_1, a_2, \ldots, a_q$ is a partition of the probability space β.

For brevity, in what follows we shall simply write 'graduated frame' instead of the words 'graduated frame over β'.

Assume that a graduated frame f is represented in the form (1). By its integral we shall mean the FR-number

$$\sum_{i=1}^{q} (\mathscr{P}(a_i) \cdot R_i),$$

which will be denoted by $E_0(f)$.

We introduce in a natural way an operation of addition of graduated frames, an operation of multiplication of graduated frames, an operation of multiplication of a graduated frame by a rational number, and an operation for constructing the absolute value of a graduated frame.

We agree to denote by the expressions

$$+_\ni, \times_\ni, \cdot_\ni, M_\ni, -_\ni$$

the operation of adding graduated frames, the operation of multiplying graduated frames, the operation of multiplying a graduated frame by a rational number, the operation of constructing the absolute value of a graduated frame, the operation of subtracting one graduated frame from another (defined in the usual way on the basis of the preceding operations), and binary operations max and min of two graduated frames.

3.2. In the following exposition we shall consider a fixed rational number d satisfying the condition $d \geq 1$. We construct an algorithm $_dN^*$ over the alphabet $A \cup \{\sigma, \tau, 0, |, -, /, \Diamond\}$ such that, for any graduated frame f representable in the form (1):

$$_dN^*(f) \simeq \left(\sum_{i=1}^{m} |R_i|^d \cdot \mathscr{P}(a_i)\right)^{1:d}.$$

We construct an algorithm $_d\rho$ over the alphabet $A \cup \{\sigma, \tau, \Box, 0, |, -, /, \Diamond\}$ such that, for any graduated frames f_1 and f_2,

$$_d\rho(f_1 \Box f_2) \simeq {}_dN^*(f_2 -_\ni f_1).$$

$_d\rho$ is a metric transforming the space of graduated frames into a con-

SOME PROBLEMS IN THE CONSTRUCTIVE PROBABILITY THEORY 91

structive normed space, denoted below by $\mathcal{T}_d$. The algorithm for computing the norm in $\mathcal{T}_d$ coincides as a function with $_dN^*$. This algorithm will be denoted by $_dN$.

It is easily seen that $\mathcal{T}_2$ is a constructive Hilbert space.

By the characteristic representation of an element b of the p-set β we shall mean the word $b\tau\overset{\circ}{C}(b)\,\sigma 0\,|\,\tau 0$. The algorithm over the alphabet $A \cup \{\sigma, \tau, 0, |, -, /\}$ transforming any element of B into its characteristic representation will be denoted by χ_0.

3.3. The standard normed FR-completion of the space $\mathcal{T}_1$ will be denoted by $\mathcal{T}$. Points of the space $\mathcal{T}$ will be called integrable FR-constructs over β.

Remark. We note that integrable FR-constructs over the probability space K are summable FR-constructs on 0 Δ 1. The concept of summable FR-construct on 0 Δ 1 is a natural constructive analogue of the concept of Lebesgue-summable function on the interval [0, 1] (cf. Šanin (1962), §14). The concept 'integrable FR-construct' suggests itself as a constructive analogue of the concept 'random variable possessing a mathematical expectation'. Thus, the terms 'integrable FR-construct over β' and 'random variable over β, having a mathematical expectation' will be considered synonymous.

Below, for brevity we shall often write simply 'integrable FR-construct' instead of 'integrable FR-construct over β'.

We introduce the symbol $\oplus$ as a new separating symbol intended for the construction of FR-constructs in the space $\mathcal{T}$. The metric in the space $\mathcal{T}$ will be denoted by $_1\bar{\rho}$, and the algorithm for computing the norm in this space by $_1\bar{N}$.

It follows that the algorithm M is a Lipschitz operator from $\mathcal{T}_1$ into $\overset{\ni}{\mathcal{T}}$; moreover, 1 is a Lipschitz coefficient of this operator. Hence one can construct an extension of the operator M to the space $\overset{\ni}{\mathcal{T}}$. We also construct the standard extensions to the space $\overset{\ni}{\mathcal{T}}$ of the operators $\overset{\ni}{+}, \overset{\ni}{\cdot}, \overset{\ni}{-},$ and M, considered together with appropriate Lipschitz coefficients. The resulting operators will be denoted, respectively, by

$$+_{\mathcal{T}},\ \cdot_{\mathcal{T}},\ -_{\mathcal{T}},\ M_{\mathcal{T}}.$$

Instead of this notation we shall use the simpler notation $+, \cdot, -, M$, if this does not cause ambiguity.

The algorithm E_0 is a locally bounded linear operator from $\mathcal{T}_1$ into the space of all real duplexes, and 1 is a Lipschitz coefficient of this operator. In the space $\mathcal{T}_1$ applying to the algorithm E_0 and the number 1 the algorithm for the standard extension of Lipschitz operators, we obtain a linear Lipschitz operator from $\mathcal{T}$ into the space of all real duplexes, and 1 is a Lipschitz coefficient of this operator. The resulting operator will be denoted by E. Let f be any integrable FR-construct; the duplex $E(f)$ will be called its integral.

Applying to the word

$$\{\chi_0\}\, [1]\, \{R\},$$

which is the complete cipher of a uniformly continuous singular operator, the algorithm for the standard extension of complete ciphers (cf. Šanin (1962), Section 10.4), we obtain an operator χ mapping the space of events over β into $\mathcal{T}$.

It is not difficult to show that the operators $\max_{\mathfrak{s}}$ and $\min_{\mathfrak{s}}$ are binary Lipschitz operators from the space $\mathcal{T}_1$ into the space $\mathcal{T}$, and, moreover, 1 is a Lipschitz coefficient of these operators. Hence one can construct extensions of these operators to the space $\mathcal{T}$, which will be denoted by max and min.

Let x be a real duplex. By $\overset{\circ}{\mathsf{C}}(\Lambda)\, \sigma x$ we mean the integrable FR-construct

$$\{X\} \oplus \{R\},$$

where
$$\forall n(X(n) \simeq \overset{\circ}{\mathsf{C}}(\Lambda)\, \sigma x(n)),\ \forall n(R(n) \simeq \overline{x}(n)).$$

Below, for brevity we shall often simply write x instead of $\overset{\circ}{\mathsf{C}}(\Lambda)\, \sigma x$.

For points of the space $\mathcal{T}$ we shall introduce an order relation $\leqslant_{\mathcal{T}}$. Let f_1 and f_2 be points of the space $\mathcal{T}$. We shall say that the point f_1 is majorized by the point f_2, and we shall write $f_1 \leqslant_{\mathcal{T}} f_2$ if $f_2 - f_1 = {}_{1\rho}M(f_2 - f_1)$.

It is easily seen that the relations just introduced are transitive.

We shall say that an integrable FR-construct f over β is z-bounded, where z is a duplex, if $M(f) \leqslant_{\mathcal{T}} z$.

We define in a natural way an operation of multiplying an integrable FR-construct f by a graduated frame g.

SOME PROBLEMS IN THE CONSTRUCTIVE PROBABILITY THEORY

It is clear that the operation of multiplying a graduated frame by a fixed graduated frame g is a Lipschitz operator from the space $\mathcal{T}_1$ into the space $\mathcal{T}$.

Finally, fixing a duplex z, one can define an operator multiplying any integrable FR-construct f by any z-bounded FR-construct g. We use the operation of multiplying an integrable FR-construct by graduated frames. It is clear that the operation of multiplying a fixed integrable z-bounded FR-construct g by graduated frames is a Lipschitz operator from the space $\mathcal{T}_1$ into the space $\mathcal{T}$, with z as a Lipschitz coefficient of this operator. Hence one can construct an extension of the operator to the space $\mathcal{T}$, which we shall denote by $\underset{z}{\times}$.

3.4. Let m be an event over β, f an integrable FR-construct, and e a real duplex. We shall write

$$m \succ \{f > e\}, \quad \text{if } f \underset{1}{\times} \chi(\mathsf{C}(m)) \underset{\mathcal{T}}{\leqslant} e;$$
$$m \succ \{f < e\}, \quad \text{if } f \underset{1}{\times} \chi(\mathsf{C}(m)) \underset{\mathcal{T}}{\geqslant} e;$$
$$m \prec \{f \geqslant e\}, \quad \text{if } f \underset{1}{\times} \chi(m) \underset{\mathcal{T}}{\geqslant} e;$$
$$m \prec \{f \leqslant e\}, \quad \text{if } f \underset{1}{\times} \chi(m) \underset{\mathcal{T}}{\leqslant} e;$$

and

$m \approx \{f > e\},\quad$ if $\ m \succ \{f > e\}\ $ and $\ \forall m'(m' \succ \{f > e\} \supset (m' \supseteq m));$
$m \approx \{f < e\},\quad$ if $\ m \succ \{f < e\}\ $ and $\ \forall m'(m' \succ \{f < e\} \supset (m' \supseteq m));$
$m \approx \{f \geqslant e\},\quad$ if $\ m \prec \{f \geqslant e\}\ $ and $\ \forall m'(m' \prec \{f \geqslant e\} \supset (m \supseteq m'));$
$m \approx \{f \leqslant e\},\quad$ if $\ m \prec \{f \leqslant e\}\ $ and $\ \forall m'(m' \prec \{f \leqslant e\} \supset (m \supseteq m')),$

where m' is a variable for events over β.

We shall say that a probability space β satisfies property S^* if there is a realizable sequence α of elements of the p-set β such that

$$\forall ij (i \neq j \supset \alpha(i) \lambda \alpha(j)),$$

and, for any event m over β possessing the property

$$\forall n (m \supseteq \bigcup_{i=0}^{n} \alpha(i)),$$

an event m' over β is quasi-realizable such that

$$(\rho(m \underset{B}{\square} m') > 0) \,\&\, \forall n(m \supseteq m' \supseteq \bigcup_{i=0}^{n} \alpha(i)).$$

3.4.1. Assume that a probability space β satisfies the condition S^*. Then one can construct an integrable FR-construct f over β such that $\neg(m \approx \{f \leqslant 0\})$ for any event m over β.

Remark. An example of an integrable FR-construct f over the probability space K for which every Lebesgue-measurable set m satisfies the condition $\neg(m \approx \{f \leqslant 0\})$ was shown to the author by Demuth (1967).

Analogous examples are given by certain continuous summable FR-constructs having 'bad' maximal regulators of continuity, and introduced in Slisenko (1967, §4).

By inessential modifications of the example constructed in the proof of Theorem 3.4.1, one can prove the theorems obtained from 3.4.1 by replacing the symbol $\leqslant$ by any of the symbols $\geqslant$, $<$ or $>$.

3.5. We shall say that a sequence α of integrable FR-constructs over β converges to an integrable FR-construct f over β almost everywhere if for any natural numbers k and l there are realizable a natural number N and an event m over β such that, for any natural number n, if the condition $n \geqslant N$ is fulfilled, then, first,

$$m \succ \{|\alpha(n) - f| > 2^{-l}\},$$

and, second,

$$\mathscr{P}(m) < 2^{-k}.$$

The conceptual origin of this definition is the theorem of Egorov (1911) which is proved in classical mathematics and asserts the equivalence of convergence almost everywhere and almost-uniform convergence.

The concept of 'convergence almost everywhere' which has just been introduced is essentially different from the concept introduced in Demuth (1967) and denoted by the same term. In particular, the former does not have a 'pointwise character'.

The presence of 'uniformity' in convergence almost everywhere is essential, since, as the following theorem shows, Egorov's theorem does not hold in constructive mathematics.

3.5.1. One can construct a sequence of graduated frames over K which, first, converges to 1 at every point of the constructive segment $0 \, \Delta \, 1$ (with the possible exception of rational points), and, second, does not converge almost everywhere to 1.

Remark. The theorem remains true if in its formulation the words 'graduated frames over K' are replaced by the words 'uniformly continuous functions on $0 \triangle 1$'.

3.6. In this section we shall prove a theorem establishing a connection between convergence of a sequence of graduated frames with respect to the integral metric and convergence almost everywhere.

3.6.1. Let f be an integrable FR-construct over β. Then the sequence α such that $\forall n(\alpha(n) - f(\overline{f}(n)))$ converges to f almost everywhere.

COROLLARY. Let $\overline{f}$ be an FR-construct which is summable on $0 \triangle 1^1$. Then one can construct a sequence of graduated frames which converges almost everywhere to f.

Remark. In particular, this theorem permits us to associate with every summable FR-construct f a sequence of rational-valued functions with a finite number of 'steps' on intervals with rational end-points, which constructively converges almost everywhere to f.

4. WEAK AND STRONG LAWS OF LARGE NUMBERS IN CONSTRUCTIVE PROBABILITY THEORY

4.1. Instead of the concept of a 'random variable possessing a mathematical expectation' (synonym: 'integrable FR-construct') we shall use the abbreviated expression 'random variable'.

We shall say that a sequence $\mathfrak{A}$ of random variables is a sequence of pairwise integrally-independent random variables if an algorithm R exists such that for any natural numbers i, j, n, k, q if $0 \leq i < j \leq q$ and $n \geq R(q \square k)$ then

$$|E(\mathfrak{A}(i)(n)) \cdot E(\mathfrak{A}(j)(n)) - E(\mathfrak{A}(i)(n) \times \mathfrak{A}(j)(n))| < 2^{-k}.$$

By a *dispersion* of the graduated frame f we mean the duplex

$$E((f - E(f)) \times (f - E(f))).$$

4.2. We say that a sequence $\mathfrak{A}$ of random variables over β obeys the weak law of large numbers, if algorithm R_1 and algorithm R_2 exist such that for any natural numbers n, m, k, l and element b of the p-set B if $n \geqslant R_1(k \square l)$, $m \geqslant R_2(k \square l)$ and

$$b \approx \left\{ M\left(\frac{1}{(n+1)}(\mathfrak{A}(i)(m) - E(\mathfrak{A}(i)(m)))\right) \geqslant 2^{-l}\right\}$$

then $\mathscr{P}_0(b) < 2^{-k}$.

4.2.1. Let α and α' be sequences of random variables such that

$$\forall n (E(M(\alpha(n) - \alpha'(n))) = 0)$$

and α obeys the weak law of large numbers. Then the sequence α' obeys the weak law of large numbers.

The following constructive analogue of the Chebyshev (1867) theorem on the weak law of large numbers is valid.

4.2.2. Let α be a sequence of pairwise integrally-independent random variables and for any natural numbers n, k

$$\mathscr{D}_0(\alpha(k)(n)) \leqslant C,$$

where C is a natural number. Then sequence α obeys the weak law of large numbers.

4.3. We say that a sequence $\mathfrak{U}$ of random variables obeys the strong law of large numbers if algorithm R_1 and algorithm R_2 exist such that for any natural numbers j, β, k, l, m, n and element b of the p-set B if $m \geqslant n \geqslant R_1(k \square l)$, $\beta \geqslant R_2(m \square k \square l)$ and

$$b \approx \left\{ \sup_{j=n}^{m}\left(M\left(\frac{1}{(j+1)} \sum_{i=0}^{j} (\mathfrak{A}(i)(\beta) - E_0(\mathfrak{A}(i)(\beta)))\right)\right) \geqslant 2^{-l}\right\},$$

then $\mathscr{P}(b) < 2^{-k}$.

4.3.1. Let α and α' be sequences of random variables such that

$$\forall n (E(M(\alpha(n) - \alpha'(n))) = 0)$$

and α obeys the strong law of large numbers. Then sequence α' obeys the strong law of large numbers.

Let there be given a finite list of graduated frames $f_1, \ldots, f_q$ ($q > 1$) and for every i ($1 \leq i \leq q$) let the graduated frame f_i have the form

$$a_1^i \tau a_2^i \tau \ldots \tau a_{t_i}^i \sigma R_1^i \tau R_2^i \tau \ldots \tau R_{t_i}^i$$

where $R_1^i, R_2^i, \ldots, R_{t_i}^i$ are distinct rational numbers, $t_i \geq 1$ and the list of elements $a_1^i, a_2^i, \ldots, a_{t_i}^i$ of the p-set B is a partition of the initial probability space β. Let an element b of the p-set B be constructible from elements a_j^i of the p-set B ($1 \leq i \leq q$, $1 \leq j \leq t_i$) by the operations of union, intersection and complementation. In this case we write

$$[b \leftarrow f_1, f_2, \ldots, f_q].$$

We say that a sequence α of random variables is a sequence of induced-independent random variables if algorithm R of type $(HH \to H)_0$ exists such that for any distinct natural numbers V_1, V_2 taken from an arbitrary interval $0 \, \Delta \, q$ and elements b_1 and b_2 of the p-set B, if $n \geq R(q \, \Box \, k)$ and

$$[b_1 \leftarrow \underbrace{\alpha(0)(n)}, \underbrace{\alpha(1)(n)}, \ldots, \underbrace{\alpha(q)(n)}],$$

$$[b_2 \leftarrow \underbrace{\alpha(0)(n)}, \underbrace{\alpha(1)(n)}, \ldots, \underbrace{\alpha(q)(n)}]$$

then

$$|E(\chi(b_1) \times \underbrace{\alpha(V_1)(n)} \times \chi(b_2) \times \underbrace{\alpha(V_2)(n)}) -$$
$$- E(\chi(b_1) \times \underbrace{\alpha(V_1)(n)}) \cdot E(\chi(b_2) \times \underbrace{\alpha(V_2)(n)})| < 2^{-k}.$$

4.3.2. Let a sequence $\mathfrak{A}$ of random variables be a sequence of induced-independent random variables. Then sequence $\mathfrak{A}$ is a sequence of pairwise integrally-independent random variables.

The following constructive analogue of the strong law of large numbers holds.

4.3.3. Let a sequence α of induced-independent random variables be such that for any natural numbers i, n, m, k if $m > n \geq R_1^0(k)$, $i \geq R_2^0$ ($n \, \Box \, m \, \Box \, k$), then

$$\sum_{j=n+1}^{m} (j+1)^{-2} \underbrace{\mathscr{D}_0(\mathfrak{A}(j)(i))} < 2^{-k},$$

where R_1^0 and R_2^0 are algorithms with values in natural numbers substitutions. Let the sequence F of rational numbers be such that for any natural numbers j, l

$$\mathscr{D}_0(\mathfrak{A}(j)(l)) < F(j).$$

then sequence $\mathfrak{A}$ obeys the strong law of large numbers.

4.4. Using the dispersion concept, we can (weakening theorem 4.3.3) prove the strong law of large numbers in a simplest form analogous to the formulation of the Kolmogorov theorem in classical probability theory.

4.4.1. Let a sequence $\mathfrak{A}$ of induced-independent random variables possessing dispersion be such that the series

$$\sum_{j=0}^{\infty} \frac{\mathscr{D}(\mathfrak{A}(j))}{(j+1)^2}$$

converges constructively. Then the sequence $\mathfrak{A}$ obeys the strong law of large numbers.

Here the expression $\mathscr{D}(f)$ denotes the dispersion of f.

Faculty of Mathematics of Leningrad State University,
Leningrad

NOTE

[1] The concept of a summable FR-construct is a natural constructive analogue of the concept of Lebesgue-summable function introduced by Šanin (1962).

BIBLIOGRAPHY

Chebyshev, P. L.: 'About Mean Quantities', *Matem. sb.* **2**, Moscow, 1867 (Russian).
Demuth, O., 'The Lebesgue Integral in Constructive Analysis', *Zap. Naucn. Sem Leningrad. Otdel. Mat. Inst. Steklov.* (LOMI) **4** (1967) 30–43 (Russian). MR 40 # 7112.
Egorov, D. F.: 'Sur les suites de fonctions mesurables', *Comptes Rendus, Acad. Sci., Paris* **152** (1911) 244–246.
Glivenko, V. I., *Course on the Theory of Probability*, ONTI, Moscow, 1939 (Russian).
Goodstein, R. L., *Constructive Formalism. Essays on the Foundations of Mathematics*, University College, Leicester, 1951, MR 14, 123.

Kolmogorov, A. N., 'Algèbres de Boole métriques complètes', *Ann. Soc. Polon. Math.* **20** (1948) 21–30 (appendix).

Kolmogorov, A. N., 'Three Approaches to the Definition of the Concept of the "Amount of Information"', *Problemy Peredaci Informacii* **1** (1965) 3–11; English transl., *Selected Transl. Math. Stat. and Prob.*, vol. 7, Amer. Math. Soc., Providence, R. I., 1968, pp. 293–302. MR 32#2273.

Kossovsky, N. K., 'Some Questions in the Constructive Theory of Normed Boolean Algebras', *Trudy Mat. Inst. Steklov.* **113** (1970) 3–38 = *Proc. Steklov. Inst. Math.* **113** (1970) 1–41.

Markov, A. A., 'On Integration in Boolean Algebras', *Sixth Scientific Session, Abstracts of Lectures*, Gos. Univ., Leningrad, 1949 (Russian).

Martin-Löf, P., 'On the Concept of Random Sequence', *Teor. Verojytnost. i Primenen.* **11** (1966) 198–200.

Šanin, N. A., 'Constructive Real Numbers and Constructive Function Spaces', *Trudy Mat. Inst. Steklov.* **67** (1962) 15–294; English transl., *Transl. Math. Monographs*, vol. 21, Amer. Math. Soc., Providence, R.I., 1968. MR 28#30.

Slisenko, A. O., 'The Construction of Maximal Continuity Regulators for Constructive Functions', *Trudy Mat. Inst. Steklov.* **93** (1967) 208–249 = *Proc. Steklov Inst. Math.* **93** (1967) 269–317.

Specker, E., 'Nicht konstruktiv beweisbare Sätze der Analysis', *J. Symbolic Logic* **14** (1949) 145–158. MR 11, 151.

KEITH LEHRER

EVIDENCE AND CONCEPTUAL CHANGE*

In this paper we shall formulate a rule of evidence that is based on subjective probabilities and is sensitive to conceptual change. Whether a sentence is selected as evidence on our account will be determined by the subjective probability of the sentence and of the sentences with which it competes for evidential status. Both the subjective probability of a sentence and what sentences it competes with will depend on the conceptual and semantic relations among sentences of the language. When these relations are altered, new sentences may be added to evidence and old ones deleted.

We shall assume the existence of an assignment of antecedent subjective probabilities to sentences of languages having a finite number of atomic sentences.[1] The extension of our methods to infinite languages will not be discussed.[2] Our selection of evidence will depend only on these subjective probabilities and the logical or semantic relations between sentences. The subjective probabilities will be a set of ratios assigned to sentences in such a way as to be coherent, that is, to conform to the calculus of probability. The assignment is relative to a person and a time. A subjective probability assignment may be construed as a degree of belief a person has in a sentence at a time or as his estimate of the chance of the sentence being true. In our discussion we shall use the probability functor '$p_i(e)$' to mean 'the probability of e at time t_i,' and suppress reference to the subject for the sake of convenience. What is critically important, is that the subjective probability of a sentence may shift in time, the degree of belief or the estimate may change from one time to another. It is this sort of shift that is influenced by conceptual and semantic relations.

Our rule for evidence will be based on the further assumption that a sentence competes for the status of evidence with some sentences and not others. In deciding what sentences a sentence competes with for the status of evidence, we cannot without circularity, appeal to evidence. Instead, we shall rely on a conception of competition that will remain viable no matter what our evidence is. To that end, we propose that a sentence competes with every sentence with which it *could* conflict in the light of evidence.

Are there any sentences with which a sentence cannot conflict no matter what the evidence? To answer this question, we require a clearer idea of when one sentence conflicts with another. Consider any two sentences, p, and q. Suppose that we conclude that if p is true, then q is false. Whether this conclusion is derived from logic alone or from evidence as well, it affirms *conflict* between the two sentences. But if q is a logical consequence of p, then we cannot conclude that if p is true, q is false. Thus the logical consequences of a sentence are the only ones with which it cannot conflict. Hence we adopt the following definition of competition: e competes with d for the status of evidence if and only if d is not a logical consequence of e.[3] A sentence must compete for the status of evidence with all statements except its logical consequences.

The foregoing considerations provide the basis for the formulation of a selection rule for evidence sentences. The rule tells us that a logically consistent sentence may be admitted to evidence if and only if the sentence is more probable than those with which it competes. More formally, the rule is as follows: Letting '$E_i(e)$' mean 'e is admitted to evidence at time t_i,' we obtain a formal statement of the rule as follows:

E. $E_i(e)$ if and only if it is not the case that $e \vdash \ulcorner p \ \& \sim p \urcorner$ and for any s such that it is not the case that $e \vdash s$, $p_i(e) > p_i(s)$.

We shall now attempt to justify the rule.

The rule has four features that make it suitable as a rule of evidence. First, the rule is conjunctive, that is, if $E_i(e)$ and $E_i(d)$, then $E_i(e \ \& \ d)$. Second, it guarantees that there will be some sentence of total evidence, that is, there will be some sentence T such that $E_i(T)$ and for any e, if $E_i(e)$, then $T \vdash e$. Third, the rule is consistent, that is, sentence T and the set of evidence sentences do not logically imply a contradiction. Fourth, no false sentence will be as probable as any true sentence selected as evidence, that is, if $E_i(e)$ and e is true, then, for any s, if s is false, then $p_i(e) > p_i(s)$.[4] The last feature is important if we wish to say that the probabilities explain the truth of a sentence selected as evidence. If a false sentence could be more probable than a true sentence selected as evidence, it would be implausible to contend that the probabilities provided any explanation of the truth of the evidence in question.

Further justification for rule (E) may be obtained by showing that the rule directs us to select as evidence a sentence having a maximum of expected utility and other sentences of positive utility.[5] Here we are concerned with the utility of selecting a sentence as evidence. There are two possible outcomes of selecting a sentence as evidence, the outcome of selecting a true sentence as evidence and the outcome of selecting a false one. Letting '$EU_i(e)$' mean 'the expected utility of selecting e as evidence at time t_i', '$UT_i(e)$' mean 'the utility of selecting e as evidence at time t_i when e is true' and '$UF_i(e)$' mean 'the utility of selecting e as evidence at time t_i when e is false', we have the following formula for calculating expected utility:

$$EU_i(e) = p_i(e) \, UT_i(e) + p_i(\sim e) \, UF_i(e).$$

To complete the analysis of expected utility, we need a rule for ascertaining the utilities of selecting sentences as evidence.

The negative utility of selecting a false sentence as evidence may be measured in terms of the probability of a strongest competitor of it. That probability is the measure of loss or disutility. On the other hand, the positive utility of selecting a true sentence as evidence may be measured as the probability of the denial of a strongest competitor. Hence, if we let e^* represent any arbitrary strongest competitor of e, that is, a competitor of e having as high a probability as any competitor, we obtain the following equalities:

$$UT_i(e) = p_i(\sim e^*)$$
$$UF_i(e) = - p_i(e^*).$$

What we lose when we select a false sentence as evidence is equal to the probability of being right had we selected some strongest competitor instead, and what we gain when we select a true sentence as evidence is equal to the probability of being right in denying a strongest competitor the status of evidence.

For finite languages of the sort we are considering, it is possible to provide a formula for finding a strongest competitor for any given sentence. To do so, we need to appeal to the basic partition of a language which is a set P of sentences of the language. The sentences of a basic partition P are logically inconsistent in pairs and the disjunction of all such members is a logical truth. Moreover, every statement of the language is logically equivalent to either a member of the set, a disjunction of members, or a conjunction of members. We can obtain a strongest competitor by appeal to the members of such a set P as follows:

$p(e^*) = p(d)$ where d competes with e and is a disjunction of all members of P (in alphabetical order) except one member, m_i, such that for any other member, m_j, if $m_j \vdash e$, then $p(m_i) \leqq p(m_j)$.

This method of specifying the probability of a strongest competitor of a sentence does not assign any value to contradictory sentences because such sentences logically imply every other and hence do not compete with any.

With these equalities specified, we discover that a rule directing us to select as evidence all those sentences whose expected utility is positive will yield the same results as rule (E). The usual principle of rationality is to select an alternative with a maximum of expected utility. The rule to select all those sentences having a positive expected utility includes the directive to select one having a maximum expected utility but does not limit selection to those. The departure from the usual application of utility theory is justified for two reasons. First, the usual applications are in the context of action, where the alternatives are logically incompatible and only one can be chosen. In such cases we obtain all the expected utility we can by selecting a maximal alternative. But in the present application, the alternatives are not mutually incompatible, more that one can be chosen, and hence it is reasonable to follow a rule directing us to obtain all the positive utility we can in our selection. Second, not only is it possible to adopt more than one alternative, that is, select more than one sentence as evidence, but the set of sentences selected will be logically consistent. Hence by selecting all sentences as evidence which have a positive utility, we gain all the expected utility we can, and at the same time insure the mutual consistency of the sentences we select.

The results of following rule (E) may be expressed in terms of the members of the basic partition P. If we assume that all sentences are logically equivalent to members, disjunctions or conjunctions of such members, and if we also note that a sentence is selected as evidence by our rule if and only if every logically equivalent sentence of the language is also selected, then by specifying which members, disjunctions, and conjunctions of members are selected as evidence we shall have completely specified which sentences are selected. Conjunctions of members are logically contradictory and are never selected by rule (E). A member of the partition P is selected if and only if it is more probable than its denial. A disjunction of members is selected if and only if it is more probable than the denial of its least probable disjunct. A disjunction of all members, and hence any other logical truth of the language, is selected because it only competes with sentences that are not logical truths. The probability of one is assigned to logical truths but not to other sentences of such languages.

With the foregoing characterization of rule (E), we are now in a position to explain how the selection of evidence may be influenced by conceptual change and semantic shifts. There are two kinds of conceptual change. The first consists of the adoption of some fundamentally new theory or law, and, generally, the rejection of some formerly accepted theory or law, without any change in the meaning of any terms of our vocabulary. The second kind of conceptual change also consists of some change in the theories and laws we accept, but such changes are so radical as to require changes in the meaning of terms used to formulate the theories and laws in question. We shall now illustrate how conceptual change of either kind can alter what we select as evidence according to rule (E).

Radical conceptual change necessitates changes in the semantic status of sentences and in the semantic relations between sentences. Whether a sentence is logically true or logically contradictory depends on the semantics of the language expressed in the meaning rules. Any change in the meaning of words entails that some sentences will be logically true or logically contradictory that were not so before or vice versa. Some sentences will be logical consequences of others that were not so before or vice versa. Such radical conceptual change will therefore alter the relation of competition; a sentence may compete with another with which it did not compete before conceptual change or vice versa. As a result, a sentence that was more probable than its competitors may turn out to be less probable than some new competitor subsequent to conceptual change. On the other hand, a sentence less probable than some competitors prior to conceptual change may cease to have those sentences as competitors subsequent to conceptual change.

Conceptual change, whether requiring semantic change or not, may alter the subjective probabilities assigned to sentences of the language and consequently the selection of evidence. A sentence might become more probable than its competitors as a result of the adoption of some theory or law and hence be selected as evidence, or on the contrary, it might become less probable and be rejected. The way in which subjective probabilities may shift has been investigated by other authors and need not be elaborated here.[6] It will suffice for our purposes to note that the semantic change which renders a sentence logically true requires shifting the probability of the sentence to one while a semantic change which renders a sentence logically contradictory requires shifting the probability to zero. Conceptual change that does not alter the semantics of a language will only shift the probabilities of sentences that are neither logically true nor logically contradictory. Sentences of a finite language that are neither logically true nor logically contradictory will be assigned some value greater than zero and less than one. The probability of such sentences may be shifted to any other coherent set of values inside that interval. Consequently, any such contingent sentence that is selected as evidence prior to conceptual change may subsequently be rejected as a result of shifting probabilities, or vice versa, even if conceptual change does not alter the meaning of words. More radical conceptual change that requires semantic shifts may lead to the selection or rejection of any sentence of the language as evidence. Fundamental changes in the way in which we conceive the world require a complete reevaluation of what sentences we should take as evidence in cognitive inquiry.

The preceding explication of the manner in which the selection of evidence by rule (E) may be influenced by conceptual change is a first step in the construction of an overall theory of inductive inference. Having specified a rule for the selection of evidence, the next step would be to specify a rule of inductive inference based on such evidence. Such rules have been proposed elsewhere. They are based on the conditional probabilities of sentences relative to those sentences selected as evidence and on the utility we assign to inductively inferring a sentence that is correct (or incorrect) from the evidence.[7] The need for such a rule of inductive inference arises because the objectives of inductive inference are somewhat different from those of evidence selection. The selection of evidence is strongly guided by the objective of avoiding error and the choice of a utility function should express that objective. On the other hand, induction is strongly guided by the objective of explaining the evidence in an informative way and thus requires a choice of a different utility function. Hypotheses that explain what is described by the evidence in a highly informative way may reasonably be accepted as working hypotheses subject to further testing and investigation. Such inquiry may lead, not only to the rejection of those hypotheses, but to a shift in subjective probabilities. That shift in turn may alter the selection of sentences as evidence and consequently the sentences inductively inferred. Inductive inference and the selection of evidence are mutually influential. Inquiry spins in the constant flux of conceptual, evidential, and inferential innovation.

University of Rochester

NOTES

* Research for this paper was supported by a Grant from the National Science Foundation. (Paper read at the Congress under the title 'Induction and Conceptual Change'.) This paper is printed here by kind permission of the publisher and editors of *Philosophia*, where it has already appeared in vol. 2, no. 4 (Oct. 1972).

[1] For a collection of articles on subjective probability, see *Studies in Subjective Probability* (ed. by H. Kyburg and H. Smokler), John Wiley and Sons, New York, 1964. For a unified approach, see Richard C. Jeffrey, *The Logic of Decision*, McGraw-Hill, New York, 1965. I argue that finite languages are adequate to deal with problems concerning length and other concepts defined in infinite languages, in 'Induction, Rational Acceptance, and Minimally Inconsistent Sets,' forthcoming in *Minnesota Studies in the Philosophy of Science*. Available from author on request.

[2] The extension of such languages is discussed in my paper 'Induction and Conceptual Change', *Synthese* **23** (1971) 206–225.

[3] In discussion Bar-Hillel noted that the proposed competition is not symmetrical. If q is a logical consequence of p but not vice versa, them q competes with p but not vice versa. This asymmetry may be justified as follows. Suppose that r is such that from the empirical evidence we could conclude that if q is true then r is false, though q and r are logically consistent with each other. Now suppose p is the statement $(q \& r)$. Then q is a logical consequence of p, but from the evidence we could conclude that if q is true, then p is false because r is false. Thus q competes p, we can conclude from evidence that if q is true than p is false, but p does not compete with q because we cannot conclude that if p is true, then q is false.

[4] Proofs of these claims are to be found in my 'Justification, Explanation, and Induc-

tion', in *Induction, Acceptance, and Rational Belief* (ed. by M. Swain), D. Reidel, Dordrecht, 1970, pp. 127–134.

[5] Others who have applied rules of expected utility to epistemic and cognitive problems are the following: Carl G. Hempel, 'Deductive-Nomological vs. Statistical Explanation', in *Minnesota Studies in the Philosophy of Science*, Vol. III, (ed. by H. Feigl and G. Maxwell), University of Minnesota Press, Minneapolis, 1962, pp. 98–169, Jaakko Hintikka and J. Pietarinen, 'Semantic Information and Inductive Logic', in *Aspects of Inductive Logic* (ed. by J. Hintikka and P. Suppes), North-Holland Publishing Company, Amsterdam, 1966, pp. 96–112, Risto Hilpinen, 'Rules of Acceptance and Inductive Logic', *Acta Philosophica Fennica* 21, North-Holland Publishing Company, Amsterdam, 1968, and Isaac Levi, *Gambling with Truth*, Alfred A. Knopf, New York, 1967.

[6] I examine some such shifts in 'Induction and Conceptual Change', cited above.

[7] Such rules are proposed by Hempel, Hintikka, Pietarinen, Hilpinen, and Levi, in sources cited above. I propose such a rule in 'Induction, Rational Acceptance, and Minimally Inconsistent Sets', cited above and in 'Belief and Error', forthcoming in a book edited by M. Gram and E. Klemke.

ILKKA NIINILUOTO

EMPIRICALLY TRIVIAL THEORIES AND INDUCTIVE SYSTEMATIZATION

In recent discussions about the role of theoretical terms in inductive systematization, a claim has frequently been repeated to the effect that a theory having no non-tautological deductive observational consequences cannot achieve any inductive systematization among observational statements. This claim has been made by Bohnert (1968, p. 280), Hooker (1968, p. 157–158), Stegmüller (1970, pp. 423, 428–429), and Cornman (1973), for the purpose of showing that the examples given by Hempel (1958, pp. 214–215), (1963, pp. 700–701), and Scheffler (1963, pp. 218–222) do not suffice to prove the indispensability of theoretical terms for inductive systematization. The purpose of this note is to criticize this claim in the form it is made by Stegmüller, who has given, as far as I know, the only explicit argument for it. Elsewhere (see Niiniluoto, 1973) I have tried to discuss more generally the notoriously vague notion of 'inductive systematization achieved by a theory' and some main problems connected with it.

To get an idea of the importance of the problem at hand, let us see how it arises. Let T_1 be a simple theory with two axioms

A1. $(x)(Mx \supset O_1 x)$
A2. $(x)(Mx \supset O_2 x)$,

where 'M' is a theoretical predicate, while 'O_1' and 'O_2' are (possibly complex) observational predicates. Now Hempel and Scheffler argue that from '$O_1 b$' we may induce 'Mb' by A1 and from 'Mb' then deduce '$O_2 b$' by A2. Therefore T_1 achieves inductive systematization between observational sentences '$O_1 b$' and '$O_2 b$'. However, the Craigian transcribed theory of T_1 and the Ramsey-sentence of T_1 are both logically true and cannot provide any inductive connection between '$O_1 b$' and '$O_2 b$'. Hence, the Craigian and Ramseyan substitutes of theories do not in general preserve inductive systematization.

The crucial premise in this argument is that a theory like T_1 can achieve inductive systematization. This has been denied by reference to two different kinds of counterarguments. The first refers to some properties of inductive inference (see Bohnert (1968, p. 280), Lehrer (1969), and Cornman (1973)). The second, in turn, points out that T_1 "has no deductive empirical content" (Hooker), is "immune to observational confirmation" (Bohnert) or "empirically trivial" (Stegmüller), and therefore does not achieve any systematization in the first place. If so, there is no systematization to be preserved, and the blame cannot be put on the methods of elimination of theoretical terms suggested by Craig and Ramsey.

If these counterarguments were true, then we would be left where we started: in the course of the debate, since 1958, whether inductive systematization is preserved in the elimination of theoretical terms, most of the theories which have reached any attention have been such that they cannot achieve inductive systematization. These two types of counterarguments do not seem to me conclusive, however. To be sure, the first one raises serious problems concerning very general features of inductive inference and shows that Hempel's argument, as it was cited above, cannot be true. But it seems, nevertheless, that we can, by means of a more definite sense of 'inductive systematization', give support to the claim that a theory like T_1 can, after all, achieve some inductive systematization. This question will be discussed in another paper of mine (Niiniluoto, 1973). The second counterargument will be the topic of this note.

The following terminology will be used in the sequel. Let T be a deterministic theory stated in a language L. Let L_B be the 'observational' sublanguage of L. Then T achieves *deductive systematization* (with respect to L_B) if and only if (i) $(T \& e) \vdash h$, but (ii) not $e \vdash h$ for some factual (i.e. nontautological) statements e and h in L_B. Let $\mathscr{I}$ be an inducibility relation characterized by a set of inductive rules. Then T achieves *inductive systematization* (with respect to L_B) if and only if (i) $(T \& e) \mathscr{I} h$, but (ii) not $e \mathscr{I} h$ (and not $(T \& e) \vdash h$) for some factual statements e and h in L_B. O_T is the set of the factual observational *deductive consequences* of T, i.e.

$$O_T = \{h \text{ in } L_B \mid T \vdash h \text{ and not } \vdash h\}.$$

Similarly, I_T is the set of the factual observational *inductive consequences* of T, i.e.

$$I_T = \{h \text{ in } L_B \mid T\mathcal{I}h, \text{ not } T \vdash h, \text{ and not } t\mathcal{I}h\}$$

where t is a tautology in L_B.

The nature of the inducibility relation $\mathcal{I}$ cannot be discussed in any detail here. Among possible candidates for it are all inductive acceptance rules, the positive relevance criterion ($e\mathcal{I}h$ iff $\mathcal{P}(h, e) > \mathcal{P}(h)$), or the high probability criterion ($e\mathcal{I}h$ iff $\mathcal{P}(h, e) > k$, where k is a constant). Practically all claims concerning inductive systematization are relative to the properties of the relation $\mathcal{I}$. In the sequel, it will suffice to refer to the following ones:

(C1) If g is probabilistically independent of h relative to e (i.e. $\mathcal{P}(h, e \& g) = \mathcal{P}(h, e)$), then $e\mathcal{I}h$ iff $(e \& g) \mathcal{I}h$.

(C2) If $(e \& g) \mathcal{I}h$, then $e\mathcal{I}(g \supset h)$.

(C3) If $e\mathcal{I}(g \supset h)$, then $(e \& g) \mathcal{I}h$.

The high probability criterion is easily seen to satisfy (C1) and (C2). The positive relevance criterion satisfies (C1) and also (C2) in some cases (e.g., if $\mathcal{P}(g, \sim h) \geq \mathcal{P}(g, e)$). Here (C1) is an inductive analogue of the principle of adding premises in deduction. (C2) is an analogue of the Deduction Theorem (which might be called 'The Induction Theorem'). The converse of (C2), that is (C3), is not true for the positive relevance or high probability criteria. In fact, it should not be satisfied by any reasonable inductive rule (cf. Niiniluoto, 1973).

Stegmüller (1970, p. 423) says that a theory T is empirically trivial, if T itself or its Ramsey-sentence is logically true. In particular, T is empirically trivial, if it has no deductive observational consequences. Moreover, empirically trivial theories (like T_1 above) cannot achieve inductive systematization.

To give a more detailed argument, Stegmüller considers three possibilities (pp. 428–429):

(1°) The theory T has no factual deductive observational consequences, i.e. $O_T = \phi$.

(2°) The theory is not empirically 'confirmable', 'corroborable' or 'testable' (empirisch bestätigungsfähig, prüfbar).

(3°) The theory T does not achieve inductive systematization (kann auch nicht als Mittel dafür verwendet werden, gewisse empirische Aussagen mit Hilfe anderer zu stützen).

And he claims that the following two theses can hardly be questioned:
THESIS I. (1°) implies (2°).
THESIS II. (2°) implies (3°).

As Stegmüller notes, Thesis I corresponds to Bohnert's claim that the theory T_1 is immune to observational confirmation. Besides this, we may note that Hooker (1968, p. 158) has argued that it is "not entirely implausible" to accept a thesis which amounts essentially to the claim that (1°) implies (3°).

Let us note first that (1°) is equivalent to

(4°)   The theory T does not achieve deductive systematization.

(Because of The Deduction Theorem, (4°) is equivalent to the condition: if not $\vdash (e \supset h)$ then not $T \vdash (e \supset h)$. This follows immediately from (1°). Conversely, if (1°) does not hold, then $T \vdash h$, for some h such that not $\vdash h$. But then $(T \& e) \vdash h$, for some e such that not $e \vdash h$.)

If the Theses I and II were correct, we could now conclude that (4°) implies (3°), i.e. that if the theory T achieves inductive systematization it achieves also deductive systematization. This is not very plausible, and it gives us a reason to suspect at least one of the Theses I and II. I shall argue below that by defining the notion of 'testability' in a suitable way we can make either of these theses true but not both of them at the same time.

Stegmüller does not give any explication of his crucial notion of 'Prüfbarkeit', nor does he make use of the extensive discussions about confirmability, testability, and corroborability between the Carnapians, the Hempelians, and the Popperians. There are readily at hand two possibilities to define this concept:

(A)   There is at least one factual observational statement which is positively or negatively relevant to T.
(B) T has at least one factual deductive observational consequence which is testable.

The definition (A) is suggested by the (positive) relevance criterion of qualitative confirmation and likewise by the proposals of metrical degrees of confirmation or degrees of evidential support (see the summary of them in Kyburg, 1970, p. 169). The definition (B) reduces the testability of a theory to the testability of its observational consequences. No general definition of the latter is attempted (nor needed) here. The definition (B) may, for instance, be interpreted as requiring that T has empirical content in the sense of Popper (1959), i.e. that the set of potential falsifiers of T is not empty. The definition (B) is obviously stronger than (A), because statements which are relevant to T need not belong to O_T. That is, (B) implies (A), but not conversely.

These definitions give us two interpretations of (2°), namely

($2°_A$) $\mathscr{P}(e, T) = \mathscr{P}(e)$ for all factual observational statements e.
($2°_B$) T has no testable deductive observational consequences.

(The third obvious possibility, viz. of requiring that T has no factual deductive observational consequences whatsoever does not interest us here, because it would make (2°) equivalent to (1°)). If the theory T is not confirmable or disconfirmable in the sense (A), then it is not testable in the sense (B), i.e. ($2°_A$) implies ($2°_B$). If $O_T = \phi$, then ($2°_B$) is trivially true. Hence, (1°) implies ($2°_B$). Every statement e in O_T (such that $\mathscr{P}(e) < 1$) is positively relevant to T. Hence, ($2°_A$) implies (1°), but not conversely.

From ($2°_A$) it follows that for all factual statements e and h in L_B, $\mathscr{P}(h, T \& e) = \mathscr{P}(h, e)$ (i.e. T is independent of h relative to e). The fact that T does not achieve inductive systematization (3°) is equivalent to the condition: if $(T \& e) \mathscr{I} h$ then $e \mathscr{I} h$ (or $(T \& e) \vdash h$). From $(T \& e) \mathscr{I} h$, ($2°_A$), and C1 we can conclude $e \mathscr{I} h$. Hence, assuming (C1), ($2°_A$) implies (3°). There is however no way of deriving (3°) from ($2°_B$).

It is interesting to note that while a theory achieves deductive systematization if and only if it has factual deductive observational consequences the same does not hold for inductive analogues of these notions. That is, (3°) is not equivalent to (5°), where

(5°) The theory T has no factual inductive observational consequences, i.e. $I_T = \phi$.

The reason that we cannot derive (5°) from (3°), or conversely, is that (C3) fails. However, (5°) is implied by ($2°_A$), assuming (C1). (To see this, it suffices to note that $I_T = \phi$ is equivalent to the condition: if $T \mathscr{I} h$ then $t \mathscr{I} h$ or $T \vdash h$).

The results that have been obtained can be summarized in Figure 1, where the arrows → indicate the implications that are found to hold:

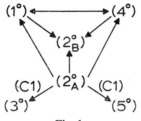

Fig. 1.

We have found no support for the thesis that the claim of 'empirical triviality' in the sense of (1°) of a theory is relevant to its possibility of achieving inductive systematization. This is reflected by the fact that Stegmüller's thesis (1°) → (3°) does not hold in our diagram. Instead, we have concluded that a theory which is empirically trivial in the strong sense (A) cannot have factual deductive or inductive consequences, neither can it achieve deductive or inductive systematization. In other words, a theory having inductive consequences or achieving inductive systematization is not trivial in the sense (A) of confirmability – even if it would lack deductive observational consequences.

National Research Council for the Humanities,
Helsinki, Finland

BIBLIOGRAPHY

Bohnert, H. G., 'In Defense of Ramsey's Elimination Method', *Journal of Philosophy* 65 (1968) 275–81.

Cornman, J. W., 'Craig's Theorem, Ramsey-Sentences, and Scientific Instrumentalism', forthcoming in *Synthese* 25 (1973).

Hempel, C. G., 'Implications of Carnap's Work for the Philosophy of Science', in *The Philosophy of Rudolf Carnap* (ed. by P. A. Schilpp), La Salle, Ill. 1963.

Hempel, C. G., 'The Theoretician's Dilemma: A Study in the Logic of Theory Construction' (1958), reprinted in *Aspects of Scientific Explanation and Other Essays in the Philosophy of Science*. The Free Press, New York, 1965.

Hooker, C. A., 'Craigian Transcriptionism,' *American Philosophical Quarterly* 5 (1968) 152–63.

Kyburg, H. E. Jr., *Probability and Inductive Logic*, Macmillan, London, 1970.

Lehrer, K., 'Theoretical Terms and Inductive Inference', *American Philosophical Quarterly* (Monograph Series, Studies in the Philosophy of Science, Monograph no. 3), Oxford 1969.

Niiniluoto, I., 'Inductive Systematization: Definition and a Critical Survey', forthcoming in *Synthese* **25** (1973).
Popper, K. R., *The Logic of Scientific Discovery*, Hutchinson & Co., London, 1959.
Scheffler, I., *The Anatomy of Inquiry*, Alfred A. Knopf, New York, 1963.
Stegmüller, W., *Theorie und Erfahrung. Probleme und Resultate der Wissenschaftstheorie und analytischen Philosophie*, Vol. II, Springer-Verlag, Berlin-Heidelberg-New York, 1970.

TOM SETTLE

ARE SOME PROPENSITIES PROBABILITIES?

1. This paper concerns presuppositions of the two extant propensity theories of probability: Peirce's and Popper's. Of course the frequency theory, the logical theory, and the subjective theory are more popular. By contrast with these theories, propensity theory digs deeply below the phenomenal surface, resembling in this respect what Bunge calls representational theories.

Indeed, propensities may be thought of as rooted in just those structures, characteristics and relations to which deep explanatory theories draw attention. This points to the main philosophical presuppositions of any theory of propensities – and noticing it helps us avoid certain confusions. For instance, failing to recognize propensities as deep, even occult, hence only indirectly testable, resembling in these respects field theories in physics, may easily mislead thinkers into confusing propensities with their chance outcomes or into confusing the sense of propensity with how propensities might be measured or tested.

2. The question to which this paper is addressed is: 'Are some propensities probabilities?' It is a deliberate inversion of the more customary kind of question which asks what kind of property probability is – for example: 'Is probability a propensity?' This customary kind of question challenges thinkers to *compare* rival theories of probability on the mistaken tacit assumption that there is only one proper theory – this in spite of all of Carnap's influence, though it was probably Popper who first pointed out that there may be many theories corresponding to various interpretations of the calculus of probabilities. By contrast, my aim here is to examine just one permissible theory of probability, the propensity theory, and to elucidate that theory by exposing a general background of philosophical presuppositions against which it might with advantage be set. Hence, I discuss PROPENSITY IN GENERAL and on condition of some propensities being probabilities, I focus on Popper's theory.

I choose to focus on Popper's theory rather than Peirce's, because Peirce's system is of absolute probability while Popper's is of relative; and I do not think it is possible in general to give an adequate statement of propensities of physical systems, whether the propensities be probabilities or not, without reference to the external conditions of the system, that is, to its perhaps virtual environment. The conventional ellipsis of ordinary language and even of much philosophical discussion of probabilities masks a relativity of propensities which is usually made explicit in physical theories (though not always: see, for example, the classical theory of nuclear disintegration). Popper's preference for relative over absolute probability *as primitive* matches my view that the system-environment complex is primitive, perhaps with isolated systems as special cases.

Thus the first presupposition to which I draw attention is ontological:

(1)   it is REALISM of a kind.

Propensity theory presupposes a real world as a locus of the propensities referred to. And the second is epistemological:

(2)   it is that PROPENSITIES ARE NOT OBSERVATIONAL, BUT DEEP.

Theories about propensities are intended to refer to deep physical properties of system-environment complexes.

Realism is not peculiar to propensity theory of probability, of course; but the presupposition of *depth* is: the more popular theories of probability tend to presuppose positivism and to be consistent with what Bunge calls black-boxism.

3.   I deliberately emphasize propensity-in-general as an OCCULT property to draw attention to a deep and far-reaching tacit error in modern philosophy of science, namely the assumption that proper knowledge claims require to be justified or at least to be justifiable. This assumption, coupled with the traditional empiricist assumption that experience is the only legitimate source of warrant for knowledge claims about the world, led to the widespread denunciation of speculation about what lies below the surface of things. Scientific knowledge was presented as a more or less satisfactory ordering of reports of experience. Where scientists'

speculations gave deep theories, thinkers in this empirical justificationist tradition reinterpreted those theories – to their ontological impoverishment. Propensity theory clashes with this tradition: (3) if it does not *presuppose* its falsity, at least it implies it, since no statement of propensities can meet its demands.

Popper has suggested giving up justificationism, thus allowing, once more, deliberate speculation about what the world is made of fundamentally. There need be no repetition of errors such as essentialism and intellectualism, since these views were themselves justificationist, though not empiricist.

Perhaps part of the task of science is freely to propose hypotheses concerning the hidden structure and properties of physical systems and environments which may explain their observable structure and properties, and perhaps even help us to control their future behavior in some respects. The rationality, and hence the intellectual respectability, of such speculation is perhaps sufficiently safeguarded if all hypotheses are held open to criticism, including empirical criticism, where any can be devised.

4. Propensity theory's presupposition of depth suggests it is not the primary concern of statements of propensities to anticipate outcomes of particular states of affairs but rather to draw attention to latent powers of system-environment complexes, to point to potentialities or *potencies* they may possess. In a word, propensity theory presupposes (4) dynamism.

Customarily, modern philosophers of science eschew the ancient concept of efficient cause – a concept captured roughly in ordinary language by such verbs as 'push', 'pull', 'move', 'produce', 'bring about'. In this they follow David Hume who correctly declared efficient causes to be unperceivable, but who in addition, and perhaps mistakenly, invited philosophers to treat only of the perceivable, invited them, that is, to restrict their scientific discourse to what might be thought empirically verifiable.

In this tradition, natural law, which presupposed efficient causation, came to be replaced by the correlation of antecedent with subsequent conditions, giving rise to a number of problems including the problem of distinguishing accidental from natural correlations! The chronically perplexing nature of this problem perhaps suggests the error of outlawing ontology (or metaphysics) and with it the idea of efficient causation or dynamism.

In spite of the strength of the Humean tradition, the idea of dynamic cause remains a presupposition of much discourse both in the arts – in law, in medicine, in historiography construed as narrative – and in the sciences, not to mention common speech. In physics, for example, we have not been able quite to rid ourselves of the notion of *force* despite some energetic (not to say forceful and influential) analytic attempts – notably Ernst Mach's. Of course, by no means all laws in science make use of the concepts of force, or power, or some other version of dynamism, and even in those which do, the idea of drive or cause is often masked by the symbolic representations of the theories, all of which might help explain why all scientific explanation is widely supposed to be merely the subsumption of instances under generalities (what Hempel calls deductive-nomological explanation) or its weaker statistical counterpart.

Of course, some explanations in science *are* merely matters of subsumption under general laws: this is not denied. Nonetheless, explanations satisfying the deductive-nomological schema are often unsatisfactory unless they are, in addition, *causal* or *dynamical* explanations in this sense:

> The explanans refers to forces which bring about the state of affairs described in the explanandum.

An example from physics is the use of the dynamic notion of gravitational collapse to explain both the genesis and the nemesis of giant stars or stellar systems (collapsars or 'black holes').

The idea of propensity is closely linked to the idea of natural law since natural laws imply propensities. A statement of a propensity may perhaps be regarded as a conjectural ontological hypothesis (or at least an hypothesis rooted in an ontology) concerning some physical property of a system-environment complex declaring or hinting at a natural (dynamical) *relation* between the complex in certain states and certain other, perhaps virtual, states into which it might enter. Hence propensity clashes with the popular theory of scientific explanation and may be said to presuppose (or at least to imply) (5) its falsity.

Incidentally, our consideration of propensity draws attention to the possibility that scientific theories differ from accidental generalizations by virtue of their ontological rootedness.

5. Nonetheless, the gap that remains between deep theory and reality is great and perhaps unbridgeable. In particular, we may notice that law statements refer simply to one level of analysis of reality (or to one degree of resolution or focus).

Programmes for the *integration* of bodies of theory referring to varying domains of reality remain incomplete and may be incompletable, despite a few successes, such as the integration of theories in electrical and magnetic domains. Programmes for the *reduction* of law statements at some levels to law statements at other levels have met with little or no success.

While it may be said that all systems of theories intended to illuminate or explain how the world appears to us, model the real world to an extent or after a fashion, including existentialist theories of man (such as Martin Buber's), the claim of any one particular theory system (say, elementary particle physics or existentialism-cum-phenomenology) to sovereignty over all is to be resisted in view of the illumination afforded our comprehension of reality by a variety of systems arising from a variety of types of analysis and comprising a variety of law statements, some qualitative, some quantitative, some deterministic, some stochastic.

In particular, the claim made by some philosophers, oblivious perhaps of the fragmented many-layered quality of the body of scientific knowledge, the claim namely that there are only two metaphysical options regarding orderliness – strict determinism and the doctrine of sheer chance – this claim is to be resisted.

Both strict determinism and the doctrine of sheer chance are extreme and happen to be oversimplifications: both of them oversimplify the relation between law statements belonging to different levels of analysis by supposing them all reducible to one rock-bottom level which is taken to be either deterministic or stochastic. Clearly, both strict determinism which contradicts ontic probability and the doctrine of sheer chance, which undermines the lawfulness propensity theory relies upon, are inimical to probabilities as propensities. Hence we may say that propensity theory of probability, though perhaps not propensity theory in general presupposes (6) the falsity of both those ontological doctrines as well as the falsity of the reductionism which they presuppose.

Of course, even if all natural laws were deterministic, there might still be statistical laws in science owing to our ignorance or to our habit of referring to just one level of analysis or of focus at a time. Such laws would be superficial: to uncover probabilities as propensities we should have to dig more deeply. The question whether some propensities, some natural hidden potencies of some systems in some environments might be probabilities is equivalent to the ontological question whether the universe is thoroughgoingly deterministic or at least in part stochastic.

According to me, the universe is at least partially indeterministic. HENCE SOME PROPENSITIES MIGHT BE PROBABILITIES. But this is not the place to defend this point of view. Here it is my task, rather, to expose what I think are the presuppositions of a propensity theory of probability. Two final remarks.

The first concerns Professor Suppes' criticism of Popper's propensity theory. So far, Suppes correctly says, no analysis has shown that propensities are probabilities, in the sense that they satisfy the axioms of the probability calculus. Of course, such a task need not be undertaken in general: it is the task of physical theorists to show in regard to any particular stochastic physical property of a system-environment complex, whether or not that property is probabilistic in the strict sense (as it is analogously for the subjective theory the task of psychologists or sociologists to show whether or not the betting behavior of otherwise rational men obeys the probability calculus).

The second is more general: for generations we have tried in science to do without ontology (or metaphysics) despite the deep roots of science in ontology. The consequent divorce of science and philosophy has weakened both. Perhaps the introduction of propensity theory, particularly, propensity theory of probability may help build the more fruitful relation between philosophy and science for which many of us hope.

University of Guelph

KLEMENS SZANIAWSKI

QUESTIONS AND THEIR PRAGMATIC VALUE

Let S be a set. The elements of S will be called 'states (of the world)'. They are assumed to be exhaustive and pairwise disjoint (the formal expression of this assumption is easily obtained if the elements of S are interpreted as sentences).

By 'question on S' we shall mean the ordered pair $Q = \langle X, \{p_s\}_{s \in S} \rangle$, where X is an exhaustive set of pairwise disjoint events and p_s is a family of probability distributions on X, parametric with respect to S. In other words, a probability distribution p_s on X is associated with s, for each $s \in S$. The elements of X will be called 'answers'.

The above definition of 'question' seems to cover all the essential uses of the notion. Thus, for instance, Q is a categorical question on S if all the probability distributions p_s are of zero-one type.

If Q is categorical two cases are possible. (1) For each x, there is exactly one s such that $p_s(x) = 1$. This establishes a one-one correspondence between X and S. An answer x can then be interpreted as the statement that the actual state of the world is this s which makes x certain. (2) For some x, there is more than one s such that $p_s(x) = 1$. In this case X defines a partition of S, the elements of which are: $S_x = \{s : p_s(x) = 1\}$, for $x \in X$. An answer x is then interpreted as the statement that the actual state of the world belongs to S_x. The remaining case of some x being such that $p_s(x) = 0$, for all $s \in S$, can be reduced to either (1) or (2), in view of the assumption that the set S is exhaustive; for such answers are impossible under all states of the world and, therefore, can be eliminated from the set X.

A question can be 'nearly categorical' if the distributions p_s approximate the zero-one type; i.e., if they are unimodal, the modal probability being sufficiently high. It is then natural to interpret the most probable answer, under the state s, as the statement that the actual state of the world is s (that it belongs to S_x, respectively). Now, however, the answer may be false, in contrast with the categorical question. The degree of reliability of the answer is determined by the values of the modal probabilities.

The above considerations show that semantic interpretation of answers (i.e., of the elements of X), if any, is entirely determined by the relation between S and X, i.e., by the character of the probability distributions p_s. Indeed, the semantic interpretation is secondary to the probabilistic relations between S and X. In some cases such an interpretation seems formally inessential. Consider, for example, the above-mentioned case of unreliable solution of Q. We so interpret X if for each x there is an s such that $p_s(x)$ is 'sufficiently high'. It is difficult to see how this limitation imposed upon the distributions p_s could be exploited in a theory of questions. Hence, it seems advisable to consider a general case, i.e., to impose no limitations on the distributions p_s, rather than single out 'unreliable solutions of Q' as a special case.

The pragmatic value of a question Q on S can be defined, relative to a decision problem the outcomes of which depend on S. Given such a problem $U = \langle A, S, u \rangle$, where A is the set of actions, and u is a utility function defined on the Cartesian product $A \times S$, any criterion of decision-making (maximization of expected utility included) generates a valuation of the question. The valuation is based on the fact that the choice of action can, in principle, be improved (in the sense of the criterion adopted) if it is made dependent on x.

Thus, in the general case i.e., when no limitations are imposed on the character of probability distributions p_s, the value of the question Q, relative to a decision problem and (for instance) an a priori probability distribution p on S is determined as follows.

Any use that can be made of having the question Q answered is represented by a decision function d from X to A. The set D of all d is an extension of A, in view of the fact that an action a in A is equivalent to a d such that $d(x) = a$ for all x in X. The role that Q plays consists therefore in providing the decision-maker with some additional ways of behaving. The profit to be derived from such an extension is defined by the difference between the value of the optimal d in D and that of the optimal a in A. In the case under consideration, i.e., when the probability distribution p on S is given, this difference is easily shown to be

$$C(Q; U, p) = \sum_x \max_a v(a, x) - \max_a \sum_x v(a, x),$$

where $v(a, x) =_{df} \sum_s u(a, s) \, p_s(x) \, p(s)$.

$C(Q; U, p)$ is read as: the value of the question Q, relative to the decision problem U and the a priori probability distribution p. So defined, the value of a question is, of course, non-negative: if in a given case the solution of Q is useless, relative to U and p, the optimal d is identical with optimal a, which makes C equal to zero.

If the a priori distribution is not assumed to be given, a criterion of decision making (such as, for example, maximin or minimax loss) plays the role of maximization of expected utility.

Several problems arise in connection with the maximal and minimal value of $C(Q; U, p)$. A number of them have been solved. The notion of the value of a question also has a bearing on the choice of a criterion: it seems plausible to reject criteria which make all questions valueless, relative to a non-trivial decision problem (i.e., such a problem in which no dominating action exists.)

University of Warsaw

GIUSEPPE TRAUTTEUR

PREDICTION, COMPLEXITY, AND RANDOMNESS

The initial consideration is based upon a conjecture of von Neumann (1966) which we will refer to as *VNHp*. This conjecture seems to be one of the few perspicuous explorations of the notion of complexity besides the common statements about the feeling that some new phenomena will emerge given some sufficiently complex system. *VNHp* is bound both to factual issues and to epistemological issues. Von Neumann never formulated it explicitly but a fair rendering is as follows: there exists a certain level of complexity, $\bar{n}$, such that for objects of complexity lesser than $\bar{n}$ it is simpler to describe what they do (the behavior) than how they are made (the structure); whereas for objects of complexity larger than $\bar{n}$ the opposite occurs. Any object such that it is simpler to say how it is made rather than what it does will be said to have the von Neumann property (*VNP*). Thus *VNHp* is simply the statement that all but the objects of complexity lesser than $\bar{n}$ have *VNP*. The context in which these ideas appeared was a discussion about the shape a logical theory of complex automata should have and the epistemological side of them is best expounded by an example given by von Neumann (1966) himself: "It is absolutely not clear *a priori* that there is a simpler description of what constitutes a visual analogy (what the visual brain does) than a description of the visual brain (how the visual brain is made)" (parenthetical notes are mine). On the other hand, the factual side of the discussion was concerned with the possibility of a self-reproducing machine and the feeling was present, based on the assumption that a self-reproducing machine should contain a description of itself, that the complexity of such a machine should be the critical value $\bar{n}$. However, the concepts appearing in *VNHp* were never formalized. The concepts and the new outlook which have been developed in the field of computational complexity may help to make a precise formulation of *VNHp*.

The universe in which we will be interested is one in which all the objects and structures are Turing machines of the Hartmanis and Stearns (1965) type, that is multi-tape, multi-head machines, with a non-erasable one-way output tape, which will be interpreted either as the means of communication with the outside or as representing the observable quantities associated with the machine. Although, as is well known, these machines are capable of computing the partial recursive functions, I will rather regard them as potentially infinite processes with the output tape as a record of the evolution of the system. Description will be identified with Gödel number in a fixed, fully effective Gödel numbering in the sense of Rogers (1958). This Gödel numbering, in a sense, will constitute the 'language' in which 'descriptions' are meaningful. The complexity will be the 'size of machines' defined by Blum (1967) as any finite-one recursive function ρ which assigns to any machine an integer which rates its complexity. The choice of a particular ρ will depend on external circumstances of fact and will vary from field of application to field of application. However, it is requested of ρ that there exists an effective procedure to tell which machines have a given complexity (ρ-value). Thus ρ explicates the phrase: how difficult it is to describe the structure of an object. It remains to define exactly the phrase: how difficult (complex) it is to say what an object does. I introduce here a notion of prediction. It seems obvious that if one is capable of saying what a machine does one must be capable of answering questions about the object ahead of time, or of some other resource, since otherwise it might have been more expedient to simply look at the object and our alleged knowledge would have appeared singularly useless. The definition of prediction is as follows: with each machine A_i there is naturally defined the function $T_i(n)$ which is the number of operations (or amount of some other resource) from start up to the printing of the n-th symbol on the output tape. A_j predicts $A_i (A_i <_p A_j)$ iff when A_i starts on some initial configuration α and A_j is started on some initial configuration β which encodes in a fixed manner i, α, m, d, the output tape of A_j, for some k, is identical between the k-th and the $(k+d)$-th symbol with the output tape of A_i between the m-th and the $(m+d)$-th symbol for all α, almost all m and all d and $T_j(k+d) < T_i(m+d)$. Of course, if A_i stops we consider its tape as completed by a string of blank symbols as long as necessary. The behavioral complexity $\hat{\rho}(i)$ of A_i is defined as the structural complexity, that is the ρ-value, of the smallest machine, if any, which predicts it, if this is smaller than $\rho(i)$, and $\rho(i)$ itself otherwise:

$$\hat{\rho}(i) = \min{}_z [\exists j [A_i <_p A_j \& \rho(j) = z < \rho(i)] \vee \rho(i) = z]$$

A machine has *VNP* if $\hat{\rho}(i) = \rho(i)$. This implies that all its predictors, if any, are structurally no simpler than A_i itself. It is immediate by padding in the enumeration of machines that if $<_p$ is not empty then there are infinitely many machines without *VNP*. On the other hand, using a theorem of Hartmanis and Stearns (1965), it is easy to build for any machine another machine which predicts it, since prediction in this very restricted sense, only requires a linear speed-up. In this set-up VNH_p is not true because there are machines of arbitrary complexity without *VNP*, but something interesting happens instead. It is clear that all machines with the lowest ρ-value have *VNP*. A simple induction shows that there are infinitely many more and in fact not only all their predictors are no simpler but all machines in the transitive closure of $<_p$ are not. We therefore have this structure of intermingled machines, some with *VNP* some without, which extends infinitely upwards.

A parallel phenomenon has been exploited to give epistemological substance to the notion of finite random string. As is now well known, Kolmogorov (1968) and Chaitin (1970) suggested independently, albeit following different formal developments, that a random object might be one such as to be very difficult to predict. Kolmogorov especially pursued the goal, along ideas sprung from von Mises' approach to probability, to put probability theory on an algorithmic base, via a definition of random string which satisfied both the formal requests of probability theory and the intuitive explicandum which lies behind. In particular, the finite random string which has a strong intuitive and practical appeal and which could not be situated anywhere in classical probability theory, was the natural approach. We will consider only these and try to show the generality and allusive, if not explicative, power of *VNP*. Both Kolmogorov and Chaitin suggest that a good candidate to the status of finite random string would be a string such that its inherent complexity is lesser or about the same as the complexity of a device capable of predicting it. This is very similar to the *VNP* for machines with the following changes: the function ρ becomes the inherent complexity of the string and is usually taken as the length of the string itself; while the behavioral complexity $\hat{\rho}$ becomes the Kolmogorov's complexity $K_A(x) = \rho(\min_y \times [A(y) = x])$. Under the simple condition on ρ, which must rate also the complexity of pairs of strings and satisfy $\rho(x, y) < C_x + \rho(y)$ for all strings y, the main theorem of Kolmogorov assures us of the existence of an universal programming system such that the complexity computed with respect to that system is not much larger than the complexity computed with respect to any other algorithm: $K(x) \leqslant K_A(x) + c$. It is also an immediate consequence of the definitions that K has a simple upper bound $K(x) \leqslant \rho(x) + c$. It is natural to say, in this framework, that a string has *VNP* if the ρ-complexity of the smallest algorithms which compute it fall within a fixed constant of the ρ-complexity of the string itself: $\rho(x) - c \leqslant K(x) \leqslant \rho(x) + c$. This coincides with the definition of finite random string given by Martin-Löf (1969) and when ρ is interpreted as the length of the string it turns out that the vast majority of strings are random in this sense. However, the length, or the length of the code, do not seem to be the only measures of interest because: (i) for such strings as proteins, capable of very refined folded structures it would certainly be convenient to assign different complexities, possibly not monotone with length, to strings of the same length, and (ii) it may also be convenient to assign complexities to objects not obviously unidimensional. In such cases the number of objects of given complexity may be any function and the previous result does not obtain. Therefore it seems not entirely useless to observe that nevertheless for any ρ satisfying the above requirements there are in fact infinitely many strings with *VNP*. The proof is the same as the one for machines when the relation $<_p$ is replaced by the relation $<_k$ defined by $x <_k y$ iff $U(y) = x$, where U is the universal programming system associated with K.

Now this rather pervasive phenomenon has a certain allusive power which expands in various directions. (i) Simon (1962) attempted to illustrate the fact that most complex systems of interest have a hierarchical structure, and this permits their analysis by much simpler means than the systems themselves. This may be interpreted by saying that in fact most systems of interest do not have *VNP* so that they can be predicted or explained away by simpler systems. (ii) An epistemological hope might be offered by the fact that even those systems with *VNP* which, being non-hierarchical, would have been considered by Simon not knowable, might possess approximations not necessarily simpler but without *VNP*. (iii) One might surmise that a well-known object without *VNP* might suddenly acquire it by being broken and thus explain why diagnostic problems are generally very difficult. Recently, Chaitin (1970) proposed a new criterion for life which seems to fit in the *VNP* scheme. His main point is that a living organism is an object with an inherent complexity lesser than the sum of the complexities of its components. Elaborating a little upon Chaitin's idea one might think of the decomposition of the object as a deduction of a formal grammar with the given object represented by the sentence symbol, every part of the object represented by some node and the elementary components of the object as terminal symbols. The branching at each node might be taken to represent the action of some operator putting together the higher node from more elementary ones. One may well think that the physical world may involve only a finite number of types of elementary objects and of ways of mutual interaction. Now a measure of inherent complexity is imposed on every object as a Blum size. But to every decomposition of an object one may assign a complexity value equal to the sum of the complexities of the parts in which the object can be exhaustively decomposed in accord with that decomposition. Thus, besides its own inherent complexity, to every object there belongs (with the exception of the elementary components) a set of – let them be called – decomposition complexities. Objects such that their inherent complexity is no larger than the smallest decomposition complexity would naturally be said to possess *VNP* and, according to Chaitin, something of a living nature.

Istituto di Fisica Teorica
Napoli

BIBLIOGRAPHY

Blum, M., 'On the Size of Machines', *Info and Control* **11** (1967) 257.
Chaitin, G. J., 'On the Difficulty of Computations', *IEEE* IT **16** (1970) 5.
Chaitin, G. J., 'To a Mathematical Definition of "Life"', *SIGACT News* No. **4** (1970) 12.
Hartmanis, J. and Stearns, R. E., 'On the Computational Complexity of Algorithms', *TAMS* **117** (1965) 285.
Kolmogorov, A. N., 'Three Approaches to the Quantitative Definition of Information', *Int. J. of Comp. Math.* **2** (1968) 157.
Martin-Löf, P., 'Algorithms and Randomness', *Rev. Int. Stat. Inst.* **37** (1969) 3.
Rogers, Hartely Jr., 'Gödel Numberings of Partial Recursive Functions', *JSL* **23** (1958) 331.
Simon, H. A., 'The Architecture of Complexity', *Proc. Am. Phil. Soc.* **106** (1962) 467.
Von Neumann, J., *Self Reproducing Automata* (ed. by A. W. Burks), University of Illinois Press, Urbana, 1966, pp. 47–56; p. 78.

JOHN M. VICKERS

RULES FOR REASONABLE BELIEF CHANGE

Probability, Transparency, Coherence

When probability is applied to events or propositions it depends upon a concept of necessity: Every necessary proposition has unit probability and a disjunction of pairwise necessarily incompatible propositions has as probability the sum of the probabilities of the disjuncts. Thus different concepts of necessity give different concepts of probability. Truth may be thought of as the limit of concepts of necessity; in this limit there is just one probability measure, that which assigns unity to every true proposition and zero to every false proposition.

Consider a class T of transformations on propositions which includes all tautologically implicative transformations, and each member of which is truth-preserving. Such a class gives a concept of *necessity* according to which a proposition is T-necessary if it is T-implied by every proposition, and gives also concepts of T-incompatibility, T-consistency and T-equivalence.

The relativization of probability to necessity can be expressed as relativity to a class T of transformations; a T-probability measure assigns 1 to every T-necessary proposition and is additive over T-incompatible disjunctions.

DEFINITION 1. p is a T-probability measure if and only if
(i) If A is T-necessary then $p(A) = 1$
(ii) If A and B are T-incompatible then
$p(A \vee B) = p(A) + p(B)$

A second way in which probability relates to necessity is by way of necessary equivalence: We say that a measure is *transparent* for an equivalence relation if the measure is invariant under replacement of equivalent arguments. Thinking of necessity in terms of classes of transformations:

DEFINITION 2. p is *T-transparent* if and only if: If A and B are T-equivalent then $p(A) = p(B)$

Yet another way in which probability relates to necessity is by way of *coherence*[1]: A *betting function* for an agent on a collection of propositions assigns a number to each proposition in the class which gives the odds at which the agent is willing to bet on the proposition: $b(A) = m/n$ when the agent will put up the proportion m/n of a stake S on condition that he receive S if A happens and lose what he put up otherwise. Such a function is said to be *incoherent* if there is some set B of bets in accordance with the function such that the agent will *necessarily* suffer a loss should he accept all the bets in B. A function is coherent if not incoherent. If we restrict the stakes to unity, then a simple way to think of coherence is as follows:

Let p be a measure on a logically closed class Ω of propositions. We call such a class a *field* of propositions. Define the *p-cost* of a subset X of Ω to be the summation of the values assigned to members of X by p. This is the amount a bettor would put up who bet on every proposition in X at unit stake. In order for p to be coherent there must be some consistent subset of X, the cardinality of which is no less than the p-cost of X. That is to say, there must be some maximally consistent subset of Ω, call it Φ_0, and the cardinality of the intersection of Φ_0 with X must be no less than the p-cost of X. For this cardinality is just what the above described bettor would receive should Φ_0 be actualized. Similarly, since coherence requires a willingness to take either side of a bet; in order for p to be coherent there must be some maximally consistent subset of Ω, call it Φ_1 such that the cardinality of the intersection of Φ_1 with X is no greater than the cost of X. Coherence requires that there be some possibility of not losing and some possibility of not winning.

Described in this way coherence is also seen to be relative to a notion of necessity. Still operating under the simplifying assumption of unit stakes, we have:

DEFINITION 3. A measure p on a field Ω of propositions is *T-coherent* if and only if for every subset X of Ω there are maximally consistent ϕ_0 and ϕ_1 subsets of Ω such that the p-cost of X is no less than the cardinality of $(X \cap \phi_0)$ and does not exceed the cardinality of $(X \cap \phi_1)$.[2]

These three definitions give distinct views of reasonableness; involving respectively probability, coherence and transparency.

The laws of probability have traditionally, since the time of the classical theorists, been viewed as characterizing reasonable belief. In Hume, where the distinction between reasonable belief and belief *simpliciter* is for important reasons ignored, there is an argument that the laws of probability based on the principle of indifference are a *priori* laws of thought; that our concept of partial belief is essentially probabilistic.

Coherence identifies unreasonableness of behavior with a tendency to put oneself into situations in which one is necessarily deprived of what he values. To be reasonable in this sense is to be insured not against loss, but against the certainty of loss. Taking into account the relativity of these notions; to be T-reasonable is to be insured against the T-certainty of loss. Of course after the bets have been paid off, the losing bettor whose beliefs were coherent but whose wagers were unfortunate is no better off than the bettor whose beliefs were incoherent and who on that ground lost the same amount. There is a limit sense of reasonableness in which any losing bettor's beliefs are unreasonable, this is the sense in which reasonableness is identified with belief in the truth, it is T-coherence where propositions are T-equivalent just in case they have the same truth-value, where implication reduces to the material conditional, and where the class of true propositions is also the class of T-necessary and the class of T-possible propositions. For this T there is but one T-probability measure; that which assigns unity to every true proposition and zero to every false proposition. But in general, we wish to allow a larger class of coherent belief measures so as to allow, for example, the distinction of an unreasonable bettor from an unfortunate one. For this reason some smaller selected T, some weaker notion of necessity, is usually used as a basis for coherence and probability.

Transparency gives another notion of reasonableness; conformity of beliefs to the laws of logic. A man who fails to believe the logical consequences of what he believes, or who believes an inconsistency is said to be unreasonable. As our view of logic gets stronger, as we allow more laws of logic, we get in this sense stronger views of reasonableness, since stronger views of logic put stronger requirements on consistency and license more entailments than do weaker views. Reasonableness based on transparency has to do also with the extent to which a man's beliefs are public, the more transparent his beliefs are to us, the more his beliefs agree with ours.

All these views of reasonableness also relate reasonable belief with belief *simpliciter* in analogous ways, namely reasonable belief differs from belief in being based upon a larger class of transformations. All belief is T-coherent, T-probabilistic and T-transparent so long as we pick a small enough class T. To say that belief is T-unreasonable is to say that it fails of T-coherence, T-probability or T-transparency in some sense T. But to support these claims we shall in general have to suppose that it satisfies these requirements for some weaker T.

These three views of reasonableness, each relative to a class T of transformations, can be shown to be in general equivalent for all classes T of transformations which are truth-preserving and which include all tautological transformations. The two theorems which elucidate this equivalence are:

THEOREM I. *If T is truth-preserving and includes all tautological transformations, then a measure on a class of propositions is a T-probability measure just in case it is T-transparent and (at least) a tautological probability measure.*

THEOREM II. *If T is truth-preserving and includes all tautological transformations, then a measure on a class of propositions is a T-probability measure just in case it is T-coherent.*

Theorem II is a generalization of the well known standard coherence results.[3] That coherence is in each case sufficient for probability can be established by the usual methods; given a function which violates some probabilistic law one devises a set of bets which shows incoherency. The sufficiency of T-probability for T-coherence is proved in general here as a consequence of a truth, the Boundary Theorem, which has also some other applications.

First the proof of Theorem I. It is simple enough, and in two parts:

Proof of Theorem I.

I.1. Every T-probability measure is T-transparent (assuming T to satisfy the constraints of the theorem).

Proof. Suppose p to be a T-probability measure and A and B to be T-equivalent. Then $A \leftrightarrow B$ is T-necessary so

$$p(A \leftrightarrow B) = 1$$
$$p(AB \vee \bar{A}\bar{B}) = 1$$

$$p(A\bar{B}) = p(\bar{A}B) = 0$$
$$p(A) = p(AB) = p(B).$$
So p is T-transparent.

I.2. If p is (at least) a tautological probability measure and is T-transparent then p is a T-probability measure (assuming T to satisfy the constraints of the theorem.)

Proof. It must be shown first that p assigns unity to every T-necessary proposition, and second that p is additive over disjunctions of T-incompatible propositions. To see the first notice that if A is T-necessary then A is T-equivalent to $A \vee \bar{A}$ so

$$p(A) = p(A \vee \bar{A})$$

and, since p is a tautological probability measure

$$p(A \vee \bar{A}) = 1 = p(A).$$

To see the second assume A and B to be T-incompatible. Then (since T includes all tautological transformations) $A \vee B$ is T-equivalent to

$$A\bar{B} \vee \bar{A}B$$

A is T-equivalent to

$$A\bar{B}$$

and B is T-equivalent to

$$\bar{A}B$$

$A\bar{B}$ and $\bar{A}B$ are tautologically, thus T, inconsistent, so $p(A\bar{B} \vee \bar{A}B) = p(A\bar{B}) + p(\bar{A}B) = p(A) + p(B) = p(A \vee B)$.

Now to the Boundary Theorem. The general form of the theorem is:

THEOREM III.

Boundary Theorem. If p is a T-probability measure on a denumerable field Ω of propositions and X is any subset of Ω, then there are maximally T-consistent subsets Φ_0 and Φ_1 of Ω such that

$$c(\Phi_0 \cap X) \leq \sum_{A \in X} p(A) \leq c(\Phi_1 \cap X)$$

where c gives the cardinality of its argument.

The Boundary Theorem can perhaps most easily be understood by seeing it first as it applies to probability measures in the classical sense. The classical definition of probability is the ratio of favorable to possible cases. This definition has two important restrictive features: (i) It is applicable only when the number of possible cases is finite. (ii) It assumes that all possible cases have the same weight.

The theorem is first developed below for this special case and then established in general by ramifying this simple development.

Let F be a finite set of logically T-independent propositions. In what follows T is assumed to be truth preserving and to include all tautological transformations. A *constitution* of F includes for each proposition in F either it or its negation but not both. Since the members of F are logically T-independent, every constitution of F is T-consistent. Let Ω be the field generated from F by including all finite truth functions of members of F. Suppose F to have just k members. If B is any member of the field Ω then B is T-equivalent to a disjunction

$$\Lambda F_1 \vee \Lambda F_2 \vee \ldots \vee \Lambda F_n$$

(ΛF_i is the conjunction of the members of F_i) where $F_1, \ldots, F_n$ are the constitutions of F each of which T-implies the proposition B.

We can think of a (T-possible) case as the T-closure of a constitution of F. For each constitution F_i the set of all those elements of the field Ω each of which is T-implied by F_i is the case Φ_i

$$\Phi_i = \{A \mid F_i \text{ } T\text{-implies } A\} \cap \Omega.$$

If p is any T-probability measure on Ω then for each member B of Ω,

$$p(B) = \sum_{B \in \Phi_i} p(F_i).$$

The *indifference measure*, q, is that T-probability measure which assigns the same probability, $1/2^k$, to each case. The value of this measure for a proposition $B \in \Omega$ can be expressed as the ratio

$$q(B) = \frac{\text{number of } \Phi_i \text{ of which } B \text{ is a member}}{\text{total number of } \Phi_i = 2^k}$$

or, more concisely

(a) $\quad q(B) = 1/2^k \cdot c(\{\Phi_\iota \mid B \in \Phi_\iota\})$

where c gives the cardinality of its argument.

Thus if X is any subset of Ω

$$\sum_{B \in X} q(B) = \sum_{B \in X} 1/2^k \cdot c(\{\Phi_\iota \mid B \in \Phi_\iota\})$$
$$= 1/2^k \sum_{B \in X} c(\{\Phi_\iota \mid B \in \Phi_\iota\}).$$

The sum

(b) $\quad \sum_{B \in X} c(\{\Phi_\iota \mid B \in \Phi_\iota\})$

is also

(c) $\quad \sum_\iota c(\Phi_\iota \cap X).$

To see the identity of (b) and (c) notice that a case Φ_ι will contribute one to the first sum for each element of $\Phi_\iota \cap X$.

Thus $\sum_{B \in X} q(B)$ is the average of the cardinalities of the intersections of the cases Φ_ι with X:

$$\sum_{B \in X} q(B) = 1/2^k \sum_\iota c(\Phi_\iota \cap X).$$

Since this sum is an average of the quantities $c(\Phi_\iota \cap X)$ it can neither exceed nor be exceeded by all of them. Thus:

III.1. *Special case of the boundary theorem.* If q is the indifference measure on a field Ω defined as above, then for each finite subset X of Ω, there are cases Φ_0 and Φ_1 such that

$$c(\Phi_0 \cap X) \leq \sum_{B \in X} q(B) \leq c(\Phi_1 \cap X).$$

III.1. is generalized below in the following ways.
1. The restriction to indifference measures is removed.
2. The restriction to specially structured Ω, that Ω be generated from a finite collection of T-independent propositions, is removed. Ω may be any denumerable field of propositions and there may be infinitely many T-independent propositions in Ω.

3. The restriction to finite X is removed, so we consider also the generalization of III.2 where $\sum_{B \in X} p(B)$ is not finite. In this case the boundary theorem asserts that a set has a denumerably infinite T-consistent subset so long as some T-probability measure has an infinite sum over the set.

These restrictions are removed in the order stated. (1) and (2) are unproblematic; the technique differs from that used to support III.1 above mainly in that weighted averages are used in place of the simple average

$$1/2^k \sum_i C(\Phi_i \cap X).$$

The removal of the restriction to finite X is straightforward consequent upon (1) and (2). Some variation in method is required, since if Ω includes infinitely many T-independent propositions, the cardinality of the collection of maximally T-consistent subsets of Ω is non-denumerable. This means that the additivity of T-probability measures can no longer be employed, since this characteristic is restricted to denumerable collections of arguments. Thus the additivity employed in supporting III.1 is no longer possible. It is not true in general that the T-probability of a proposition is the sum of the probabilities of the T-possible cases[4] in which it obtains. For there may in general be non-denumerably many of these cases if we take them to be maximally T-consistent subsets of Ω. For this reason the argument in case (3) differs in character from those to IV.1 and in cases (1) and (2).

Cases (1) and (2) are established in the

LEMMA III.2.
Finite Boundary Theorem. If p is a T-probability measure on a field Ω of propositions and X is a finite subset of Ω, then there are maximally T-consistent subsets of Ω, Φ_0 and Φ_1, such that

$$c(\Phi_0 \cap X) \leqslant \sum_{A \in X} p(A) \leqslant c(\Phi_1 \cap X).$$

Proof. Since X is finite the difficulties inherent in handling the non-denumerable collection of maximally T-consistent subsets of Ω can be avoided by making use instead of the finite collection $\{X^j\}$ of T-consistent constitutions of X. For if Φ_i is any maximally T-consistent subset of F,

RULES FOR REASONABLE BELIEF CHANGE 137

there is some X^j, a T-consistent constitution of X, such that

$$(X^j \cap X) = (\Phi_t \cap X).$$

For each T-consistent constitution X^j of X, define the function pj on F.

$pj(A) = 1 \Leftrightarrow X^j$ T-implies A
$pj(A) = 0$ otherwise.

Clearly each pj is a two-valued probability measure, and the sum of the values of pj over the members of the set X is just the cardinality of the intersection of X^j and X

$$\sum_{A \in X} pj(A) = c[X^j \cap X].$$

In each case $X^j \cap X$ is T-closed within X, that is to say, if X^j T-implies the proposition A of X, then $A \in X^j \cap X$. Thus for each constitution X^j of X there is some maximally T-consistent ϕ_t such that

$$(\Phi_t \cap X) = (X^j \cap X).$$

In view of this, attention can be directed to the field $\Omega(X)$ consisting of all finite truth-functions of propositions in X. The analogue of (a) above is

LEMMA III. 3 If A is a proposition in the field $\Omega(X)$ then

$$p(A) = \sum_j [p_j(A) \cdot p(\Lambda X^j)].$$

Proof of the lemma. Define the function g on $\Omega(X)$ which gives for each $A \, \varepsilon \, \Omega(X)$ the collection of those constitutions X^j each of which T-implies A. Then, since for each $A \in \Omega(X)$ and each X^j either X^j T-implies A or X^j T-implies $\bar{A}$,

$$g(A) \cup g(\bar{A}) = \Omega(X).$$

Each proposition A is T-equivalent to the disjunction of all those ΛX^j which T-imply it. Since these conjunctions are pairwise T-inconsistent, the value of p at A is just the sum of its values for the ΛX^j which T-imply it.

$$p(A) = \sum_{X^j \in g(A)} p(\Lambda X^j).$$

p_j, by definition, assigns unity to just those propositions in $\Omega(X)$ which are T-implied by X:

$$X_j \in g(A) \Leftrightarrow p_j(A) = 1.$$

Thus
$$p(A) = \sum_{X^j \in g(A)} p_j(A) \cdot p(\Lambda X^j)$$

and if $X^j \in g(\bar{A})$, then $p_j(A) = 0$. So

$$\sum_{X^j \in g(\bar{A})} p_j(A) \cdot p(\Lambda X^j) = 0.$$

Hence $p(A)$ can be given as a weighted sum of the values the various p_j assign to A

$$p(A) = \sum_j p_j(A) \cdot p(\Lambda X^j)$$

which establishes the lemma.

The lemma entails first that

$$\sum_{A \in X} p(A) = \sum_{A \in X} \sum_j p_j(A) \cdot p(\Lambda X^j)$$

so
$$\sum_{A \in X} p(A) = \sum_j p(\Lambda X^j) \cdot \sum_{A \in X} p_j(A).$$

This is the analogue of the move from (b) to (c) above complicated here by the weightings $p(\Lambda X^j)$. The result is that the sum of p over X is expressed as a weighted average of the values $\sum_{A \in X} pj(A)$ and can neither exceed all these quantities nor be exceeded by all of them. In view of the relations of pj, X^j, and the maximally T-consistent Φ_ι, we conclude that for some maximally T-consistent Φ_0, Φ_1, subsets of F.

$$c[\Phi_0 \cap X] = \min_j [X^j \cap X] \leq \sum_{A \in X} p(A) \leq \max_j c[X^j \cap X] =$$
$$= c[\Phi_1 \cap X]$$

which establishes the finite boundary theorem.

The next step in generalization is to remove the restriction to finite X.

Proof of Theorem III. It must be assumed that the set T of transformations is *compact* in Ω. That is to say that every T-inconsistent subset of Ω has a finite T-inconsistent subset.

If X has a finite subset X^1 such that $p(A)=0$ for all $A \in X - X^1$, then this reduces to the finite form of the theorem. So suppose there to be no such finite set.

Since X is denumerable, we can assume an enumeration $A_1, A_2, \ldots$ of X and a corresponding sequence

$$X_0 = \Lambda, X_{n+1} = X_n \cup \{A_{n+1}\}$$

of finite subsets of X such that

(i) $\quad \cup X_\iota = X$

(ii) $\quad \lim_{\iota \to \infty} \sum_{A \in X_\iota} p(A) = k.$

We say that such a sequence of sets $\{X_\iota\}$ is a *covering nest* of X. Clearly the limit k is invariant regardless of which nest of subsets is chosen.

Two cases are distinguished according to whether the sum of p over X is finite or infinite. Suppose this sum to be finite. Then III.2 entails that $c[\Phi_\iota \cap X]$ is finite for some ι. For if no such intersection is finite then the quantity $\min_j c[\Phi_j \cap X_i]$ increases without bound as i indexes increasingly larger members of the covering nest $\{X_i\}$ and hence may be made to exceed any given finite quantity.

The assumption of compactness entails that if X intersects every maximally T-consistent subset of Ω, then some finite subset of X intersects every such set. And from this it follows that if $\min_j c[\Phi_j \cap X] = m$ then for some finite subset X^* of X, $\min_j c[\Phi_j \cap X^*] = m$.

Now to establish the lower bound part of Theorem III in case the sum of p over X is finite, assume this to be false. We argue that III.a is contradicted.

Let $\min_j c[\Phi_j \cap X] = m$. So our assumption is that $\sum_{A \in X} p(A) < m$. As we argued above, X has a finite subset X^* such that $\min_j c[\Phi_j \cap X^*] = m$. And since

$$\sum_{A \in X^*} p(A) < \sum_{A \in X} p(A) < m = \min_j c[\Phi_j \cap X^*]$$

the finite form of the theorem is contradicted.

This shows that so long as T is compact in Ω, if $X \subseteq \Omega$ then for some maximally consistent subset Φ_0 of Ω,

$$c[\Phi_0 \cap X] < \sum_{A \in X} p(A).$$

This establishes the existence of a Φ_0 as asserted in the theorem in the case in which the sum of p over X is finite. Still assuming this case, we now argue to the existence of a maximally T-consistent Φ_1 such that

$$\sum_{A \in X} p(A) \leqslant c[\Phi_1 \cap X].$$

Assume that the largest T-consistent subset of X is of size n less than k. Then for some X_ι in the nest $\{X_\iota\}$

$$n < \sum_{A \in X_\iota} p(A) \leqslant k$$

and by the finite form of the theorem there is some T-consistent subset of X_ι, and hence of X, of size no less than

$$\sum_{A \in X_\iota} p(A) > n.$$

Contradicting that n is the size of the largest T-consistent subset, and thus leading to the rejection of the assumption. Thus X has some T-consistent subset of at least size k. Since this subset has a maximally T-consistent extension, the existence of an appropriate upper bound on the sum of p over X is established, which completes the argument for the first case.

In the second case the sum of p over X is denumerably infinite. The existence of a lower bound is in this case trivial. To establish the case and the theorem it will suffice to show that X has an infinite T-consistent subset. Let $\{X_\iota\}$ be a covering nest for X. For each ι let

$$x_\iota = \sum_{A \in X_\iota} p(A)$$

then, since $\sum_{A \in X} p(A)$ is finite, as the subscript ι increases without bound, so does x_ι. Were X to have no T-consistent subset larger than some finite k, then k would be a bound on x_ι and $\sum_{A \in X} p(A)$ would also be finite.

Remark on Kyburg's lottery paradox[5] in the light of Theorem II: The lottery paradox depends upon the fact that for plausible T-measures p, there are T-inconsistent sets every member of which has a p-value arbitrarily close to unity.

Consider, for example, a finite lottery of k tickets. Suppose T to be such that it is not T-possible for any two tickets to win, nor T-possible for all k tickets to lose. Let p be a T-measure which assigns $1/k$ to each proposition asserting that a given ticket loses.

Now the sum of the p values of the set of assertions that each ticket loses is $k - 1$ and thus Theorem II guarantees that there is some subset of this set of size $k - 1$ which is T-consistent. Theorem III also implies that *every* subset of size $k - 1$ is consistent; that it is consistent to assert of any $k - 1$ tickets that each of them loses.

The lottery paradox provides examples of sets of propositions not all of which can consistently be accepted. Theorem III gives in some cases largest subsets of these sets, all the members of which can be consistently accepted.

Carnap's [6] approach to the problem of characterizing reasonable belief was to look for constraints in addition to the laws of probability which could be placed on measures of belief. He assumed a clear distinction between the necessary and the contingent, and saw the problem as one of varying the class of measures without varying this distinction. The present approach is contrary to this: The distinction between the necessary and the contingent is seen as variable within certain limits, and the class of probability measures is narrowed by strengthening the concepts of necessity upon which they are based.

When we turn to the problem of characterizing reasonable ways of changing beliefs through time, it is natural to think of reasonable change in terms of successive increases in the class of transformations for which belief is transparent. We say that such change is *convergent*. This conforms to a principle due to Peirce, though not quite in this form, that reasonable belief change should converge in the limit to truth.

Restriction to convergent belief change is also a generalization of Carnap's principle that belief in A at a time should be equal to belief in A conditional upon E at an earlier time, where E is the total observed in the interim. This principle depends again upon a fixed distinction between the necessary and the contingent, to distinguish what can from what cannot be observed. Convergence is what remains of this principle when this distinction is no longer assumed.

Claremont Graduate School, California

NOTES

[1] For a guide to the considerable literature on this subject see the introduction to Kyburg and Smokler (1964).
[2] The restriction to bets of unit stake is innocuous in view of the weakness of the concept of *utility*. The measurement of utility is invariant only up to the assignment of a zero and a unit, and thus, if wagers are placed in utility, any collection of them is equivalent to some collection with unit stakes.
[3] As found in Ramsey (1950), de Finetti (1964), Shimony (1955), and Lehman (1955). for example. See Note 1 above.
[4] Strictly speaking this can be understood only when probability is defined for sets of propositions. A serviceable definition is to take the conjunctive probability of a set to be the infimum of the probabilities of finite conjunctions of members of the set.
[5] See, for example, Kyburg (1961).
[6] See Carnap (1962).

BIBLIOGRAPHY

Carnap, Rudolf, *Logical Foundations of Probability*, 2nd ed., Univ. of Chicago Press, Chicago, 1962.

Finetti, Bruno de, 'Foresight: Its Logical Laws, Its Subjective Sources' in *Studies in Subjective Probability* (ed. by H. E. Kyburg, Jr. and H. E. Smokler), John Wiley & Sons, Inc., New York, 1964.

Kyburg, Jr., Henry E., *Probability and the Logic of Rational Belief*, Wesleyan Univ. Press, 1961.

Kyburg, Jr., H. E. and Smokler, H. E. (eds.) *Studies in Subjective Probability*, John Wiley & Sons, Inc., New York, 1964.

Lehman, R. Sherman, 'On Confirmation and Rational Betting', *Journal of Symbolic Logic* **20** (1955) 251–262.

Ramsey, Frank P., *The Foundations of Mathematics and Other Logical Essays* (ed. by R. B. Braithwaithe), Humanities Press, New York, 1950.

Shimony, Abner, 'Coherence and the Axioms of Confirmation', *Journal of Symbolic Logic* **20** (1955) 1–28.

PART III

LANGUAGE
(Section XI)

TEUN A. VAN DIJK

MODELS FOR TEXT GRAMMARS*

1. INTRODUCTION: TOWARDS TEXTUAL LINGUISTICS

1.1. One of the major recent developments in linguistics is undoubtedly the rapidly growing interest for the elaboration of a theory of texts and 'text grammars'. What first seemed to be merely a series of scattered and rather marginal attempts to extend current sentence grammars, now turns out to be a wide-spread and even well-programmed 'movement' to establish a more adequate type of grammar.[1]

It is the aim of this paper to discuss briefly the main ideas underlying this tendency, especially with respect to their methodological status within the theory of linguistics, and to present a hypothetical framework for the construction of explicit text grammars. In this early state of the investigation it will be specifically necessary to consider the possible heuristic and theoretical models which could serve the elaboration of such grammars. For historical surveys about work done in discourse analysis, text grammars, and related domains we have to refer to the literature mentioned in the notes. Similarly, for the details of the descriptive, theoretical, and methodological arguments, in favour of textual linguistics.[2]

The motivations for the empirical and theoretical interest in the description of the structures of discourse have been provided both within linguistics and within such neighbouring disciplines as anthropology, poetics, and social psychology (content analysis). Although a great part of valuable descriptive work and also some models have been provided rather early in these last disciplines, we will here be exclusively concerned with the proper linguistic aspects of text structure. This restriction also has methodological reasons: the description of any type of discourse should be based on an explicit knowledge of its general or specific linguistic properties as they are formalized in text grammars and their (meta-) theory.

1.2. The current linguistic approaches to the study of texts can roughly be divided into two directions, although the interest of recent suggestions precisely has to be sought in the attempt to combine these different orientations into one grammatical theory.

First of all, there is the wish to describe discourses or texts in their own right, and not merely as a collection of sentences. In that case it is the text and not the sentence which is considered to be the abstract basic unit underlying the empirical performance unit of the utterance. Harris' 'discourse analysis' can be considered to be an early descriptive attempt in this direction, although it merely provides some rather superficial characteristics of textual structure. Theoretically more adequate studies of this type have recently been provided, especially in Germany.

The second series of studies more closely relate to the central topics discussed in current generative theory. They may be considered as contributions to the extension of sentential grammars, in order to provide more adequate descriptions of the different phenomena characterizing the relations between connected (embedded, coordinated or subsequent) sentences: anaphorical relations, definite articles, relative clauses, tense consecution, modal adverbs, focus, topic and comment, presupposition and entailment, and so on. We see that the need felt here for the description of discourse directly derives from the theoretical problems encountered in the description of sentence structures.

The problem at issue is the question whether textual structures can be adequately accounted for by such 'extended sentence grammars'. That is, can a generative sentence grammar in principle specify, in a consistent and relatively simple way, sequences of sentences with their structures and the relevant grammatical relations and conditions that hold between related sentences in a coherent text? If it does, is there a type of grammar which performs the same task in a more adequate way? And if it does not, which type of new grammar will be necessary to account for the empirical facts, i.e. structures and relations in coherent texts?

1.3. Such questions cannot be solved on aprioristic grounds, and require extensive methodological and empirical research. There must be adduced explicit reasons, firstly, to take the text, and not the sentence, as a basic abstract object for grammatical description and, secondly, there must be decisive arguments for the assumption that extant sentence grammars cannot possibly account for all relevant textual structures.

The first set of reasons is of socio-psychological nature and has often been given now. Some of them may be listed here:

(a) The observable manifestations of language as a psychological and social system, viz. utterances, are tokens of texts rather than tokens of sentences.

(b) Native speakers are able to process (produce, receive, interpret) such utterances as coherent wholes and not merely as a sequence of (unrelated) sentences.

(c) Native speakers are able to recognize different types of relations between the sentences constituting the text of which the utterance is a performance realization. This fact enables them to differentiate between grammatical and less grammatical texts.

(d) Native speakers are able to recognize different relations between utterances/texts which do not directly depend on the respective sentence structures alone: paraphrases, abstracts, question/response, intertextual reference, etc.

(e) Native speakers do not process texts/utterances verbatim, e.g. by literal recall and by programming sentence structures, but by the formation of (underlying) plans.

These few socio-psychological facts have to be explained by an adequate theory of language, i.e. they are part of the empirical domain of linguistic theories. Not treating them, therefore, may be motivated only by explicit practical reasons of, say, feasibility with respect to the actual state of the theory, not by unprincipled and arbitrary reduction of the domain. Although the set of empirical phenomena which a theory is purported to account for is pragmatically determined by the conventions in a given discipline, we might, with e.g. Sanders (1969), speak of the 'natural domain' of a linguistic theory. Thus, in order to provide the formal model for the linguistic abilities listed above, a theory of language has to specify a type of grammar which does not have the sentence as its natural domain, since relevant linguistic properties would then not be accounted for. Similar arguments have earlier been brought forward, by generative grammar itself, against all types of 'word' – based grammars.

Once adopted the text, or rather the infinite sets of possible discourses and their underlying formal (textual) structures, as the natural domain of grammatical theory, it is necessary to stress that this object is not (only) accounted for by a theory of performance. That is, the rules and conditions determining the ideal production and reception of texts are systematic and not ad hoc or probabilistic. They are therefore part of linguistic competence.

The second set of reasons that have led to the elaboration of text grammars presupposes these empirical facts and is of a more formal nature. We must prove that S-grammars are inferior in weak and/or strong descriptive (and explanatory) adequacy with respect to T-grammars. Although a strict proof comparing different types of grammars can be given only for completely formalized systems, sufficient arguments can be adduced for the descriptive inadequacy of S-grammars. The important fact is that T-grammars not only describe intersentential and textual relations but also purport to yield more satisfactory descriptions of the traditional phenomena described by S-grammars. This means that extant generative grammars would not simply constitute a sub-set of rules (and categories) of future T-grammars, but would themselves first be highly modified. It is possible that only rather few strictly intrasentential aspects would be accounted for by the S-component of a T-grammar. Not only the well-known phenomena of cross-reference enumerated earlier would fall under the special textual rules and constraints, but also many rules of stress and intonation, many properly syntactic formation and transformation rules (e.g. topicalization), and probably all semantic and lexical rules. That is, many of the traditional 'context'-free or (sententially) 'context'-sensitive rules of the grammar turn out to be 'text'-sensitive.

1.4. As was remarked above, S-grammars and T-grammars can be explicitly compared only if this last type is worked out in detail (and if both are formalized). Given the empirical facts and the theoretical assumptions mentioned in the previous paragraph, we may adopt the hypothesis of Sanders (1969) that S-grammars are not 'natural theories' of specific languages, because the infinite set of sentences is not a natural domain since it is included in the infinite set of the well-formed texts of a language. S-grammars thus satisfy the set of reducibility conditions with respect to T-grammars: all their interpreted elements, their axioms, and their rules are proper subsets of those of T-grammars.

In many of the attempts mentioned earlier this proof of proper inclusion has not been provided. That is, although existing generative S-grammar has been extended on several points, these extensions, e.g. by complementary derivational constraints (rules, conditions), need not be inconsistent with the axioms and the other rules. If an S-grammar specifies, by rule schemata and/or recursive (initial) elements, sequences of sentences and the set of conditions determining their linear coherence, they are at least equivalent in weak generative capacity with postulated text grammars, simply by reducing texts to linearly ordered sequences of sentences. The proof to be given, then, must show that:

– T-grammars describe these conditions in a more consistent and a more simple way, and/or,

– T-grammars describe structures which cannot possibly be described by the axioms, categories and rules of extended S-grammars.

The first point is not wholly of a meta-theoretical nature alone. If we adopt the hypothesis that a generative grammar in principle is a formal model of underlying psychological abilities (competence), its relative simplicity is somehow related to the mental capacities of native speakers. It is clear that extremely complex conditions (say of intersentential coherence in a longer text) must in principle be processable by humans if the grammar should provide an adequate model. Now, it can be shown that any linear account of the (e.g. dependency) relations between sentences in a coherent text leads to rules of an as yet unmatched complexity, far beyond the limits of human linguistic competence. As an informal example we may mention the set of relevant semantic relations holding in a text: a speaker will normally be able to process (produce/ recognize/memorize, etc.) these relations between immediately subsequent sentences (and their respective deep structures), but in processing longer texts he is unable to memorize all these relevant relations in order to establish the global coherence of the text as a whole. Clearly, a process of abstraction (formation of plans) is involved as modelled by macro-rules and macro-structures postulated in text-grammar.[3] This argument is probably most decisive for the elaboration of an entirely different type of grammar, excluding the simple 'extension' of S-grammar, as advocated by the mentioned linguists.

We arrive here at the second part of the demonstrandum: can (extended) S-grammars describe such macro-structures? If not, they are also on this point inferior in strong generative capacity. Indeed, whereas it is rather misleading to use one initial symbol (S) for the derivation of sequences of sentences, it seems impossible to use the sentential initial symbol for the derivation of very abstract underlying and sentence-independent macro-structures. The categories (and rules) involved here can therefore not possibly be derivable from it, hence $G'_s \subset G_T$. The general task of T-grammars is thus twofold:

– they must specify sentential structures and the linear (surface) coherence relations/conditions holding between them in a coherent text (thus providing more adequate descriptions also of intra-sentential phenomena).

– they must specify macro-structures defining global coherence by the introduction of specific axioms and categories, the formulation of their formation and transformation rules, and the rules relating macro- or deep-structures with sentential ('surface') structures.

Before we consider the models for these two specific components of T-grammars, let us make some hypothetical remarks about their possible form, rules and categories and their respective descriptive tasks.

2. Some properties of T-grammars

2.1. One of the most striking aspects of the description of the relations between sentences in a coherent text is the close interdependency of many superficially rather diverse phenomena. Indeed, stress and intonation seem to be dependent on the relations of identity and opposition between semantic representations (van Dijk, 1972a). The same is true for the differentiation between topic and comment, focus and syntactic order (Dahl, 1969). The conditions determining pronominalization, relativization, definitivization are closely related, if not identical, and also depend on underlying semantic representations of subsequent sentences or clauses.

Clearly, S-grammars can, and did, describe such conditions for complex and compound sentences, especially with respect to pronominalization.[4] It turns out that similar conditions hold for structurally (i.e. syntactically) independent subsequent sentences which, while not dominated by one S-symbol, are not accounted for by such grammars.

We will therefore consider the relations between clauses in complex or compound sentences as special cases of the more general textual relations between sentences, or rather between their underlying 'propositions' or 'logical forms', be these described in syntactic or semantic terms. This implies that any complex sentence can be paraphrased by a set of simple sentences. The converse, however, is not always the case: not any well-formed sequence of simple sentences can be reduced to a complex sentence with the same meaning, as has often been assumed. Thus, an *S*-grammar not only would arbitrarily exclude grammatical sequences of sentences from linguistic description, but at the same time lacks some important features of generalization.

2.2. An example may be found in the conditions determining pronominalization. These are generally based on the reputedly complex notion of 'referential identity', a term which in syntactically based grammars may be made explicit only by such devices as indices, anaphoric features, or the introduction of (bound) variables.[5] We will see in the next section that at this point models from mathematics and predicate logic lie at hand, but we will restrict ourselves here to an informal description. Although many pronouns denote the same referent as their antecedent(s), there are many cases of pronominalization where referential and lexico-semantic relations of identity, membership or inclusion may underlie the generation of pronouns.

Conversely, there are cases where pronominalization is impossible although the condition of referential identity is fulfilled, for example in cases of semantic contrast and possible anaphorical ambiguity. The general rules for pronominalization as formulated by Langacker (1969) are based on the syntactic properties of the primacy-relations 'preceding' and 'commanding', defined for clauses, or rather for *S*-dominated sub-graphs of one derivational tree. According to these rules pronominalization may take place (roughly) in forward direction and backwards if and only if the antecedent (postcedent) commands the pronoun, i.e. if it is part of a sentence in which the sentence containing the pronoun is also contained. These rules provide correct predictions but lead to a paradox (the Bach-Peters-Kuno paradox) when the *NP*'s to be pronominalized contain relative clauses with pronouns having identical reference with pronouns contained in (the relative clause of) the antecedent or post-cedent, thus leading to infinite regress in mutual substitution of noun-phrases, like in such sentences as *The pilot who shot at it hit the Mig that chased him.*[6]

Recent discussions have made clear that in such cases intricate structures of definite descriptions are involved at different levels of depth, and that unambiguous reference can be made only to immediately dominating *NP*'s with their own descriptive 'scope'.

These conditions can probably be formulated in a much simpler and more general way within a *T*-grammar. It turns out, for example, that the notion of (syntactic) dominance in fact is probably derivable from the relation of preceding, defined for subsequent 'sentences' (propositions) underlying the surface structure of a text. Thus, (restrictive) relative clauses are derivable from preceding sentences $S_{i-1}, S_{i-2}, \ldots$ where the most deeply embedded S_x has the lowest index of order. When order is not reflected in the depth of embedding we obtain ambiguous or semi-grammatical sentences like the one mentioned above. At the same time we are thus able to specify the temporal, causal and other semantic relations between the clauses of complex sentences. A sentence like *The pilot who hit it shot at the Mig*, is ungrammatical because it is derived from the ungrammatical sequence *The pilot hit the Mig, The pilot shot at the Mig*, where order (temporal succession) is linked with the relation between cause and effect. The definite description of $N(P)$'s, then, is a linear process of progressive specification, narrowing the possible referential scope of the (pro-) noun, in general to a single individual (or single class). However, natural language has some freedom in the application of these rules.

2.3. Most interesting of this type of textual description of pronouns (and complex sentences) is the fact that the conditions for definitivization are practically identical. Besides the pragmatico-referential condition stating that the referent denoted by the *NP* is supposed by the speaker to be known to the hearer, we have the syntactico-semantic conditions of preceding, equivalence, inclusion or membership relations between lexemes or semantic representations in subsequent sentences of the text. This establishment of so-called 'discourse referents' (Karttunen, 1969) however, is not determined by the existence of connected lexemes in previous sentences alone. The fact that in certain cases negation, counterfactual and intentional operators or predicates do not lead to definitivization, demonstrates that 'identification' or 'existence' of referents denoted by the lexemes, depend on the entire semantic representations of S_1, $S_2, \ldots, S_{i-1}$ and their relations with S_i[7]. That is, identification, in general, is possible only within the same (possible) world, as it is specified by previous semantic representations of the text, which in current grammar are accounted for by presupposition (Lakoff, 1968). The general rule for definitivization, then, is that in modally coherent sequences (i.e. sequences interpretable within the same possible worlds) unidentified indefinite *NP*'s may be followed only by unidentified definite *NP*'s (qualifying or semantic definite descriptions) and identified indefinites by identified definites (properly referential definite descriptions), a rule which only a *T*-grammar or a corresponding text logic is able to formulate (for detail, cf. Van Dijk, 1972b).

2.4. This conclusion can be drawn also from a description of the mechanisms underlying irregular stress assignment, the identification of focus and the differentiation between topic and comment. In all these cases we must consider the structure of and the relations with semantic representations of preceding sentences. Like for definitivization and pronominalization, the identification of a topic is based on the 'definite description' of (pre-) lexical elements in preceding propositions. Any syntactic (superficial) description of these and related phenomena in isolated sentences leads to unsatisfactory results (cf. Drubig, 1967; Dahl, 1969).

2.5. It might be asked at this point already if the global remarks made above do not suggest a possible form of part of the *T*-grammar. Indeed, if syntactic formation rules are determined by semantic structures, it would be more difficult to devise a surface (micro-structural, linear or intersentential) component of a text grammar which is not semantically based. The relations between all elements of a sentence to previous sentences are based on the set of semantic relations. The selection of conjunctions and temporal, local and modal adverbs is a particularly clear example of this semantic determination: their selection restrictions are never intra-sentential.

2.6. There is a still stronger argument for the semantic basis of sentential and intersentential structures. Our main hypothesis was that a grammar also has to specify macro-structures in order to explain the ability of speakers to operate with (longer) texts. These macro-structures underlie the set of semantic representations of the sentences of the text and therefore are themselves of semantic or logico-semantic nature. Syntactic categories in the traditional linguistic sense do not play a role at this level. Already for reasons of simplicity, then, sentence deep structures in *T*-grammars will be semantic since they are formed under 'control' of the macro-structure, i.e. they are derived from them by a specific type of one-many transformations, mapping macro-semantic representations into the linear sequence of sentential (micro-) semantic representations. The character of these rules is still fully obscure and can be made explicit only if we know more about the nature of macro-structures.[8]

The need for macro-structures is intuitively easy to understand. Without such 'global' underlying patterns, the formation of semantic representations in sentences would be fully 'unorientated', determined only by the structure of preceding semantic representations. The choice of coherent lexemes is directly dependent on this process of semantic formation, and is part of the superficial (stylistic) realization of the global underlying structure. Rules for macro-structures, then, are not only psychologically (cognitively) necessary but also (grammatically) indispensable: a text grammar of purely superficial and linear character would be hopelessly complex, if it could be constructed at all. Since texts can be interpreted, memorized, paraphrased as global 'wholes', these wholes should have the form of (global) semantic representations, i.e. of a phrase marker of some type (Lakoff, 1971). It is therefore impossible that textual structure consists of an amalgam of 'summed up' sentential SR's: $SR_1 \cup SR_2 \cup \ldots \cup SR_n$, i.e. be $\bigcup_{i=1}^{n} SR_i$. Neither the converse seems to hold, although intuitively more plausible: we do not simply sum up what is common to all the semantic representations of the sentences of the text: $\bigcap_{i=1}^{n} SR_i$. Such sets of SR's are unordered and therefore do not represent a semantic phrase marker (which is partially ordered). We therefore need rules specifying functional relations within sentences and between sentences, i.e. we will need sets of ordered pairs, triples..., e.g. an identity relation between the semantic subjects (e.g. agents, patients) of subsequent sentences, or sets of relations determining the actions of such actants and their interrelationships. These sets and relations may be properly generated by global underlying structures having the form of a phrase marker. Thus, if the propositional function $g(x, y, z)$ is assumed to render the global structure of a textual semantic representation (e.g. verbalized as follows MAN TRAVELS WITH WOMAN THROUGH ITALY) it has to generate the cartesian product $G \times A \times B \times C$ of sets like {*man, he, Peter, husband,*...}, {*to travel, to go, to walk, to drive*...}, {*woman, she, Mary, wife,*...}, {*Italy, town, country, roads, houses, people,*...}, of which the respective elements are related in the form of sentential phrase markers, e.g. *walk* (*he, her, Perugia*), etc.

We notice that the relations between deep or macro-structures of texts and surface sentential structures are not well-defined (nondeterminate): the formation of SR's and the selection of lexemes is rather free, be it globally within the framework ('plan') generated by the base rules. The status of the rules relating macro- and micro-structures has still to be defined. They cannot simply be 'meaning-preserving' transformations in the traditional sense. They do not change the structure of a string, but map macro-lexicoids and their relations into sets of micro-lexicoids and their relations, i.e. textual phrase markers onto sentential ones. Macro-structures, thus, provide the global trans-derivational constraints in textual derivation.

The hypothesis that textual deep structure is a simple or complex (macro-) semantic phrase-marker, or perhaps a partially ordered n-tuple of phrase-markers, $\langle P_1, P_2 \ldots P_n \rangle$, where P_1 underlies P_2, P_2 underlies P_3 and P_{n-1} underlies P_n, leaves us with the task to identify terms (categories) and relations (functions, rules) defining these underlying structures.

The theoretical elements of the semantic language describing macro-structure may represent e.g. sets of persons, sets of (in-) animate objects, sets of characteristics (states, attributes), sets of actions, processes and events, further time and place points and sets of modal (and perhaps pragmatic or 'performative') categories, where the relations are defined by the sets of actions, processes and events. It is clear that at this level of abstraction we do not have categories determining linear structure (i.e. immediate constituents) but an abstract relational structure between semantic terms. In this respect text deep structures are analogous (isomorphous) with (semantic) sentence deep-structures. If this rather bold hypothesis is true, our system would be characterized by a highly important generalization: textual base rules would be of the same type as sentential base rules. In that case we may use S-grammars as partial models for T-grammars, as we shall see in the next section. At the same time it would provide a relatively simple competence model: the rules learned to produce textual macrostructures are partially identical with those learned to produce textual micro-structures (sentences). This hypothesis merits extensive theoretical elaboration, because it will elucidate the thorny problems of the relations between textual deep and surface structures, and the status of sentences within text grammars. We can then try to attack the problems concerning the global derivational constraints (transformations) transducing macro-structures into microstructures.

3. Models for t-grammars

3.1. Having outlined above a rough sketch of the components of postulated texts grammars, we will now consider different models for an adequate description of structures and relations in texts, in view of explicit notation and of future formalization.

Let us first briefly state which types of models we have in mind here, since the concept of model, more than ever, is becoming so general that we cannot use it without serious ambiguities.[9]

3.2. The role of models in our research can be understood only if we begin to recall the reasons which have led to the construction of text grammars. First of all, we wanted to provide a linguistic description of the structure of a formal entity, called 'text', manifesting itself empirically as a phonetic or graphical utterance (discourse). The pragmatic motivation is that such descriptions may lead to the understanding of the functions of texts in the different systems of communicative interaction in society. Secondly, text grammars have been postulated to resolve properly linguistic (or psycholinguistic) problems, e.g. about the structure of sentences and their interrelations as regular (ideal) manifestations of language systems or idealized psychological abilities. Given the fact that hardly any explicit knowledge has been acquired about the linguistic structures of texts, we intend to construct a theory of such objects. Working within linguistics, we therefore adopt the series of current methodological procedures. That is, we consider texts as abstract realizations of a system of rules (of a natural language) and will try to make explicit these rules in a grammar, as a specific type of (partial) theory, viz. an algorithm. We therefore have to recognize the different modelling capacities of such types of explicit theories in linguistics: the grammar being a model of the system of a natural language, or of idealized abilities of a native speaker to produce and to interpret grammatical and appropriate (acceptable) utterances. A text grammar, like any grammar, is thus itself a theoretical model of conceptualized (abstract, ideal) systems.

However, within linguistics we already have some knowledge about S-grammars, but we do not know what T-grammars will look like. Because there are no a priori reasons why T-grammars would be of a fully different type, we might use heuristically a specific type of S-grammar, e.g. a semantically based generative-transformational grammar (e.g. as outlined in Lakoff, 1971, and Rohrer, 1971) for the hypothetic construction of them. In that case we make use of the very general principle determining analogical models: texts sentences are both linguistic objects constructed by phonemes, morphemes, syntagms clauses, etc., and since texts consist of one or more sentences their theory will very likely be partially identical with a theory of sentences. However, not only this inclusion relation but also the form of the grammar itself, its components, categories and rules, may serve as heuristic tools. An example is the distinction between deep and surface structure and the isomorphy between sentential and textual semantic representations (deep structures). We here meet the intricate reduction relations between two theories: is it possible that, once constructed a T-grammar in analogy with S-grammar, we may reduce S-grammars to T-grammars, because these alone formulate rules for all the relevant properties of sentences?

Besides this global analogical and theoretical model provided within linguistics proper by a generative (S-)grammar, we may use different formal models. We may want to use mathematical and logical languages to make explicit specific structures and relations described by text grammars. Since linear coherence is partly determined by relations of lexico-semantic inclusion we may use e.g. different set-theoretical models. A predicate logic (standard, modal, intensional) may be used for the formalization of different identity relations between noun-phrases in subsequent sentences and for the formalization of the relational structure of macro-semantic representations. The (part of the) grammar itself, then, is a realization model of the formalized system. Similarly, generative grammars themselves are considered as realizations of the theory of semi-groups or of semi-Thue-systems, which a formalized theory of grammar may take as analogical models.

3.3. Linguistic Models

3.3.1. The first model, then, is taken from within linguistics proper: a (modified) generative sentence grammar is said to form not only a part of a T-grammar but in principle represents at the micro-level the form of a T-grammar designed for macro-structures. However, the heuristic value of such a model may not obscure the differences in the categories and rules used in T-grammars. The major difference, as we saw, is the fact that, with respect to classical transformational grammar and its (revised or extended) standard theories, textual deep structures are identified with their macro-semantic structure. There are no syntactic base rules in T-grammars (other, of course, than the formal semantic formation rules, i.e., the 'syntax' of the semantic language). Generative semantics has designed some hypothetic models for semantically based (S-) grammars and in this, still very tentative, linguistic domain, a T-grammar will look for models for its base component.[10] For the surface (inter-)sentential component we might in principle use syntactically based grammars, but for reasons of coherence and simplicity we use the same model for deep and surface structure components. The surface component will no longer retain us here because it is virtually identical with an extended form of S-grammar. Note however that many problems are of intersentential nature and therefore their description is given for sequences of sentences, not for isolated sentences. Initial rules (initial rule schemata) of the type $\# S \# \rightarrow (\& \# S \#)^n$ ($n \geqslant 1$), where '&' represents any connective element (included the zero-connection (parataxis)), may generate these sequences. The conditions of concatenation are partially identical with those for the combination of clauses in complex or compound sentences. These are, in fact, reduced to sequences, which will probably imply a re-introduction of the traditional generalized transformations. Embedding, clearly, is thus a result of rather late transformations.

3.3.2. The main theoretical problem, then, is a possible linguistic model for macro-structures. First of all, any syntactico-semantic or semantic model using syntactic *PS*-categories like *NP*, *Pred P*, *VP*, etc. will of course be irrelevant at this level of abstraction. There are only two sets of related proposals: the actant-model of Tesnière-Greimas-Heger and the different types of case-grammar proposed by Fillmore, Halliday and Chafe, which seem to be possible candidates for primary conceptualization of macro-structures. In the first system (cf. Tesnière, 1959; Greimas, 1966) we find a differentiation of such opposite categories as subject-object, destinateur-destinataire, adjuvant-opposant, between which the relations are defined intuitively as 'roles' in the sentence (or text, as with Greimas). 'Verbs' (predicates, attributes) are normally said to be static or dynamic, representing states and actions (processes, events), respectively. No rules are given for the introduction/definition of these categories and for their interrelations. Heger (1966) proposes a recursive 'modèle actantiel', which is more complete and explicit. A predicative functor is analyzed as a n-place relator and a set of n-actants. He compares some of his categories with those of Fillmore (1968), who describes a modalized proposition with the aid of the following categories: Agentive, Instrumental, Dative, Factitive, Locative and Objective. Additional cases, he says, will surely be needed (Fillmore, 1968, 24–5). Notice that the status of these cases is not wholly clear: their definition is intuitive and semantic, their use however is clearly syntactic, because they dominate 'case-*NP*' categories. We will not go, here, into the deficiences of this system for syntactic description, but will only retain its possible merits for the establishment of semantic deep structure categories. Chafe (1970) takes different verb-types (states, processes, actions) as functional nucleus of a proposition and relates them to such categories as patient, agent, experiencer, beneficiary, instrument, complement, location. Halliday (1970) finally uses such 'participant' roles as actor, goal, beneficiary, instrument, circumstantial.

3.3.3. In these proposals a number of categories are common, certain others may be translated or reduced to others. However, the importance not so much lies in their simple listing, even when used for the semantic ('logical') characterization of 'parts of speech' in sentences, but in the type of their relations. A result in most approaches is the centrality of the verb-predicate, which actually determines the roles. In *John hits Peter* and *John hears Peter*, agent and patient are different, although mapped on identical syntactic structures. In Chafe's derivations the roles have relations of dominance: agent for example normally dominates instrumental.

The problems of derivation are serious here. Instrumentals may appear without (apparent) agents (like in *Peter was killed by a bullet*, or *The bullet killed Peter*) or agents may be related with instrumentals without patient (or goal) (like in *John killed with a knife*), etc. The hierarchical structure, thus, seems rather weak, and may perhaps simply be represented as a set of relations where the relation is defined by the actantial category with respect to the verb/predicate like in the following informal and incomplete derivational steps.

(1) (i) proposition → predicate (actant$_1$, actant$_2$, ..., actant$_n$)
    (ii) predicate → {action, ...}
    (iii) actant$_i$ → {agent, patient, ...}

However, it is possible that certain categories are 'integrated' in the verb/predicate (like in *to strangle, to stabb*) which would suggest introduction of complex action or event categories, which requires further analysis.

Having thus briefly indicated some possible linguistic models for our list of primitive semantic terms (which need further definition, reduction, etc.) we now clearly need models for the explication of their apparent relational structure. We will see in the next paragraphs that a modified relational logic naturally can serve as a provisional model.

3.3.4. We finally should notice that there is a close interrelation between these types of functional models and the theory of narrative in poetics. Actually, many suggestions for the form of macro-structures in texts derive from the recent developments in the structural analysis of narrative (cf. Ihwe, 1972, van Dijk, 1972a, 1972c). We do not elaborate this topic here, but it is interesting to see that informal suggestions from a more particular theory may serve as a model for the empirical scope and the form of linguistic grammar (cf. van Dijk, *et al.*, 1972).

3.4. *Logical Models*

3.4.1. The use of logical models in present-day linguistics is extensive but has not been without methodological problems. Like for all formal models, it is necessary to be clearly aware of the fact that we have to represent linguistic structures not logical structures. Although simplification, idealization and reduction are inevitable in any process of theory formation and especially in formalization, the use of logical models may not obscure relevant linguistic facts. The use of classical propositional and predicate calculi often obscured the role of such important linguistic categories as articles, adverbs, conjunctions, mood, time, aspect, etc. and often neglected the hierarchical structure of the syntax of the sentences represented. The introduction of the more powerful (i.e. precisely 'weak' and 'flexible') modal systems obviously has greater promises for linguistic description. The relatedness between grammar and logic, recently re-emphasized by Lakoff (1970), is evident and trivial for any formalized theory. Moreover, since logic originated in systems for reasoning in natural language it is also clear that even modern mathematical logic will still represent some idealized structures of natural language. However, modern propositional logic which only deals with propositional variables and their truth values, will hardly be able to yield a model for the linguistic relations between (meaningful) sentences and their underlying 'logical forms'. The deductive rules of propositional logic, then, have value only when the variables are substitutable by constants within a truth-functional frame. This is why all types of predicate logic have until now, together with set theory, dominated as heuristic, descriptive or formalizing models in linguistics, since they make explicit the internal structure of propositions.[11]

3.4.2. Even in modern logic, however, there is a striking fact which seems to temper any optimism about its use in text-grammar: both propositional and predicate logical systems are of the 'sentence'-type. That is, rules are formulated for the formation of well-formed (complex) propositional formula and rules for their deductive relationships. The basic unit in the formal sciences, thus, is the proposition and not the (macro-)structure over propositions, which is the proof, derivation or similar structures. At the level of syntactical calculus this does not matter, since here only the rules of direct (linear) deduction apply. However, as soon as we want to give some semantic (intensional or extensional) interpretaion of a structure, we deal with conditions of completeness, coherence and consistency, for which the proofs, of course, are not given for isolated propositions but for sequences (systems) of interrelated propositions (cf. van Dijk, 1972b).

Although logical reasoning is remote from reasoning in natural language we thus have some clues for the demonstration of the superficial (intersentential) coherence of texts: relations of equivalence and inference also relate the sentences in a discourse, be it mostly in an inductive and rarely in a deductive way. Formal (logical) semantics, however, is restricted to the latter type of relationships, using for example relations between explicit meaning postulates (as they are informally provided by the lexicon, or represented by feature matrices of semantic primitive categories). The linear coherence of a discourse like: *Everybody at the office has gone out for lunch; Johnson must also have lunch now*, can be demonstrated logically (by a rule of instantiation) if we insert the presupposition *John is a man of the office*. We see that the sentences preceding a sentence S_i in a coherent text serve as a set of premises for grammatically continuing it; that is, this sentence has to be, ideally, consistent with S_1, $S_2, \ldots S_{i-1}$. Most types of discourse, however, have only occasional instances of such (deductive) reasoning, and even then it is partially implicit such that entailments and presuppositions are left for the reader in order to reconstruct textual coherence. It might be asked to what extent these may be considered as parts of the logical form of textual surface structures and hence of a theory of competence (grammar) and not only of performance. Normally the relations between sentences of a discourse are established by relations of identity (like in pronominalization) equivalence (definite noun phrases and pronouns, synonyms, paraphrases), membership and inclusion between (pre-)lexical elements. A text like *John was walking in town; The streets were empty*, is grammatical because $street \in town$, which can be made explicit by lexical rules in the following way: if x is a town, then x has streets; there is a town, therefore it has streets. In general, then, the lexical rule is the premiss representing the universal statement, whereas the previous sentence(s) and their consequences form the particular conditions. The sentence in question is grammatical if it (or its presupposition) follows as a conclusion by rule (e.g. modus ponens). This conclusion not only permits the regular introduction of a noun (argument) but also its definite character, which can only be explained by the implicit presence of the presupposition following from the lexical rule and the preceding sentences of the text (cf. Dahl, 1969). Presuppositions, as we see, can be described only in text grammars because only these specify explicitly, as logical forms of preceding sentences, the conditions for the truth valueness and well-formedness of a given sentence. A complex sentence like *John knows Mary is ill*, can thus be derived from the text *Mary is ill, John knows that*, where the presupposition is simply a preceding sentence. The sentence *Mary is ill* is not presupposed by *John pretends that Mary is ill*, where S_2 is under the scope of the 'world creating' (hence 'discourse creating') locutionary verb of S_1, which does not imply the truth of S_2. The same holds for the more complex sentence, quoted by Lakoff (1970, 181): *Max pretended that he realized that he was sick*, which in fact is ambiguous since it can be derived from the following underlying texts: (i) *Max was sick; Max pretended: Max realized that*, or from (ii) *Max pretended: (Max is sick; Max realized that)*. In the first case, *Max was sick* is a presupposition, and in the second case it is not, because it falls under the scope of *to pretend*. Similarly, for such predicates as *to believe* for which a belief logic may provide underlying topical forms.

3.4.3. Similar remarks may be made for sequences of modalized sentences. Notice that modal verbs or adverbs, represented by operators of different types, may have a whole sequence of sentences as their scope. We will not further study these logical models here, since – as far as they are relevant for relations between sentences – they have already been studied elsewhere. Their role in textual grammar, as was stated above, is obvious since only these grammars specify coherent sequences of sentences.

Notice further the possible role of systems of action logic for the description of textual surface structures. Intersentential relations not only are subject to rules for (existential) truth values and for modal and temporal coherence (where e.g. tense logic will be of importance) (cf. Rescher, 1968), but also to rules relating the 'actions' represented in the semantic structures. In the scope of predicates indicating permission it will not be possible to generate predicates expressing obligation, and so on. Such a model will be of particular interest for the description of narrative texts (cf. van Dijk, 1972c).

3.4.4. Let us briefly return to the structure of macro-representations and their possible logical models. Since we here deal with abstract relations between actants, a predicate logic may be used. In treatises of logic such use is often made for the symbolization of sentence structures. Similarly, current generative semantics makes extensive applications of such notations. Our hypothesis seems to be trivial: textual deep structures are supposed to be formalizable in the same way. Some aspects and problems will be mentioned briefly here.

In general, underlying macro-structure is considered to be a (macro-) proposition modified by performative and modal operators. The proposition itself is formed by an n-place predicate and a series of (bound) variables, where the relation is expressed by the order of the arguments. However, this order is purely linear and does not indicate hierarchical relations between the arguments. Traditionally, this order mirrors roughly the syntax of some natural languages: subject, object, indirect object, complements (see Reichenbach, 1947). Obviously, these surface categories and their order are irrelevant for abstract semantic description. We therefore need a strict semantical syntax relating the different categories like agent, patient, object, instrument with the respective argument-variables. Otherwise it is impossible to make explicit such paraphrases as *X receives Y from Z* $\simeq$ *Z gives Y to X*, with the following structure: Action (Agent, Object, Patient) implying that *receive* and *give* must be semantic converses, and that the categories are replaced by different variables or constants to generate the two different verbs:

Action (Agent, Object, Patient) $\simeq$ Action (Agent, Object, Patient)
h (x, y, z) $\simeq \tilde{h}$ (z, y, x)

Note that when the roles are not specified like in $h(x, y, z)$, $h(z, y, x)$, the paraphrase does not necessarily hold: the following paraphrase is ill-formed:

John goes with Peter to Paris $\neq$
x g y z
*Paris comes with Peter from John
z g y z

The rules of formation of the semantic language (be they for sentences/propositions or macro-structures) therefore require explicitation of ordering relations by specifying e.g. Argument$_1$ as Agent, Argument$_2$ as Object, etc. There is a great number of unresolved theoretical problems here (e.g. about the number of the primitive categories and their possible hierarchical structure or mutual presupposition), which we will not treat.

We only want to show that current predicate logic is probably too poor to represent the character of the ordered pairs, triples, ... underlying semantic propositions. We might formalize functional roles by coordinate specification of the following type:

$$h(x, y) \text{ \& Action } (h) \text{ \& Patient } (x) \text{ \& Agent } (y)$$

or by introducing them as a type of functional operators, e.g. as follows:

$$Ag(x) \, Pat(y) \, [h(x, y)].$$

Clearly, these suggestions are fully speculative and ad hoc, and only indicate some possibilities for introducing names (labels, nodes) of relational functions (roles) into the representation (cf. Petöfi, 1971, 1972a,b). For instance, introducing them as operators does not remind in any way of the use of quantifiers (or even scalar predicates) limiting the domain of validity for the variables to a certain constant class of Agents, Patients, etc. The problem is not easy to solve because logical theory itself is hardly explicit about the status of the different argument places of a relation. Its differentiation, e.g. as given by Carnap (1958: 72f), of first, second, ... domain of a relation, is based on such intuitive definitions as 'the individual who bears the relation to...', 'the individual to whom the relation bears', ... etc. Difficulties arise here for 3-place relations, but also for such relations as *to get, to suffer from*, etc. Carnap's definitions of the first member of a relation e.g. by the predicate expression $mem_1 (R)$, as $mem_1 (H) (x) \equiv [(\exists y) Hxy]$, and of $mem_2 (R)$ as $mem_2 (H) (x) \equiv [(\exists y) Hyx]$, do not seem to be of any help to define the functions (roles) of the arguments with respect to the (n-place) predicate. The functional roles agent, patient, etc. must be defined (or introduced) independently of the other arguments, and only with respect to the predicate. Thus we would have, when taking 'to be an agent of H' as a predicate expression (where Agent is a functor), something like: Agent $(H)(x) \equiv [(\exists H) \varphi(x, H)]$, where φ is a higher predicate like 'to accomplish', 'to instigate', 'to be the origin of'. Clearly, such definitions belong to the metatheory of semantics, and we simply may accept agent as a primitive. This result is somewhat poor, but logical theory (as far as we know) has not provided more explicit models for the characterization of different types of relations. Reichenbach (1947:229) employs the α-method for the derivation of some specific second-type formula (like $(\exists f) f(x)$). The 'α', then, is a relation of the second type to define $f(x)$ as follows: $\alpha[f, x]$, where α is merely a logical constant expressing that f and x have a mutual relation. Similarly, the relations agent, patient are to be defined as constants combined with n-place predicates and individual variables. Perhaps we are more close to the aims of our description here because we would like to have a closed, rather small, class of actants. They are constants of semantic theory and a logical model will have to account for this status. We will not go further into these intricate matters here.

3.4.5. After these remarks about the possible structure of the macro-proposition underlying a simple text (several macro-propositions would underlie complex or compound texts), it may be assumed that other logical aspects of (macro-) sentences may be modelled in a similar way. First of all, we will need a set of quantifiers to bind the variables of the proposition. Besides the existential and universal operators, we will surely need operators for (in) definite descriptions. (e.g. eta- and iota-operators for identified indefinite and definite descriptions, and epsilon- and lambda-operators for unidentified (qualifying) (in) definites, respectively; see van Dijk, 1972a, Chapter 2 and especially 1972b.) Intuitively, a whole text may 'speak about' a definite (unique, identified) individual, as determined by intertextual or pragmatico-referential presuppositions, or an indefinite individual, like for example in narrative texts (cf. *Once upon a time there was a king...*, the initial formula of fairy tales). Note that indefiniteness for the whole text does not imply indefiniteness for the same individual as realized in the sentences of the text. Once introduced as a discourse referent, an individual is properly identified at the level of sentence structure.

It is possible that other restrictions are somehow related with the operators, for example fictitious or intentional existence, without which the truth value of many texts (novels, dreams, and other counterfactuals) would pose some problems. These must be introduced in the macro-representation, since they have the whole text as their scope. In the same way we perhaps should introduce time- and place-operators as semantic correspondences for the pragmatic operators of time and place. We here briefly meet the problem of the relations between the semantic and the pragmatic components of grammatical description.[12]

At the same level we may specify the set of what can be called the 'typological' constants (or variables?) such as 'assertion', 'question', 'imperative', which also have the entire text as their scope. These are related with the type of speech act in which they are embedded. At the formal semantic level they may figure either as operators (constants) or as explicit 'hyper'-propositions. Dominated by these we find a set of modal operators, with the nuclear proposition as their scope: 'factual', 'possible', 'probable', 'necessary', etc., for which a rich series of logical models can now be provided. The implications of these modal operators and rules of usage are interesting for a theory of text grammars. For instance, a text having 'probable' (PROB) or fictitious (FICT) as its macro-operator may have modally necessary sentences, but their truth value is restricted to these probable or fictitious worlds, and do not have any interpretation in the object (zero level) world of the speech act producing the text. This very interesting (logical) conclusion is particularly valuable for the textual description of e.g. literary texts, where sometimes apparently 'real' individuals (names of wellknown, historical persons or towns) are realized. When placed under the scope of a fictitious operator they thereby automatically acquire the status of fictitious persons and places. We see that the lengthy ontological discussion about 'reality' in literature may be founded on a sound logical basis.

3.5. *Mathematical Models*

3.5.1. After this (too) short discussion of some logical models for the description of semantic (macro-)structures, let us now turn to the possible use of mathematical models for text grammars. These, of course, are closely related to the logical models, and what was said about properties and relations may of course also be reformulated in terms of set theory. We will use mathematical models especially for the description of textual surface structures, i.e. for the characterization of linearly ordered sequences of sentences and their properties. As for the application of logical models we have to underline that formalizing at this level presupposes preliminary knowledge of the linguistic properties to be investigated. We do not normally construct a calculus without having already some interpretation in mind, although conversely, many mathematical systems may somehow receive a linguistic interpretation. Nevertheless, we are interested in those systems which produce some insight into abstract structures, not merely in trivial translations by the substitutions of variables by linguistic units of some kind. More specifically, we will be interested here in the properties of textual structures and their grammar which do not directly derive from our knowledge of sentential structures and grammars. This kind of investigation, clearly, is still in an embryonic stage. It is impossible to review here the whole set of models already used in mathematical linguistics for the description of (intra-)sentential structures, e.g. of phonemic and syntactic systems of definition, identification, context-free and context-sensitive concatenation,[13] etc. It is well-known that models for semantic systems are still rare, with the exception of very clear-cut sub-systems as those of kinship-terms, number-names, etc. This faces us with serious problems since our hypothesis was that textual coherence is based on sets of semantic micro- and macro-relations. The following survey, then, has merely an explorative character and is neither mathematically nor linguistically explicit. It merely matches some hypothetical textual structures with possible mathematical systems.

3.5.2. Proceeding from the most general properties to more particular ones, we may first try to find some characterization of the language generated by the hypothetical text grammars, especially in relation with the existing S-languages.

Traditionally, a natural language has been characterized as a subset L of the set $\Sigma(V)$ of all possible strings over a finite vocabulary V, being the lexicon of that language; $\Sigma(V)$, then, is a (free) semigroup based on the binary operation of concatenation. At this general level both L_S, being the set of sentential strings, and L_T, being the set of textual strings, are subsets of $\Sigma(V)$. However, we know that not any combination of grammatical sentences (i.e. of members of L_S) are admissible texts of a language. Thus L_T is a subset of the semi-group $T(L_S)$ defining the set of all possible sentence concatenation of a language L_S into sequences (cf. Palek, 1968). Clearly $L_S \subset L_T$ is true because any sentence of L_S is by definition a possible element of the set of texts consisting of one sentence, which is a subset of L_T. The grammar G_S is the formal device separating L_S from $\Sigma(V)$ by enumerating and formally reconstructing any of the elements of L_S and not the other strings of $\Sigma(V)$. Similarly, for a grammar G_T with respect to L_T and $T(L_S)$.

As is well-known, an (S-)grammar for a natural language can be defined as a quadruple $\langle V_N, V_T, R, S \rangle$ of a non-terminal vocubulary (V_N), a distinct terminal vocabulary (V_T), a set of rules (R), and a specific initial element $S (S \in V_N)$. Similarly, a text grammar may be defined as the quadruple $\langle V'_N, V'_T, R', T \rangle$, where $T \in V'_N$, and where $V'_N \cap V'_T = \emptyset$. One of the tasks of a theory of grammar is to compare these two systems. The hypothesis is that the following relations hold: $V_N \subset V'_N$, $V_T = V'_T$, and $R \subset R'$, i.e. the two grammars have the same lexicon, whereas the sets of theoretical symbols and rules of S-grammars are included in those of T-grammars. We may construct an intermediate system G'_S, or Extended S-grammar, as a quadruple $\langle V''_N, V_T, R'', S \rangle$ where $R'' - R$ is the set of rules (conditions) defining the grammatical relations between the sentences of G'_S in a sequence, and where $V''_N - V_N$ is the set of non-terminal symbols used to define these relations. We should demonstrate further that $G_T \neq G'_S$, e.g. by proving that $V''_N \neq V'_N$ and $R' \neq R''$.

3.5.3. Of course, the tasks formulated above are not easily accomplished, because they require that we introduce categories and that we construct rules which are necessary for the enumeration and description of the elements of L_T, and which do not belong to G_S or to G'_S.

Palek (1968) was the first who used a mathematical model to demonstrate that G_S and G'_S are not identical, e.g. by proving that G'_S necessarily contains at least a set of elements (for describing cross-reference) which is not included in V_N (or R) of G_S. He states that this set is only a part of the necessary extension of G_S: there are other means for the description of coherence between the sentences of a text. He first constructs a system for the identification of the terms of cross-reference by (a) defining different occurrences of the same naming unit in subsequent strings, and by (b) relating these with a system of denotata, consisting of unique individuals or classes of individuals, with the aid of a specific denotative function ψ. The mechanism of cross-reference can thus be reduced to a description of the set of relations between the denotata entering into this function. The system R of relations consists, first of all, of the four basic types of relations $R_=, R_{\neq}, R_{\in}, R_{\subset}$, representing, respectively, identity, difference, membership and inclusion relations between the occurrences of a (same) naming unit as referential descriptions of the corresponding set theoretical relations between the denotata. Thus, each set of possible relations in a sequence of occurrences of a naming unit is a matrix of possible state descriptions (with the restrictions defined by the grammar), viz, the relation between the first occurrence and the second, between the second and the third, ... etc., and between the second and the first, the second and the second, the second and the third, etc.... between the nth and the nth. In a similar way, we define other referential sequences of a given string. Once established this system of referential relations, each occurrence of a naming unit is associated with its referential characterization (its place in a relation), and the corresponding linguistic means ((in)definite articles, prepositions, etc.). Since a string of sentences thus receives a more elaborate structural description, G'_S is superior in strong generative capacity with respect to G_S (provided, of course, that G_S does not possess a system of cross-reference).

3.5.4. From this very rough characterization of an example of the mathematical description of the mechanism of cross-reference, i.e. of one of the components of a text grammar, we saw that essentially set theory and its theorems served as the basic model, especially the properties of the relations of identity, difference, membership, and inclusion. Further, some notions from matrix theory were necessary to define the possible state descriptions of denotata and corresponding occurrences of naming units. Already Harris (1952) in his early approach made use of matrices for the description of discourse structure. Finally, function theory was used to map sets of denotative relations onto sets of relations between linguistic units.

Note that Palek's models in principle can also be used for the definition of coherence on a semantic basis. In that case the referential description will be matched with a description of the relations between lexemes and groups of lexemes. The same set of relations will then probably be necessary and defined over configurations of semantic features: identity, difference, inclusion, membership, and further undoubtedly intersection. Such a description is necessary in an adequate grammar because the relation between sentences are only rarely based on identical naming units, which are surface realizations of semantic identities.

3.5.5. Let us consider briefly some other branches of mathematics which might supply models for text grammars. Recall that we asked of a G_T that it produces descriptions of macro-structures underlying the sentences of surface structure. In this respect G_T has to be of stronger generative capacity than G'_S which only assigns superficial relational structure to a string. The base-grammar generating these macro-structures was sketched earlier with the aid of a logical model. The problem at issue now is the elaboration of the system relating macro-structure with micro-structures. Which types of rules are necessary for this mapping and what are their mathematical properties? These rules must somehow relate any element of the macro-structure M with a corresponding element in each sentence (micro-structure) S_i of the surface structure. That is, the model we are looking for must establish a one-many relationship between M and Σ. Recall that Σ is an ordered n-tuple of sentences. These sentences, or rather their deep structure representations, are each isomorphous with M. The function ϕ mapping M onto (into) Σ thus might relate the first element of M with all the first elements of S_i, the ith element of M with the ith element of each S_i, = the nth element of M with the nth element of each S_i. Since each macro-lexicoid of M can be considered as a class abstraction we first substitute the corresponding class of lexemes for this lexicoid, e.g. as follows: $F_1 \rightarrow \{x: F_1(x)\}$. From this class are selected the terminal elements of the S-derivations. The type of transformations corresponding to the one-many relations could be called poly-transformations (or P-transformations) since they have a single structure as input and an ordered n-tuple of isomorphous structures as output. Surface structures, then, are 'realizations' of the abstract deep structure model. Its set of functional classes of actions, agents, patients, etc. represents the abstract macrocontent, which is particularized and distributed at the surface level.

The model informally outlined here is however too strong and restricted to a specific subset of texts. A more complex model is needed which will more clearly indicate that M is only a global derivational constraint upon the formation of Σ. Not any S-formation is directly guided by M, but only the sets formed by the predicates and argument-types of all the sentences of the text.

3.5.6. Leaving the still obscure domain of macro-structures and its relations with surface-structure, we will finally return to some mathematical models for the description of intersentential structures in texts.

Superficially studied, then, a text is an ordered n-tuple of sentences ordered by the relation of preceding: *Prec* (S_{i-1}, S_i), defined over unanalyzed sentences. (It might be more appropriate to localize this description somewhat deeper, e.g. on the level of semantic representations. In that case we would define the text as a morpho-syntactic realization of an ordered n-tuple of propositions). This relation is irreflexive, asymmetric and transitive. Its converse, the relation of 'following' will be less interesting, since, as we saw earlier, the whole linguistic mechanism of cross-reference (pronominalization, definitivization, topic-comment, etc.) is based on the structure of preceding propositions and their elements. Let us stress again that we here describe a linear surface level and its corresponding notion of intersentential coherence. Although, as was demonstrated by Palek (1968), Kummer (1971a, b, 1972) and others, a processual description of the subsequent 'states' of the texts in its sentences is very important, it is not sufficient to describe texts with the aid of such linear, finite state models. The determination of semantic content does not depend merely on the set of transition rules $S_1 \xrightarrow{t} S_2 \xrightarrow{t} ..., \xrightarrow{t} S_{i-1}$, as a sort of composition function $f_1 \circ f_2 \circ f_3 \circ ... \circ f(S_i)$, but is globally programmed at the abstract underlying macro-structure level. At the surface level this implies that $S_{i+1}, S_{i+2}, ..., S_n$ may retrodetermine the semantic structure of the propositions $S_1, S_2, ... S_i$, which were at first either ambiguous or unspecified. The semantic relations holding in the ordered n-tuple, then, seem therefore essentially symmetric: if sentence S_i is dependent on S_j, then S_j is dependent on S_i. Of course, this is true only with respect to the general theoretical property of dependency, not of specific semantic relations between elements of the respective sentence.

3.5.7. The notion of dependency is obviously of particular importance in T-grammars: surface structures depend on deep structures, sentences depend on other sentences and their elements and structures upon elements and structures in those other sentences. Mathematical models for dependency grammars are well-known, and we might profit from them to state textual relations. Note that these are probably not simply described by (trans-)formation rules as in generative S-grammars. Proceeding from functional (analytic) calculi to algebraic systems, categorial (or dependency) grammars have been modelled especially on Lesniewski's/ Ajdukiewicz's Associative syntactic calculus, where primitive categories and derived categories analyze a string until an S-symbol is reached. As was demonstrated by Chomsky (1963), such models are far too complex and descriptively inadequate; especially textual dependencies would involve extremely complicated notations (when there are no recursive elements). Such a system would demonstrate more clearly than a *PS*-grammar, however, not only how elements in a sentence are functionally related, but also how elements in different sentences are related, especially when we use indices (or some related device) to identify the categories. A (transitive) verb in S_i may thus formally depend (e.g. as a consequence) on a verb in S_{i-j} (where $i \geqslant j$):

$$[n \backslash S_{i-j}] \backslash n \ldots [[n \backslash S_{i-j}]/n] \backslash [[n \backslash S_i]/n].$$

The copying procedure here is a formal way of representing (parts of) presuppositions, but might be dispensed with in a *T*-grammar. In a similar way, we may represent dependency relations between sentences in a text. The trivial (primitive) symbol would then be *t* (for text). A premiss, condition or cause could be rendered as $[t/s]$, and a consequence as $[t/s] \backslash t$, where the sequence $[t/s]$, $[t/s] \backslash t$ resolves in a specific text (an implication) (cf. Geach, 1970). Here we cannot investigate the further formal properties of such dependency systems, but any satisfactory *T*-grammar should include a device which somehow registrates such dependencies. Simple coordination by a rule-schema does not produce this structure, although intersentential connectors can be introduced as (logico-)semantic constants. For the dependency-relations between other elements we may investigate how propositional phrase markers can be combined into complex phrase markers when these relations are explicitly given.

3.5.8. The other types of algebraic models have already been discussed: since a textual language is also generated by the binary operation (of concatenation) over a terminal vocabulary V_T, they can be considered as free semi-groups. More specifically, generative grammars are semi-Thue-systems defined as ordered quadruples $\langle V_A, V_T, S, P \rangle$. Text grammars have the same form but only a different starting symbol and a different set of rules (productions). The theory of abstract automata, as we saw, will define a text in a processual way, by specifying its consecutive states. The rules of transition would probably have a semantical and relational character, i.e. when one of the lexemes, has an identity, inclusion, membership, relation or when state S_{i-1} is a necessary and or sufficient condition for S_i. These relations determine the ordered set of (cumulative) definite (identifying) descriptions of the individuals.

3.5.9. Finally, texts may be considered as matrices, that is as ordered m-tuples of n-tuples, e.g. for m sentences having each n elements. Clearly not every concrete sentence will have n elements, but there are n possible theoretical 'places' for the elements of a proposition of which some may be empty. These places, for example, are the predicates and arguments of a relational actant-structure. In this way it is easy to compare explicitly the structures of the subsequent sentences of a text. Note that such matrices may be constructed simply by the rules of grammar and a recursive device which will construct, m times, similar rows of a vector (the sentence).

3.5.10. The last group of models we may consider are of geometrical nature, although of course closely related with the previous ones. Besides the already mentioned theory of graphs – as representations of complex (partial) order relations (e.g. of macro-phrase-markers, of complex sentential phrase-markers, of relations of linear dependency, of one-many mappings, etc.), we first encounter all possible topological models here, when we conceive of texts as (semantic) spaces. Typical concepts of general topology as 'neighbourhood', 'environment', 'coherence', immediately recall terms from intuitive text theory. When a topological structure over a vocabulary V_T is defined such that any element l of V_T can be assigned a system of subsets of V_T (the 'contexts' of S), and a topological space as the relation between S and V_T, we may say that a text is such a topological space. Such a text is said to be coherent if it cannot be divided into two (non-void) open subsets. Thus, as soon as an element of a text cannot be assigned a context (e.g. a lexical relation defined over sets of features) the text is no longer coherent. Similarly, a text is coherent when all pairs of coherent subsets intersect non-vacuously. We here meet the familiar concept of linear coherence: the concatenation condition for textual surface structures. Similarly, other notions of topology like compactness, distance, overlapping, etc. can serve the study of semantic surface relations between sentences. Although the application of topology is still embryonic, if not plainly metaphorical, in this domain, already intuitive connections with traditional concepts as 'semantic field' suggest interesting possibilities for a future theory (for examples cf. Edmundson, 1967).

3.5.11. For the description of similar problems we may try to elaborate models from the theory of (semi-)lattices. Thus, properties of feature specification, like inclusion of lexeme l_i by lexeme l_j, can be considered as partially ordering the universe of the lexicon, and therefore defines also the possible relations between lexemes in subsequent sentences in a text. If we thus take the lexical surface structure of a text as a partially ordered set T and the sentence (or sequence) as a set S, we may say that S has an upper bound in T, if for all lexemes of S there is a lexeme in T which is identical or superior to these lexemes; that is if there is a synonym (having the same number and order of features) or a hyperonym including these lexemes. Similarly, $S \subset T$ has a lower bound in T if the lexemes of S are all hyperonyms of a lexeme in T. If these respective lexemes of the text are immediately hyperonymous (or hyponomous) to those of S, they can be considered as least-upper-bounds (supremum) and greatest-lower-bounds (infimum) of the lexemes of S. At first sight these characterizations do not seem to model the intricate system of linear coherence. However, first of all the system of macro-structure formation is probably highly dependent on such notions as (semantic) supremum or infimum. The process of abstraction, based on the lexical units of the text, will tend to delete lexemes of sentences as soon as a supremum is generated. Conversely, polytransformational processes will normally particularize macro-structure at the level of infimum-lexemes. Further, this model might enable us to measure linear coherence between the sentences in a sequence by considering the degrees of particularization or generalization of their connected lexemes. The closer the semantic connection (i.e. the smaller the difference in feature specification) the higher will be the linear coherence of a text. This coherence reaches its maximum when all the lexical elements of all sentences have an identical lexeme in all connected sentences (complete redundancy).

Since linear coherence is normally defined for pairs, (S_1, S_2), $((S_1, S_2), S_3)$, ..., we may consider a text as a lattice, where each pair has a supremum and an infimum, viz. $S_i \cup S_{i+1}$, when S is considered as a set of lexical elements. It is not possible here to work out in detail these lattice-models for a theory of semantic coherence in texts. We merely wanted to suggest their possible relevance (cf. for application, especially phonological, Brainerd, 1971).

4. Concluding remarks

4.1. As has become clear from the rapid sketch of some logical and mathematical models for theory construction in text grammar, only the main branches of these formal disciplines could be mentioned. Further research will have to specify the details of their possible relevance. In many respects, of course, this relevance is shared with their role for S-grammar, although within the simple sentence (or its underlying proposition) notions as coherence, consequence, etc. cannot be defined.

4.2. We have to stress that the models that were programmatically considered are models for a grammar, not for a theory of textual performance. Performance models acquire theoretical relevance only when based on systematic models. Actually, most of extant work done in the theory of texts and its models belongs to the theory of performance, even such that texts ('messages') were simply considered identical with utterances and therefore beyond the scope of grammar. We have to localize here all inductive (probabilistic) models for probability, stochastic processes, information (redundancy), frequencies and their distribution, etc., which were normally worked out for (empirical) sets of discourses (corpuses). From our theoretical discussion it was obvious that any linear model, e.g. a Markov chain, cannot be considered to be a possible candidate for a grammar: the complex dependency structure in sentences, and a fortiori in texts, is not based on transition probabilities. In general the number of possible texts over a vocabulary V_T is theoretically infinite. Conditional probabilities for strings of, say, more than 15 words, i.e. for nearly any text, are therefore not computable: no corpus would be large enough or sufficiently representative to obtain the relevant data. Information, based on probabilities of elements, can therefore not be measured for meaningful texts, and its technical aspects (as measure for sign-(un) expectedness) is hardly relevant for an adequate theory of textual performance. Only global conclusions (hypotheses) may then be made. We are thus left with the traditional statistical techniques for the description of frequencies, distribution, type/token ratio, variance, etc., of discourses, in order to characterize them stylistically with respect to a given corpus. This work and models are well-known and need not be treated here. We must repeat, however, that their theoretical relevance, as studies of the use of the system, derives from the insight into the properties of the system. Our claim is that only explicit *T*-grammars provide such an insight.

University of Amsterdam

NOTES

* This paper is published by kind permission of Mouton, The Hague, Holland, as it is scheduled to appear in *Linguistics*. The present version is identical, besides some minor corrections, to the one presented during the Congress. Some important points, e.g. the logic of definitivization, pronominalization and presupposition, have received detailed attention in van Dijk (1972a, 1972b), to which I refer for the general framework as well.

I am indebted to Asa Kasher for some useful critical remarks.

[1] This movement takes shape, for instance, in a research group in textual linguistics at the University of Constance (Western-Germany) having a very complete program of research, publications and symposia. For information cf. the note in *Linguistische Berichte* 13 (1971) 105–106, and in *Foundations of Language* (forthcoming) and *Poetics* 3 (1972) 128–130.

[2] A first survey is given by Hendricks (1967). More complete and systematic is Dressler (1970). For detail, cf. van Dijk (1972a), Ihwe (1972), Isenberg (1972), Petöfi (1971).

[3] We cannot go into detail about the psychological implications of our claims. The role of plans (or TOTE-units) has however generally been recognized. We consider them as relevant a fortiori for the cognitive processing of macro-structures. Cf. Miller *et al.* (1960).

[4] The literature on this subject is overwhelming at the moment. Cf. e.g. the contribution in Reibel and Shane (eds.) (1969) and the work of Lakoff and Postal. For fuller bibliographical details, cf. van Dijk (1972a).

[5] Concrete proposals of these different possibilities have been made by Palek (1968). Karttunen (e.g. Karttunen, 1968), Isenberg (1971), Bellert (1970), and Kummer (1971a),

[6] The discussion of this problem is extensive at the moment. Cf. e.g. Mönnich (1971) and Karttunen (1971) followed by Kuroda (1971).

[7] The relations between the logical theory of (definite) description and the conditions of pronominalization and definitivizations are regularly commented upon in much recent work. Cf. the work of Kummer (e.g. Kummer, 1971b, 1972), Wallace (1970), Karttunen (1971) and many others.

[8] The form and type of *T*-grammars is a problem about which few research until now has been done. Cf. the work of the group mentioned in Note 1, especially Rieser (1971), Ihwe (1971, 1972), Petöfi (1971, 1972a, 1972b), van Dijk (1972a). Cf. also Lang (1971).

[9] We may refer selectively to the work done about the theory of models by Apostel (1960), Chao (1962), Black (1962), Hesse (1966), Bertels and Nauta (1969), Nauta (1970).

[10] We suppose to be known here the recent discussions about the status of 'generative semantics', its claim and proposals as made by McCawley, Lakoff, Postal, Ross, a.o. and the partially justified criticism by Katz, Chomsky (cf. e.g. Chomsky, 1970)

and others. We will not enter into these discussions here. We have taken a semantically based model because it is the one which suits best our theoretical purpose.

[11] The use of predicate logic has become current practice particularly since the development of generative semantics. The only systematic application of it has been made by Rohrer (1971). In text grammar especially by Kummer (1971a, b, 1972).

[12] The important recent attention for the role of pragmatics in a semiotic theory like linguistic theory (and grammar) cannot be dealt with here. Note recent proposals about a 'communicative competence', containing the systematic rules regulating the ability to use texts/utterances 'appropriately'. See Wunderlich (1970), Stalnaker (1970), and van Dijk (1972a, chap. 9, 1973). Related with this topic are the 'typological' aspects of texts as defined by different types of speech acts. Philosophy of language would here supply valuable, although rather informal, models.

[13] For an early survey, cf. Chomsky (1963); Further see Gross and Lentin (1967), Marcus (1967), Schnelle (1968), Edmundson (1967) and, recently, Brainerd (1971).

BIBLIOGRAPHY

Apostel, Leo, 'Toward the Formal Study of Models in the Non-Formal Sciences', *Synthese* **12** (1960) 125–161.
Bellert, Irena, 'Conditions for the Coherence of Texts', *Semiotica* **2** (1970) 335–363.
Bertels, Kees and Nauta, Doede, *Inleiding tot het Modelbegrip*, de Haan, Bussum, 1969.
Black, Max, *Models and Metaphors*, Cornell U.P., Ithaca, 1962.
Brainerd, Barron, *Introduction to the Mathematics of Language Study*, Elsevier, New York, 1971.
Carnap, Rudolf, *Introduction to Symbolic Logic and its Applications*, Dover, New York, 1958.
Chafe, Wallace L., *Meaning and the Structure of Language*, University of Chicago Press, Chicago, 1970.
Chao, Yuen Ren, 'Models in Linguistics and Models in General' in *Logic, Methodology and Philosophy of Science* (ed. by E. Nagel, P. Suppes and A. Tarski), Stanford U. P., Stanford, 1962, pp. 558–566.
Chomsky, Noam, 'Formal Properties of Grammars' in *Handbook of Mathematical Psychology*, vol. II, (ed. by R. D. Luce, R. R. Bush and E. Galanter), Wiley, New York, 1963, pp. 323–418.
Chomsky, Noam, 'Some Empirical Issues of the Theory of Transformational Grammar', Indiana University Linguistics Club, mimeo, 1970.
Dahl, Östen, *Topic and Comment*, Almkvist & Wiksell, Göteborg, 1969.
Dijk, Teun A. van, *Some Aspects of Text Grammars. A Study in Theoretical Linguistics and Poetics*, Mouton, The Hague, 1972a.
Dijk, Teun A. van, 'Text Grammar and Text Logic', Paper Contributed to the Symposium 'Zur Form der Textgrammatischen Basis', Konstantz, September 3–9, 1972b. To appear in János S. Petöfi and Hannes Rieser (eds.) *Studies in Text Grammar* D. Reidel, Dordrecht, Forthcoming Paper, 1972c.
Dijk, Teun A. van, 'Grammaires textuelles et structures narratives' in *Structures narratives* (ed. by C. Chabrol), Larousse, Paris (forthcoming).
Dijk, Teun A. van, Ihwe, Jens, Petöfi, Janos, S., Rieser, Hannes, *Zur Bestimmung narrativer Strukturen auf der Grundlage von Textgrammatiken*, Buske Verlag, Hamburg, 1972 (Papers in Text Linguistics 1).

Dressler, Wolfgang, 'Textsyntax', *Lingua e Stile* **2** (1970) 191–214.
Drubig, Bernhard, *Kontextuelle Beziehungen zwischen Sätzen im Englischen*, Kiel University, Unpublished M.A. dissertation.
Edmunson, H. P., 'Mathematical Models in Linguistics and Language Processing' in *Automated Language Processing* (ed. by H. Borko), Wiley, New York, 1967, pp. 33–96.
Fillmore, Charles, 'The Case for Case' in *Universals in Linguistic Theory* (ed. by E. Bach and R. T. Harms), Holt, Rinehart and Winston, New York, 1968, pp. 1–88.
Geach, Peter, 'A Program for Syntax', *Synthese* **22** (1970) 3–17.
Greimas, A. J., *Sémantique structurale*, Larousse, Paris, 1966.
Gross, Maurice and Lentin, André, *Notions sur les grammaires formelles*, Gauthier-Villars, Paris, 1967.
Halliday, M. A. K., 'Language Structure and Language Function' in *New Horizons in Linguistics* (ed. by J. Lyons), Penguin Books, Harmondsworth 1970, pp. 140–165.
Harris, Zellig S., 'Discourse Analysis', *Language* **28** (1952) 1–30.
Heger, Klaus, 'Valenz, Diathese und Kasus', *Zeitschrift für romanische Philologie* **82** (1967) 138–170.
Hesse, Mary, *Models and Analogies in Science*, Notre Dame U.P., Notre Dame, 1966.
Ihwe, Jens, 'Zum Aufbau von Textgrammatiken'. Paper delivered at the 6th Linguistic Colloquium, Kopenhagen, August 11–14, 1971.
Ihwe, Jens, *Linguistik in der Literaturwissenschaft*, Bayerischer Schulbuch Verlag, Munich 1972.
Isenberg, Horst, 'Überlegungen zur Texttheorie' (1968) in *Linguistik und Literaturwissenschaft. Ergebnisse und Perspektive*, vol. I (ed. by J. Ihwe), Atheneum Verlag, Frankfurt a.M., 1971, pp. 150–172.
Isenberg, Horst, 'Texttheorie und Gegenstand der Grammatik', Berlin (DDR), mimeo 1972.
Karttunen, Lauri, 'What do Referential Indices Refer to?', MIT, mimeo 1968.
Karttunen, Lauri, 'Discourse Referents', paper delivered at the International Conference on Computational Linguistics, Sanga-Säby (Sweden) 1969.
Karttunen, Lauri, 'Definite Descriptions with Crossing Reference', *Foundations of Language* **7** (1971) 157–182.
Kummer, Werner, 'Referenz, Pragmatik und zwei mögliche Textmodelle' in *Probleme und Fortschritte der Transformationsgrammatik* (ed. by D. Wunderlich), Hueber, Munich, 1971a, pp. 175–188.
Kummer, Werner, 'Quantifikation und Identität in Texten' in *Beiträge zur generativen Grammatik* (ed. by A. von Stechow), Vieweg, Braunschweig, 1971, pp. 122–141.
Kummer, Werner, 'Outlines of a Model of Discourse Grammar', *Poetics* **3** (1972) 29–55.
Kuroda, S. Y., 'Two Remarks on Pronominalization', *Foundations of Language* **7** (1971) 183–188.
Lakoff, George, 'Pronouns and Reference' (mimeo), 1968.
Lakoff, George, 'Linguistics and Natural Logic', *Synthese* **22** (1970) 151–271.
Lakoff, George, 'On Generative Semantics' in *Semantics. An Interdisciplinary Reader* (ed. by L. A. Jakobovits and D. A. Steinberg), Cambridge U.P., London, 1971, pp. 232–296.
Lang, Ewald, 'Über einige Schwierigkeiten beim Postulieren einer Textgrammatik', Berlin (DDR), mimeo, 1971.
Langacker, Ronald W., 'On Pronominalization and the Chain of Command' in Reibel and Shane (eds), pp. 160–186.

Marcus, Solomon, *Introduction mathématique à la linguistique structurale*, Dunod, Paris, 1967.
Miller, G. A., Galanter, E., and Pribram, K. H., *Plans and the Structure of Behavior*, Holt, Rinehart and Winston, New York, 1960.
Mönnich, Uwe, 'Pronomina als Variablen?' in *Probleme und Fortschritte der Transformationgrammatik* (ed. by D. Wunderlich), Hueber, Munich, 1971, pp. 154–158.
Nauta, Doede, *Logica en Model*, De Haan, Bussum, 1970.
Palek, Bohumil, *Cross-Reference. A Study from Hyper-Syntax*, Universita Karlova, Prague, 1968.
Petöfi, Janos S., *Transformationsgrammatiken und eine ko-textuelle Texttheorie. Grundfragen und Konzeptionen*, Frankfurt, 1971.
Petöfi, Janos, S., 'Zu einer grammatischen Theorie sprachlicher Texte', *Zeitschrift für Literaturwissenschaft und Linguistik* 5 (1972a) 31–58.
Petöfi, Janos, S., 'The Syntactic-Semantic Organization of Text Structure', *Poetics* 3 (1972b) 56–99.
Reibel, David A. and Shane, S. A., (eds.), *Modern Studies in English. Readings in Transformational Grammar*, Prentice Hall Inc., Englewood Cliffs, 1969.
Reichenbach, Hans, *Elements of Symbolic Logic*, MacMillan, London, 1947.
Rescher, Nicholas, *Topics in Philosophical Logic*, D. Reidel, Dordrecht, 1968.
Rieser, Hannes, 'Allgemeine Textlinguistische Ansätze zur Erklärung performativer Strukturen', *Poetics* 2 (1971) 91–118.
Rohrer, Christian, *Funktionelle Sprachwissenschaft und transformationelle Grammatik*, Fink, Munich, 1971.
Sanders, Gerald A., 'On the Natural Domain of Grammar', Indiana Linguistics Club, mimeo, 1969.
Schnelle, Helmut, 'Methoden mathematischer Linguistik' in *Enzyklopädie der geisteswissenschaftlichen Arbeitsmethoden* (4, Methoden der Sprachwissenschaft), Oldenbourg, Munich, 1968, pp. 135–160.
Stalnaker, Robert, 'Pragmatics', *Synthese* 22 (1970) 272–289.
Tesnière, Lucien, *Eléments de syntaxe structurale*, Klincksieck, Paris, 1959.
Wallace, John, 'On the Frame of Reference', *Synthese* 22 (1970) 117–150.
Wunderlich, Dieter, 'Die Rolle der Pragmatik in der Linguistik', *Der Deutschunterricht* 4 (1971) 5–41.

WALTHER L. FISCHER

TOLERANCE SPACES AND LINGUISTICS

1. Relative Systems – Models

1.1. The general aspect of our contribution is the problem of an adequate representation of a concrete relative system $\mathfrak{E}$ by an abstract relative system $\mathfrak{A}$.

1.2. A *'relative system'* or *'relatum'* $(X; \sigma)$ is to be understood as a set X of entities – called *'elements'* – structured by a set σ of attributes of the entities – called *'properties* of' and *'relations'* between' the given entities.

Such a system will be called *'concrete'* iff the meaning (the ontology) of the entities and attributes is known, *'empirical'* iff the elements of $(X; \sigma)$ are given by sensory perception, and *'abstract'* iff the relatum is purely mathematical.

1.3. Iff moreover there is an application φ from $\mathfrak{E}$ into $\mathfrak{A}$

$$\varphi: \mathfrak{E} \to \mathfrak{A}$$

such that the abstract relatum $\mathfrak{A}$ is a homomorphic image of $\mathfrak{E}$, we call '$\mathfrak{A}$ an *abstract*' or *'mathematical model'* of $\mathfrak{E}$'.

1.4. To demonstrate an abstract system $\mathfrak{A}$ to be a model of an $\mathfrak{E}$ we have to prove the conditions, the properties and relations, stated in the abstract system $\mathfrak{A}$ to hold true for the elements and their interrelations in the concrete system $\mathfrak{E}$.

More precisely, to show that $\mathfrak{A}$ is a model of $\mathfrak{E}$ we have to prove that the chosen application $\varphi: \mathfrak{E} \to \mathfrak{A}$ has the property that elements of $\mathfrak{E}$ and $\mathfrak{A}$ which correspond to each other by φ have corresponding attributes under φ too.

2. Abstract Systems and Language

2.1. Precisely this is also one of the main tasks of mathematical linguistics: to test *whether a certain abstract system is relevant with respect to a sector of the reality of language*, whether the abstract laws proposed and assumed as a description of what occurs in language (as a model) are fulfilled in the applicated real phenomenon.

2.2. In the following we will consider a very simple abstract structure: the concept of an *equivalence structure*.

2.3. Equivalence structures are often used as models for linguistic phenomena, e.g. in the theory of synonymy, in defining paradigmatic and syntagmatic structures, etc. But, as easily shown in most of these cases, it cannot be proved whether the concrete relation fulfills all conditions of an equivalence relation.

3. Equivalence Structures

3.1. The concept of an '*equivalence structure*' is defined as a system of the form
$$(X; =),$$
where X is a given set of elements and '$=$' denotes an 'equivalence relation on X'.

3.2. As is well known, an '*equivalence relation* "$=$" on a set X' is characterized by the following conditions:

$=$ is a subset of $X \times X$ such that for every $a, b, c \in X$
(1) $a = a$ – *Reflexivity*
(2) $a = b$ implies $b = a$ – *Symmetry*
(3) $a = b$ and $b = c$ implies $a = c$ – *Transitivity*

3.3. With every equivalence relation '$=$' on X there is uniquely related a '*partition*' of the set X. Every equivalence relation on X induces a unique partition of the set X into a system $(X_i)_i$ of subsets $X_i \subset X$ of X – the X_i called '*classes*' – such that
(a) the totality of $(X_i)_i$ 'covers' X:
$$\bigcup_i X_i \subset X$$

(The system $(X_i)_i$ is a *'covering* of X')
(b) every two different classes are mutually exclusive (disjoint), i.e. they have no elements in common:

$$X_i \cap X_k = \phi \quad \text{for every} \quad i \neq k.$$

The X_i's are by definition the subsets of X consisting of precisely those elements of X which are equivalent to each other.

The system $(X_i)_i$ of classes induced by an equivalence relation '=' on X is called *'quotient set'* and is denoted by $X/=$.

3.4. Passing from $(X; =)$ to $X/=$ means methodologically forming a *concept by definition of abstraction*.[1]

3.5. *Example*: Let X be a set of empirical given entities and the relation '=' the relation of 'equal in colour'. Forming the classes of $X/=$ means distributing the set X of elements such that the X_i's consist in equally coloured entities. By forming $X/=$ we get the different 'colours' occurring as visible properties of the elements of X, for every colour corresponds uniquely to one of the classes X_i; the colours are represented by the X_i's.

The X_i's are individuals on a higher ontological level than the elements of X itself – the level of the X_i's represents a degree of abstractness higher than the level of the elements of X.

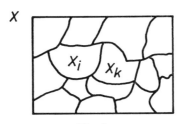

Fig. 1.

4. THE RELATION OF SYNONYMY

4.1. Now let us consider one of the relations in linguistics which is generally thought to be described (or defined) by equivalence relations: let us consider the relation of *synonymy*.

4.2. Several definitions of synonymy have been proposed by various authors. In certain cases the definition of synonymy is based upon:

I. *a geometric model*
via a function of distance measuring the semantic similarity of lexical units (words).

II. *a model of order*
via the rank of similarity of lexical units (and ultimately again via a distance function).

III. *a logical model*
via the bilateral implication of sentences in which the synonymous items occur;
or via the set of sentences implied by a sentence.

In the models I and II synonymy of two items (words) is given iff the distance of the two items (words) equals zero.

In model III 'x is synonymous to y' iff x is a hyponym of y and y is a hyponym of x.

4.3. In all of these definitions, equivalence is the very centre (kernel) of the conceptual abstract structure which is a model of the linguistic phenomenon.

But – and this is our question – is it possible to prove the validity of the three laws of reflexivity, of symmetry and of transitivity which characterize any equivalence?

4.4. In testing the conditions of the three laws in the reality of natural languages we find that the validity of at least the transitive law cannot be verified.

The reason of this negative statement is to be found in the fact of polysemy. Take, for instance, the three words,

$$x = \text{solid} \quad y = \text{dense} \quad z = \text{stupid};$$

while the words x, y and likewise the words y, z are to be considered as synonyms, the words x, z are certainly not synonymous to each other.

In other words, polysemy is the reason – or one of the reasons – why the transitive law is generally not fulfilled by the relation of synonymy.

Synonymy is not an equivalence relation – synonymy cannot be represented within an abstract model by an equivalence structure.

4.5. In place of an equivalence relation we therefore propose to introduce a weaker concept than that of an equivalence relation: the concept of '*tolerance relation*'.

5. Tolerance spaces

5.1. The concept of a '*tolerance structure*' or '*tolerance space*' is defined as a system of the form

$$(X; \xi),$$

where X is a given set of elements and 'ξ' denotes a 'tolerance relation on X'.

5.2. A '*tolerance relation* ξ on a set X' is defined by the following conditions:

ξ is a subset of $X \times X$ such that for every $a, b \in X$

(1) $a\xi a$ – *Reflexivity*
(2) $a\xi b$ implies $b\xi a$ – *Symmetry*.

Unlike an equivalence relation, a tolerance relation is not necessarily transitive.

$a\xi b$ may be read as '*a indistinguishable from b*'. It should be recalled that the notion of a tolerance relation was first introduced by the logician R. Carnap in 1928[2] under the term 'Ähnlichkeitsrelation'[3] and was rediscovered by the mathematician E. C. Zeeman in 1962.[4] The mathematical theory of tolerance spaces has in the interim been further developed by (Fischer, 1964–1971[5]; Arbib, 1966–67[6]; Zelinka, 1968–71[7]).

In a series of papers we will shortly give a detailed exposition of the set theory, of the topology in tolerance spaces, of tolerance-continuous and tolerance-differentiable functions, of tolerance spaces and interval analysis, of tolerance spaces and threshold logic, and of tolerance rankings.

5.3. To give an idea of how equivalence and tolerance structures differ from each other we would mention some elementary propositions:

(1) Uniquely associated with every tolerance relation ξ on X there is not a partition of X, but only a *covering* of X.

The covering of X induced by ξ is given by the system $(U_x)_x$ of sets $U_x \subset X$.

$$U_x := \{x' \mid x\xi x'\}.$$

U_x is the set of all elements $x' \in X$ which are indistinguishable from the given $x \in X$.

The fact that ξ is not necessarily transitive implies that the subsets of the covering are no longer mutually exclusive (Figure 2).

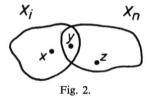

Fig. 2.

This means methodologically that a tolerance relation leads by passing from $(X; \xi)$ to the covering induced by ξ, not to a concept by definition of abstraction as is the case with equivalences, but only to a vaguely comprehended concept, an ambivalent concept.

(2) In establishing *set theory* in a tolerance space we have to remember that besides the relation of equality $A = B$ between subsets A and B of X there is a relation $A \sim B(\xi)$ – 'set A is indistinguishable from set B' –, for every tolerance relation ξ on X induces a tolerance relation $\sim$ on the lattice of subsets of X.[8]

(3) Unlike Zeeman, we have found that the *local topological structures* induced by tolerance relations are not irrelevant. The topology induced by a tolerance relation ξ on X by

$$\bar{A} := \{y \mid y \in X \text{ and } y\xi a, \text{ for } a \in A\}$$

is definitely what is called 'mehrstufig'[9], for the closure $\bar{A}$ of a set A in X is not necessarily closed.

In tolerance spaces we therefore generally find

$$A \subset \bar{A} \subset \bar{\bar{A}} \subset \bar{\bar{\bar{A}}} \subset \ldots$$

6. Synonymy and Tolerance Relations

6.1. Returning to the concept of synonymy, let us repeat that we propose to introduce tolerance relations in place of equivalence relations in defining synonymy.

6.2. Roughly speaking, the *definition of synonymy* would have the form:

'x is synonymous to y' iff $x\xi y$,
i.e. iff x is indistinguishable from y.

6.3. Although we know that not all difficulties are overcome by a definition of this kind, at any rate we are now not forced to prove the law of transitivity for a relation ξ within a concrete system. Moreover, it seems much easier to establish criteria for the indistinguishability of items than criteria for a rigid equality.

6.4. We may also establish a *topology of word fields* via the topology induced by the tolerance relation we use to define synonymy and which therefore underlies the concept of 'word field'. One of the undoubted advantages of this topology of word fields is that it is a 'topology without points', so that we do not need ideal semantic centres around which the synonyms flock together to form a topological neighbourhood.

The topology of word fields as induced by a tolerance relation is of course a multistep topology.

6.5. What we have outlined here in brief will be elucidated in detail in a book which is to appear shortly.[10]

Universität Erlangen-Nürnberg

NOTES

[1] H. Weyl, *Philosophie der Mathematik und Naturwissenschaft*, Munich 1928, pp. 9–10. R. Carnap, *Abriß der Logistik*, Vienna 1929, §20.
[2] R. Carnap, *Der logische Aufbau der Welt*, Berlin 1928, §71ff.
[3] = Relation of Similarity.
[4] E. C. Zeeman, 'The Topology of the Brain and Visual Perception', in *Topology of 3-Manifolds and related Topics*, (ed. by M. K. Fort), Englewood-Cliffs, N.J. 1962, pp. 240–256.
[5] W. L. Fischer, 'Toleranzräume', in *Archimedes* 22 (1970) 101–107. W. L. Fischer, Mimeographed Papers, 1964, ff.
[6] M. A. Arbib, 'Automata Theory and Control Theory, A Rapprochement', *Automatica* 3 (1966) 161–189.

M. A. Arbib, 'Tolerance Automata', *Kybernetika* **3** (1967) 223–233.
[7] B. Zelinka, 'Tolerance Graphs', in *Comment math. Univ. Carol.* **9** (1968) 121–131.
B. Zelinka, 'Tolerance in Algebraic Structures', *Czechoslovak Mathem. Journ.* **20** (1970) 179–183.
[8] $A \sim B(\xi) := A \subset \xi B$ and $B \subset \xi A$.
[9] Terminology following: G. Nöbeling, *Grundlagen der analytischen Topologie*, Berlin-Göttingen-Heidelberg 1954. p. 41. 'Mehrstufig' = 'multistep'.
[10] W. L. Fischer, *Äquivalenz – und Toleranzstrukturen in der Linguistik*, München 1973.

STEPHEN ISARD AND CHRISTOPHER LONGUET-HIGGINS

MODAL TIC-TAC-TOE

1. INTRODUCTION

The work we are doing began as an attempt to find a formal definition of truth for a limited class of English sentences. Guided by Tarski's definition of truth for formalised languages, and by Chomsky's conception of natural language syntax, we wanted to take a subset of English suitable for discussing a model universe, and to define in a intuitively satisfying way what it would mean for a sentence *to be true of that universe*, in terms of the syntactic structure of the sentence and the referents of the individual words. We did not feel committed to any specific grammar of English, but we were – and are – strongly prejudiced to the view that English syntax must serve some purpose in the process of communication, and we wanted our scheme to assign some function to as much as possible of the syntax we would use.

Our desire to do justice to English usage has, in fact, led us to modify our original goal. Instead of pressing the analogy between truth in English and truth in a formalized predicate language, we have chosen to view English as an *imperative* language, analogous to the languages which are used for programming computers. According to this view, English sentences are to be treated not as formulas (in the predicate language sense) but as terms whose domain of interpretation is a set of instructions to the hearer. In this setting our analogue of a truth definition becomes a way of deriving from a sentence a set of instructions for determining an appropriate response; for a well-posed question the appropriate response will be its answer, for a command it will be the execution of the command, and for a statement a modification of the hearer's state of knowledge (which will be part of the logical system upon which the set of instructions operates). This view is sufficiently general to accommodate sentences which fail to convey meaning to the hearer, for example those which incorporate erroneous presuppositions. In such a case the set of instructions for determining the response may be incomplete or impossible to carry out, and the sentence must be regarded as inappropriate. Allowing for this shift in viewpoint, our semantics has some features in common with those studied by modal and tense logicians, although our treatment of a number of locutions – in particular, those involving modal verbs – is quite different.

2. 'Must'

Consider the difference between (1) and (2) in ordinary English usage:

(1) *It is raining*
(2) *It must be raining*

Informally, we might say that it is the difference between the rain being entirely obvious to the speaker and his having to infer it from other facts. A man who comes in out of a downpour and announces 'It must be raining' is a rather comic figure; while Sherlock Holmes, when he sees a footprint and immediately says 'The man who made this has a tattoo on his left arm', is showing off. In fact, all of the English modal verbs can be used to signify that the speaker does not have direct knowledge and has to figure something out. They can, of course, be used in other ways too – 'You may go now' – but here we have in mind such usages as 'It may rain tomorrow', 'He cannot have committed the murder' and so on. Returning to (2), it would be quite unreasonable to claim that it is "stronger" than (1); if anything it is weaker, expressing less certainty on the part of the speaker. How could we capture this formally?

A formalism to deal with this use of 'must' might involve a language, a distinguished subset of 'axioms' of the language, and a method of computing a 'follows from' relation on sentences. An 'indicative' predicate would hold of those sentences which were axioms, and a 'must' predicate of those sentences, not themselves axioms, which followed from the axioms. This formalism would not, however, be expressing our intuition for 'truth', so much as for 'an appropriate thing to say'. We have, in fact, opted for formalising this notion of 'appropriateness', though not with the formalism just outlined, because it seems more likely to lead to a theory of the performance of English speakers, and because we feel that a clear intuition for what it is appropriate to say in given circumstances will extend to more sentences than those whose 'truth' is beyond dispute, making it easier to test such a formalism. It is, in any case, difficult to handle sentences other than statements within the confines of a conventional theory of truth; and a very large part of ordinary discourse consists of questions, commands and fragmentary utterances including exclamations.

3. Presuppositions

A class of sentences over which intuitions about truth tend to clash is that involving unacceptable presuppositions. Using a computational setting, and going for acceptability instead of truth, allows us to dispose of these in a rather natural way. Roughly speaking, a presupposition represents a condition which must be satisfied in order for the hearer to carry out his instructions. If the presupposition is false, he is unable to perform some step in his computation. If asked whether the present King of France is bald, he is unable to find a referent for 'the present King of France'. Notice that exactly the same view may be taken of presuppositions, whether they are embodied in statements – 'The King of France is bald', in questions, or in commands – 'Go and assassinate the King of France!'

4. Context

Another feature of English that fits naturally into a computational setting is the fact that sentences and fragments of sentences are interpreted according to context. The interpreter can be put, so to speak, into a variety of different states, and will treat the same sentence in different ways according to which state he is in. Extreme examples are the single-word utterances 'yes' and 'no', where the interpreter's state is partly revealed by the question to which these sentences are addressed; but even complete sentences which would be ambiguous in isolation are usually disambiguated by context – otherwise the speaker would not do well to employ them. Pronoun reference depends, of course, heavily on context; a pronoun may refer back to an individual who was last mentioned many sentences ago.

4. A model language user

As a means of expressing, clarifying and testing our ideas, particularly about English tenses and modal verbs, we are attempting to actually construct a model language user, in the form of a computer program. It is to be able to play a game of tic-tac-toe against a human opponent and discuss with him the progress of the game, in ordinary if simple English. At present we are concentrating on getting the program to answer questions about the game; it is not clear to us yet how to motivate it to ask questions itself.

A 'situation' for this machine consists of a board position, together with the sequence of moves leading to it. These essentially play the role of 'possible worlds'. The machine knows its own strategy, but makes no assumptions about that of its opponent. Thus it should be able to answer questions beginning 'What would you have done if...', but not, in general, questions of the type 'What would I have done if...'.

Our treatment of tense is broadly in accord with that of Reichenbach (1966). In particular, we distinguish between (i) the time at which the sentence is uttered, (ii) the time of reference, and (iii) the time of occurrence of events mentioned in the sentence. For instance, in the sentence (3),

(3)   *When I arrived, he had already left*

my arrival gives the time of reference, which is before the time of utterance and after the time of his departure. If we restrict ourselves to present and past tenses, and to sentences dealing only with moments of time, not intervals, then a good rule of thumb is that (i) the past tense will be used when and only when the time of reference precedes the moment of speech, (ii) the present tense indicates that the two moments coincide, and (iii) the presence of 'have' signifies that the time of the event precedes the reference time, while (iv) its absence indicates coincidence of these two times.

In symbols:
past tense $\Rightarrow$ time of ref $<$ time of speech
present tense $\Rightarrow$ time of ref $=$ time of speech
presence of 'have' $\Rightarrow$ time of event $<$ time of ref
absence of 'have' $\Rightarrow$ time of event $=$ time of ref

Every sentence must, in fact, be interpreted with respect to some reference time, although this time may not be mentioned in the sentence itself. Thus:

(4)   *Was he there when you arrived? No, he had already left.*

In general, a time clause in a sentence (but not one introduced by the word 'since') will indicate its time of reference; but if this has already been determined, for instance by a question to which the sentence is an answer, then the time clause can indicate the event time. An example of this is (5):

(5)   *He had left at 8 o'clock.*

Reichenbach (1966, p. 294) claims that *8 o'clock* must be the reference time in this sentence; but if (5) is the answer to (6),

(6)   *Why wasn't he there when you arrived at 8.30?*

then the '8 o'clock' in (5) is clearly the event time.

We have written a number of pilot programs, just dealing with tense, to work out our ideas in this area. One of these is designed to deal with questions about the arrival of guests at a party, and is supplied in advance with the information that Al, Bob, Bert, Charlie, David, Derek, Ed, Fred, George and Harry arrive in alphabetical order; a non-guest, Cuthbert, has not been invited and is not going to show up. Here is a sample of actual output; the time of speech was initially set to just before the party; the questions are the operator's, and the answers are the program's:

Q. Will Fred arrive before Harry?
A. Yes.
Q. Just before?
A. No.
Q. When Fred arrives will Bob have come since Cuthbert?
A. Cuthbert never will.
Q. When Bob arrives will Harry have come since David?
A. David won't have by then.

The time of speech is now moved to just after the arrival of Charlie.

Q. Has David turned up?
A. No, not yet.
Q. Has Cuthbert?
A. No, he never will.
Q. Will Bob arrive before Bert?
A. Bert already has.

The machine works by first 'parsing' the question in a primitive way to determine the time of reference, the event in question and any other time-adverbial information provided. Any essential gaps are then filled by reference to what has been said already. In the subsequent process of interpretation the reference time is first determined, and if the tense is 'inappropriate', revealing a false presupposition, then the reply has to indicate the fact. The details are somewhat complex, but the main point is to derive from the question a systematic procedure for answering it, taking account not only of the words of the question itself and the knowledge available, but also of the conversational context, and allowing for the possibility that the question may not be well posed, in which case attention must be drawn to the erring presuppositions.

Analogous problems arise with hypothetical and subjunctive clauses; they call for the construction of situations within which other sentences are tested. The construction of the 'subjunctive situation' begins from the reference point of the 'if' clause; in (7), for example, this is the present moment:

(7)   *If I go here, what will you do?*

and in (8) some point in the past:

(8)   *If I had gone here when I went there, what would you have done?*

Subjunctive reference situations can persist for several sentences, in the same way that past situations can. If (9)

(9)   *I would have done this.*

is used as an answer to (8), its reference situation is the same.

Modal verbs are connected with possible games which are continuations from the reference situation. 'May' asks whether something is true in some such possible game, 'must' whether it is true in every such game. Our provisional interpretation of 'will' is that it asks about all games which are consistent with the machine's strategy. This works well in the specific context we have chosen, but is clearly less general than it might be. The same applies to our use of 'can'. We say that a player 'can' achieve something if there is a strategy for him that leads to this result no matter what the opponent does. Thus:

(10)   *I can win.*

is taken to mean that I have a winning strategy, not necessarily that I will use it.

In general we regard 'could', 'would' and 'might' as those forms of 'can', 'will' and 'may' which appear when there is reference to a past or subjunctive situation. We accept, but do not attempt to explain, the fact that 'might' can also be used when the reference time is the present as in (11):

(11)   *Will you go here? I might.*

6. CONCLUSION

In summary, we have found it impossible to give a satisfactory account of the use of English entirely in terms of any concept which can reasonably be called 'truth'. This is because the appropriateness of a natural utterance depends heavily on the states of the speaker and hearer, and the nature of this dependence is such that it distinguishes sentences like (1) and (2) in cases where no one would want to call one 'true' and the other 'false'. And in any case it could be argued that questions and commands are likely to require for their explication basic categories additional to those of truth and falsehood. But there does seem to exist, for natural language, a useful analogue to the truth definitions offered by the formal logician, namely a set of meta-rules for constructing from a natural utterance a set of instructions to be followed by the hearer. Winograd (1971) has shown that much light can be thrown on English sentences by viewing them as representing such sets of instructions, and one of the merits of this proposal is that one can test one's intuition as to the relation between utterances and the programs which they represent by expressing the relationship in the medium of a programming language and then holding a conversation with one's theory.

University of Edinburgh

BIBLIOGRAPHY

Reichenbach, Hans, *Elements of Symbolic Logic*, MacMillan, New York, 1966.
Winograd, Terry, *Procedures as a Representation for Data in a Computer Program for Understanding Natural Languages*, MIT, Project MAC, MAC-TR-84, Cambridge, Mass. 1971.

VYACHESLAV IVANOV

ON BINARY RELATIONS IN LINGUISTIC AND OTHER SEMIOTIC AND SOCIAL SYSTEMS

Common principles of binary organization have been discovered in different sign systems. If the total number n of all the elements of the system (or of a subsystem, for instance, of a certain level of the natural language) does not exceed 10^2 ($n \leqslant 10^2$), the relations between them may be described as binary: each element is opposed to another one having the same set of binary features and differing from the first one depending on the value of one feature. In a natural language such a binary principle is characteristic of certain levels with relatively small number of elements – the total number of linguistic elements exceeding by far the above given limit. Common to the levels of phonemes (and phonological differential features), tonemes (in languages where tonemic differences are phonologically relevant), and grammemes (grammatical meanings) is the binary principle of structural organization that is valid not only for the synchronic description of a system but for its diachronic investigation as well. It is considered either as the main structural characteristics of the natural object-language itself (as in early pioneering works on binary differential features, (Jakobson, 1971)) or as the result of the classificatory use of the artificial linguistic metalanguage (Chomsky and Halle, 1968, pp. 65, 295–297).

Particularly illuminating for the evaluation of the effectiveness of this principle in the diachronic study of grammar are the results of the reconstruction of Proto-Akkadian. In the oldest reconstructed pattern of this language the binary opposition of the case of the subject (nominative) and that of the object (accusative-genitive) is parallel to the same distinction between the two verbal moods, two genders and two numbers (Gelb, 1969). The development of the system is possible due to the dichotomy in one of the members of the binary opposition; e.g., the accusative is differentiated from the genitive in the primitive case of the object, etc. The same scheme may be applied to other old Nostratic languages such as common Indo-European where the binary opposition active/inactive may be reconstructed. In the nominal paradigm, this opposition is manifested in the distinction between two cases: the active one (later nominative) and the inactive one (later accusative), two genders (animate and inanimate, still preserved in such an archaic language as Hittite), two series of verbal forms (Ivanov, 1968; cf. also Watkins, 1969, p. 66, §46). In Indo-European as well as in Kartvelian (Gamkrelidze, 1959, pp. 15 and 50–51, n. 1) the binary system of nominal classes (genders in later Indo-European) implies the existence of the pronominal category of inclusive-exclusive. The discovery of the system of binary oppositions serves to unify the synchronic description since the same (or similar) distinctions are found in nominal, pronominal and verbal paradigms on the grammatical level (in consonantal and vowel phonemes on the phonological level, respectively).

Some of the binary oppositions discovered in the grammar and in the narrow semantic fields of the lexicon are similar to those found in mythological systems and ritualistic texts. In the total lexicon of the natural language the binary relations are not given in the explicit way due to the greater number of the elements ($n \geqslant 10^4$). An analogy to such a large system in ethnology may be found in the sets of totemistic symbols, where binary relations may be traced out only in small subsets comparable to semantic fields in the lexicon.

The striking resemblance between linguistic binary oppositions and those found in mythology and rituals was given attention only in recent years, although the discovery of the latter type by Durkheim and Mauss has preceded the spread of binary theories in linguistics.

It is worth noting that the binary oppositions of phonological tones (tonemes) are interpreted in terms of binary symbolic classification in some Bantoid languages (Ivanov, 1966, pp. 260–261), as well as in old Chinese science where the classification of the low and high sounds based on the universal binary scheme of "yin-yang" distinction has preceded modern interpretation of Chinese tones (Yuen Ren Chao, 1967, p. 96).

In the simplest cases of binary symbolic classification two rows of features and/or symbols (signs) are opposed to one another. Each feature, or symbol, in such a system is in a relation of polarity to the opposed one in the other row. The difference between these systems and semantic oppositions in natural languages lies in the fact that, in mythology, every pair of binary elements may be described as a translation of the basic opposition 'good'–'bad' which corresponds to the cybernetic model of the elementary behavior, according to late M.L. Zetlin (Zetlin, 1969, pp. 11, 22, 138–9). In this model all the signals $S \in (S_1, S_2, ..., S_n)$, being perceived from the surrounding environment, are classified by the automaton either as favorable (gain, $S=0$), or as unfavorable (loss, $S=1$). Correspondingly, two rows of synonymous (from this, and only this, point of view) classes of signals are formed that may be seen exactly in such decision-making procedures in collectives as the mechanism of oracles which determines the social behavior in primitive societies; the inversion of binary opposition (connected with the formation of ternary and still more complicated systems out of them) is characteristic of the oracles as well as of other rituals. Combination of two opposites (for instance, "yin-yang") in one ritual symbol (as in the bisexual shaman) is similar to the formation of complex names of a certain category (e.g., 'size') by means of compounding the names of two possible members of the category (e.g., 'large'–'small', etc.).

It is shown that in such systems as Dagela in Africa (Pouillon, 1970) binary structure may synchronically be considered as a special case of the ternary one, with one element being manifested by a zero form or an empty set. But it would be inadequate to describe each binary system as a special case of a ternary one as it is proposed in relation to the color systems by Professor Turner (Turner, 1967, pp. 60–71). Turner's suggestion is connected with an old logician's idea (Kempe, 1890), according to which every structure (ternary one included) is perceived by the human mind as a binary one. From the logical point of view, the ternary interpretation would be as adequate as the binary one. But for most of the color systems described in Turner's study, such ternary interpretation would be redundant since the ternary member of the structure either is always represented by the zero form (as in strictly dualistic systems) or may be understood as the representative of the two members of binary opposition in the context of its neutralization (e.g., the red color in many Ndembu rituals is the archisign in a context where the opposed black and white colors should not appear). The situation is similar to the neutralization of the opposition of two phonemes in a context where an archiphoneme (to use a term of the Prague linguistic circle) appears.

It is necessary to distinguish between the elementary binary oppositions and those more complicated systems (for instance, ternary systems) that are based on them. Following Hocart's idea (Hocart, 1952, p. 31), the relation between different social (originally ritualistic) functions in ternary systems may be represented by the trees of the same type as those used in modern syntactic theory. The four-number systems of classification and of social functions are built out of the binary opposition by means of iterative (cyclical) dichotomy, as may be shown on the example of the structure of the Yoruba society where the four-member system of the main social roles and their ritualistic correlates is underlied by the binary relations of the type 'left'–'right' (Frobenius 1913, pp. 198, 280; Cassirer, 1925, p. 112; Dennet, 1968, p. 61 and others.).

Some of the main binary oppositions, and the links between them, are quasi-universal; i.e. they are (statistically) widely spread in most semiotic systems.

Thus, the association between the members 'left' and 'female' in the binary oppositions 'left'–'right', 'female'–'male', is found in most European, African, and Asiatic systems. But as an example of the rare system where 'left' is associated with 'male' the ancient Chinese "yin-yang" classification may be cited. The Chinese facts may be compared not only to Zuni mythology (Gernet, 1961, p. 70, n. 5) but to the specific abnormal position of the left hand of the ancient Chinese mythological hero Fu-Si, as interpreted by Granet (Granet, 1934, p. 263), and to the magical role of the left hand of Australian mythological heroes. Such cases of the ritual inversion between 'left' and 'right' may be compared to other examples of ritual inversion, particularly to those connected with the death and funeral among Siberian peoples.

Methodologically important is the problem of the interpretation of the formal scheme of binary symbolic classification in the social context of dual organization. It may be discussed on the example of the Ket (Yenissei Ostyak) system, where two polar rows of features and symbols are associated with two old exogamic "moieties" (Ket *qolæp* 'half, moiety') – *bogdidæy*, "the people of the fire" and *ul'dæy*, "the people of the water". With this distinction other ritualistic oppositions are bound such as that between two types of shaman (connected with the birds or with the bear, respectively) and two mythological shamans (Ivanov, 1969, pp. 112–113). Some survivals of such archaic social systems may be traced up in more developed states, such as Ancient Near Eastern societies where, for instance, the dual opposition between the king and his mother (Hittite "tawananna") is similar to that found in Ruanda, Swazi, and other

African societies having some remnants of binary structure of main social functions.

Institute of Slavonian and Balkanistic Studies,
Moscow

BIBLIOGRAPHY

Cassirer, E., *Philosophie der Symbolischen Formen*, 2 Teil (*Das Mythische Denken*), Berlin, 1925.
Chao, Yuen Ren, 'Contrastive Aspects of the Wu Dialects', *Language* 43 (1967).
Chomsky, N. and Halle, M., *Sound Pattern of English*, New York 1968.
Dennet, R. E., *Nigerian Studies of the Religious and Political System of the Yoruba*, London 1968 (2nd edition).
Frobenius, L., *Und Afrika sprach...*, Berlin 1913.
Gamkrelidze, Th. v., *Sibilant Correspondences and Some Questions of the Ancient Structure of Kartvelian Languages* (in Georgian), Tbilisi 1959.
Gelb, I. J., *Sequential Reconstruction of Proto-Akkadian*, Chicago 1969.
Gernet, J., 'L'age du fer en Chine', *L'Homme* 1 (1961).
Hocart, A. M., *The Northern States of Fiji*, London 1952.
Ivanov, V. V., 'On a Case of Semantic Archaism in Bamileke', *Yazyki Afriki*, Moscow 1966 (in Russian).
Ivanov, V. V., 'Reflexes of two Series of Indo-European Verbal Forms in Proto-Slavic', *Slavjanskoye Yazykoznaniye*, Moscow 1968 (in Russian).
Ivanov, V. V., 'Binary Symbolic Classification in African and Asiatic Traditions', *Narody Azii i Afriki*, No. 5 (1969).
Jakobson, R., *Selected Writings*, vol. I (2nd. ed.), II, The Hague-Paris 1971.
Kempe, A. B., 'On the Relation between the Logical Theory of Classes and the Geometrical Theory of Points', *Proceedings of the London Mathematical Society* 21 (1890).
Needham, R., 'The Left Hand of the Mugwe: An Analytical Note on the Structure of Meru Symbolism', *Africa* 30 (1960).
Pouillon, J., 'L'hôte disparu et les tiers incommodes', in *Echanges et communications. Mélanges offerts à C. Lévi-Strauss à l'occasion de son 60-éme anniversaire*, Paris-The Hague 1970.
Turner, V. W., *The Forest of Symbols*, New York 1967.
Watkins, *Indo-Germanische Grammatik* (hrsg. von J. Kuryowicz), Bd. III, Formenlehre, 1 Teil, Geschichte der Indogermanische Verbalflexion, Heidelberg 1969.
Zetlin, M. L., *Studies in the Automata Theory and Models of Biological Systems*, Moscow 1969 (in Russian).

ASA KASHER

WORLDS, GAMES AND PRAGMEMES: A UNIFIED THEORY OF SPEECH ACTS

One glance at the terrain of Modern Logic seems to reveal that one cannot see the wood for the trees. There are the groves of the logics of commands and of questions, the gardens of the logics of entailment and supposition, the neglected copses of the logics of promises, advices, apologies and the rest of that jungly area of speech products. But do they really fail to exhibit any kind of integrative cohesion across the whole terrain? The purpose of this paper is to answer that question in the negative, and *not* by suggesting a taxonomy of the linguistic flora and the logical fauna of our area, but rather by outlining a formal *unified* theory of speech acts.

Let us begin by having a look at the following speech situation. Alpha is an ideal speaker who means exactly what he says, and Beta is an ideal listener who understands exactly what Alpha tells him. Both belong to the same linguistically homogeneous ideal community. Each of them knows that his partner is also ideal. Suppose Alpha says to Beta: "I ask you to bring me the book right away." What, then, is the function of the expression "I ask you"?

In our terms we would say first that if the situations W_1 and W_2 differ from each other exactly in that in W_1 Beta brings the book to Alpha right away and in W_2 he does not, then Alpha *prefers* the situation, the 'possible world', W_1 over the possible world W_2. (This may be represented by the simple formula $-P_\alpha(\{p\}, \{non-p\})$.) Now, Beta is an ideal listener who understands exactly what Alpha tells him. Thus Beta knows that *Alpha says* he prefers a possible world where the book is brought to him by Beta right away, over a possible world where it is not brought to him by Beta right away, all other conditions being equal. But since Alpha is an *ideal* speaker, who means exactly what he says, then Beta not only knows that Alpha *says* he prefers one possible world over the other, but he also knows that Alpha *does prefer* one possible world where Beta brings him the book right away over another possible world where Beta does not bring him the book right away, all other conditions being equal. Hence, the price paid by Alpha to Beta for the service he asks him to render is that he gives Beta information about his (Alpha's) preferences. (The book may be pornographic, for example, and Beta may notice it on delivering it to Alpha.) Alpha pays this price *knowingly*. As an ideal speaker in a linguistically ideal community Alpha *knows* that Beta, who understands him perfectly, will get hold of some, possibly new information about his (Alpha's) preferences. From the fact that Alpha actually said what he did say it is therefore clear that he does not have only *one*, first-order preference with regard to Beta's bringing him the book right away, but he has an additional second-order preference as follows: given two possible worlds – W_3 and W_4 – that differ from each other only to the extent that in W_3 Beta brings the book to Alpha right away, Beta knows Alpha's first-order preference concerning the book and Alpha knows that Beta knows Alpha's first-order preference, while in the possible world W_4 Beta does not bring the book to Alpha right away and does not even know Alpha's first-order preference concerning the book – all the other conditions being equal, Alpha prefers W_3 over W_4.

(Symbolically: where q is the first-order preference $P_\alpha(\{p\}, \{non\text{-}p\})$,

$$P_\alpha(\{p, K_\beta q, K_\alpha K_\beta q\}, \{non\text{-}p, non\text{-}K_\beta q\}).)$$

Now, as this second-order preference is conveyed by what Alpha says, Beta, being an ideal listener, will grasp it and Alpha *knows* Beta will grasp it. Because Alpha said what he did say we thus get a third-order preference, and so on, ad infinitum. Within that infinite series of preferences we distinguish between the basic ones – let us call them '*pragmemes*' – and the derived ones. In our example we have only one pragmeme, namely the first-order preference. All the other preferences are derived from that pragmeme by the application of one single inference rule – let us call it '*the rule of the communication price*' – namely – from the premise that an ideal speaker Alpha conveyed in a speech act his preference $P_\alpha(\{p_1\}, \{p_2\})$, there follows the conclusion that in the same speech act Alpha conveyed also the following preference of his – where $q = P_\alpha(\{p_1\}, \{p_2\})$

$$P_\alpha(\{p_1, K_\beta q, K_\alpha K_\beta q\}, \{non\text{-}p_2, non\text{-}K_\beta q\}).$$

Before we start improving our theoretical framework, we put forward – still roughly – our first central contentions. First, what characterizes a linguistic function, or a kind of speech acts, is an appropriate set of preferences of certain possible worlds over some other ones. Secondly and more specifically, each of these characteristic sets of preferences is two-fold *finite*: it is finitely generated and all its elements are each finitely expressible. In other words, all the preferences conveyed by one speech act are obtained from a *finite* set of basic preferences – the pragmemes – by applying some of the finitely many inference rules – the universal rules of communication. Moreover, if in accordance with one of these preferences, one possible world is preferred over the other, then the difference between these two possible worlds is *exactly* to the extent that certain finitely many independent sentences are true in the first possible world and certain other ones, also finitely many and independent, are true in the other possible world.

The above discussion of the linguistic function of the sentence "I ask you to bring me the book right away" leads us to the conclusion that the speech acts of asking, in the sense of requesting, have characteristic sets of preferences, generated each by a single pragmeme which is a first-order preference. This is by no means true for all the speech acts. It is easily verifiable that the speech acts of threatening involve two pragmemes, which are first-order preferences, and that the speech acts of advising involve one pragmeme that is a second-order preference.

Now it is time for improvements and ramifications. Consider, first, the following list of sentences: "I ask you to bring me the book", "Please, bring me the book", "Bring me the book", "I am demanding of you that you bring me the book", "I insist that you bring me the book". Basically, a speech act in which any of these sentences is uttered conveys the same set of preferences presented in the above discussion of our first example. However, while they all share the same pragmemic structure, they are still all clearly different from each other. To express the differences between the nuances of the same linguistic function it seems appropriate to attribute a degree of intensity to each preference. Strictly speaking, we do not introduce into our theoretical framework a quantitative concept of intensity of preferences. What we do at that stage is to introduce a comparative concept – possible world W_x is preferred over possible world W_y more than possible world W_v is preferred over possible world W_u. Indeed, we assume that all our preferences are at least partially ordered.

The simplest way of introducing such a comparative concept of preference is, of course, to introduce a partially ordered classification. In other words, we introduce a family of preference relations and then we partially order its members. Now we would like to mention one problem with regard to such an ordered family of preference relations, viz. – is it a finite or rather an infinite family ? The answer seems to be roughly that only finitely many sharp distinctions can be reasonably made between different classes of preference relations, but within at least some of these classes a possible infinite number of nuances of preference relations can be found. I don't have time for more than one hint at one example – for the ideal speaker, the iterative use of the word 'very' or its counterparts in other languages produces a gradation of preferences, having upper and lower bounds but no least element.

Now, for another improvement of our framework, consider again our first example: Alpha asks Beta to bring him the book right away. If Beta were manacled and not in a position to move, then Alpha's speech act would be taken to be bizarre. If Alpha, being an ideal speaker, asks Beta to bring him the book right away, then Alpha presupposes at that context of utterance that Beta *can* bring him the book right away. Or, take another example, a little more complicated and much more interesting: On Sunday Alpha says to Beta: "I hereby promise to visit you tomorrow morning at home". Knowing that Alpha is an ideal speaker we know now that Alpha promised Beta on Sunday to visit him at home on Monday morning. Now, Alpha will not be taken to violate his promise if on Monday morning he does not come because he is dead or arrested, injured or only knows that during the night Beta's apartment had been taken over by some friends of Beta who are dangeroulsy hostile to Alpha, etc. No promise is unconditional. Every promiser has his own presuppositions with respect to those items of occurrence that are at issue in his promise.

The conclusions from this brief discussion are drawn straightaway. In our preliminary framework we related speech acts to preference relations of the following form: For any two possible worlds W_1 and W_2, if they differ from each other exactly to the extent that in W_1 it is true that p and *not-q* while in W_2 it is true that q and *not-p*, then it is conveyed by the speech act under consideration that the possible world W_1 is preferred over the possible world W_2, by Alpha, who is our steady ideal speaker. It is clear now that we have to restrict the domain of possible worlds involved. Instead of talking about 'any two possible worlds' we have to talk about 'any two possible worlds in which all the presuppositions of Alpha at the context of utterance hold.'

Thus, if Alpha asks Beta to bring him the book right away and Alpha's speech act conveys that he prefers possible world W_1 over possible world W_2, then it is true for both W_1 and W_2 that Beta has in them the capacity of bringing the book to Alpha right away. Our second example – Alpha's promise to visit Beta next morning – will show that the incorporation of presuppositions into our framework is somewhat intricate, but it will also be clear that this intricacy stems from the nature of presuppositions rather than that of the speech acts and preferences involved. Adopting Rescher's definition of a scenario as a complete possible history of the subsequent course of biographical and extra-biographical events, relevant to anything at issue in the speech act, we can say that Alpha's speech act of promise conveys a preference of possible world W_1 over possible world W_2 only if both of them are final stages of appropriate scenarios. In none of these scenarios does Alpha, for example, die before Monday morning.

Now a final remark on presuppositions. When it is conveyed by a speech act of Alpha that among each two possible worlds that differ from each other to the extent that p holds in one and not in the other, Alpha prefers one of them over the other, then Alpha presupposes at the context of utterance that in the actual world p does not hold. At least some of the performatives can be explained by the fact that Alpha presupposes that before his speech act takes place p does not hold then in the actual world and that it will hold in it after the speech act takes place.

In a nutshell, our framework includes a diversity of preferences, a realm of possible worlds, finitely expressible and appropriately restricted by presuppositions, pragmemes and rules of communication.

In passing we would like to mention two applications of this framework. Many controversial issues in the logics of commands, questions, promises and so forth can be settled by the self-evident reductions of various calculi into our framework. One of the highlights here will be the debatable issue of logical relations among commands and questions, including the problems of validity of command or question inferences. A second application is in what seems to be a novel domain of the logic of silence. Think, for example, of a situation in which a genuine question remains unanswered. We shall discuss these applications and others in detail elsewhere.

In closing, we should like to point out how our logico-linguistic framework is incorporated into an even more general framework, a *cognitive* one.

Any society may be conceived as a group of participants in one large, complicated *game*. Like any other game – chess, for instance – it is characterized by its possible situations, possible moves and possible outcomes. The possible situations for a given person are those in which he may find himself in the course of his life; his moves are of course limited by the laws of nature; his payoff function is determined by his attitude towards each of his possible situations, in terms of pleasure, honour, money, satisfaction or what have you. We assume that the ideal participant, who is our ideal speaker-listener, has the constant goal of playing – behaving – in such a manner that the expected resulting future payoff, within the range of planning, should be maximal. However, the payoff is not determined by one participant's moves solely, but by the moves of many if not all the participants. The trouble is that in our game of life we do not know what are the other participant preferences, and henceforth we have to formulate our presuppositions concerning the preferences of the other players and to play accordingly. In game-theoretical terms, we have a game with misperception or games with incomplete information. The function of communication in this game is to unveil some of the personal preferences. Whether this is a move in a strategy of conflict or a strategy of coordination, is a question to be answered on empirical grounds, for each kind of speech acts intertwined with presuppositions separately.

It is our ultimate belief that game theory – the mathematical theory of preference and decision – will play a central role in linguistic pragmatics, as model theory – the mathematical theory of possible worlds – will serve linguistic semantics. What we have done here is to show in brief how a unified theory of speech acts can be built in terms of both formal semantics and pragmatics.

Bar-Ilan University, Israel

BIBLIOGRAPHICAL NOTES

Logical wood: cf. *Logic and Foundations of Mathematics*, (ed. by R. Klibansky,) La Nuova Italia Editrice, Firenze, 1969.
Formal Semantics: cf. B. C. van Fraassen, *Formal Semantics and Logic*, Macmillan 1971; N. Rescher, *The Logic of Commands*, Routledge and Kegan Paul, 1966.

Logic of Preference: cf. N. Rescher, *Topics in Philosophical Logic*, Reidel, Dordrecht 1968, Ch. XV.

Logic and Linguistic Semantics: cf. G. Lakoff, *Linguistics and Natural Logic*, University of Michigan, Ann Arbor, 1970. (Also in *Synthese*, 22 (1971) 151–271.)

Games: cf. R. D. Luce and H. Raiffa, *Games and Decisions*, John Wiley, New York, 1957; J. C. Harsanyi, 'Games with Incomplete Information Played by "Bayesian" players I-III', *Management Science* 14 (1967–68); T. C. Schelling, *The Strategy of Conflict*, Harvard, 1960, II, and Appendices B. C.; D. K. Lewis, *Convention*, Harvard 1969.

Pragmemes: A. Kasher, *Aspects of Pragmatical Theory*, forthcoming.

JANINA KOTARBIŃSKA

ON OCCASIONAL EXPRESSIONS*

It is almost commonly admitted that occasional expressions are such as 'I', 'this', 'today', 'here', 'yes', 'no', and so on, as opposed to such as 'Warsaw', 'red', 'ring', 'or', and so on. There is also a somewhat general agreement in that the semiotic properties of occasional expressions – all or only some of them – depend not only upon the form of the expressions but on the circumstances of their use as well.

The task of this study is to grasp the specific particularities of occasional expressions whose extension has been distinguished, though only partially, on the ground of the foregoing initial assumptions.

The most current definitions attribute an occasional character to all such, and only such, expressions the meanings of which depend on some definite circumstances of use (for instance on the speaker or on the time or place of speech), respective on whichever ones. But it is easy to show that these versions of definitions are either too narrow or too large in relation to previously accepted preliminary assumptions. One can, however, try to solve the question in a different manner. One can, namely, admit that occasionality of expressions is determined not by the very fact of dependence of their meanings upon the circumstances of use but also by the *nature* of this dependence. I think that such is the direction of investigations suggested vaguely by Husserl.

The final result of our analysis is the following. The occasional expressions are characterized by two features: (1) their semiotic functions are assigned to them not only in respect to their syntactical properties, shapes or sounds, but also in respect to their pragmatical context, it is to the circumstances of their use – hence the dependence of these properties upon the circumstances of use; (2) this assignment is of such a kind that it establishes a univocal, alias functional, dependence between two variable factors: a certain type of circumstances of use and a certain type of semiotic functions.

These properties of occasional expressions determine the kind of statements suitable for definitions of such expressions. It is clear that they must be in the form of conditional statements stating that if an expression is of a given form and is used under given circumstances of use then it has given semiotic properties. Moreover, it is clear that since these definitional sentences have to establish functional relationships between variable factors, these factors must be represented in them by variable symbols, linked so that the value obtained, as the result of a substitution, from the first of the two symbols (namely the symbol occurring in the antecedent) determines univocally the value of the second. A simple example of such sentence: (x, y). If x is an expression of the form 'I' and x is spoken by the person y, then x denotes the person y; or: (x, y). If x is an expression of the form 'today' and x is spoken on the day y then x denotes the day y.

It may be well to remark that in the case of such type of definitional sentences the univocality of the relation between the semiotic properties of occasional expressions and the circumstances of their use is warranted by the rule of substitution. The reason for this is that in any such sentence the expression which characterizes the semiotic properties of the defined term includes the same variable which occurs in the antecedent and characterizes the circumstances of use, whereas the rule of substitution allows only to substitute the same expressions for one and the same variable in a given formula.

One can say that the definability by sentences of this type is one of the characteristic peculiarities of occasional expressions.

It might be added that such definitions can be considered as a particular case of partial (or conditional) definitions introduced by Carnap. The difference here is that the definitions in question are formulated in meta-language and not in object-language; and that the criteria of applicability they lay down are limited not in the sense that they permit to decide about some objects only whether these do, or do not, fall under the defined term, but in the sense that they are applicable only to some tokens of the defined term, namely of those which are used in the situational context indicated in a given definition.

These are some of the specific features of expressions at issue. These fundamental features result in various derivative characteristics. It follows, for example, that the semiotic functions are not ascribed to all expressions of the same form, but only to those which satisfy certain specified conditions (stated in the antecedent of the respective definitional sentence); that the denotations of occasional expressions change according to a certain constant principle, common to all specimens of the same form; and that, under the assumption that all changes of denotations are always reflected by modifications of the respective meanings, these expressions are essentially multivocal, but that multivocality, though unavoidable, is harmless enough, for in each particular case of use the expressions have but one meaning univocally determined by the circumstances accompanying their use.

In the literature of the subject these characteristics are, by and large, accepted. But just by and large. For there is a view quite opposed to the standpoint outlined above. I have in mind the view developed most thoroughly by Bar-Hillel and sketched only in a very general way by Ajdukiewicz and Black. It is worthy to be carefully considered.

The analysis of the concept of expression plays here the main role. In both logical and linguistic papers the usage is to give the name 'expression' either to definite inscriptions or sounds (*tokens*) – i.e. to physical objects of a determined shape or sound – or to classes (*types*) of such objects distinguished with respect to their graphical or phonic shape. Now, because of the peculiar character of occasional expressions, the authors mentioned before are inclined, with regard to such expressions, to revise this purely syntactical notion of expression. It is namely proposed to consider as an expression (the adjective 'occasional' should probably be added) not a linguistic token itself, i.e. not an inscription, or a sound, itself, but a certain complex whole composed of the elements of two kinds: of an inscription or a sound (a token), *and* of certain circumstances of use. Bar-Hillel defines an expression, so understood, as an ordered pair composed of a definite inscription or sound (token), and the accompanying pragmatic context.

On the ground of this theory occasional expressions are univocal by definition. If different tokens of a given occasional word are accompanied by different circumstances of use of a given kind, then we have to do with several expressions which differ from each other as to their component elements, and not with several specimens of the same expression which differ from each other in their semiotic functions. For example: in each time when different persons use a token of the form 'I', a series of different expressions come into being, and each of them is composed, as to its elements, of a given token of the shape 'I' and the person who used that particular token. The word 'I' does not change its denotation simply because it has no denotation at all: it is not an expression. And the expression 'I' does not change its denotation either, because occasional expressions which differ from each other by their denotations have, by this very fact, different component elements, and therefore, they are not particular cases of the same expression 'I'.

The fact that this theory preserves the constancy of meanings of occasional expressions is often considered as its advantage because of the conformity with the traditional ways of understanding the term 'meaning'. But even if this opinion is right, the conformity in question is here achieved at the cost of a far-reaching inconformity with the well established use of the term 'expression' and many other related terms, so that, all in all, the price to be paid seems exorbitant. Take a few examples.

The expressions conceived as ordered pairs are, of course, abstract entities. This concerns not only expressional types, but also the particular individual expressions. Hence, no expression can be spoken, read or heard. And, on the other hand, the occasional tokens used for the purpose of communication do not denote and do not mean anything at all. Sentence-shaped inscriptions in which such tokens occur are, accordingly, neither true nor false, and so on. These and other consequences certainly do not encourage us to adopt a view which leads to them.

One more point should be stressed. The definition of an occasional expression conceived as an ordered pair composed of a definite token and the accompanying circumstances of use gives no indications at all as to what kind of circumstances is the second element of such a pair (when its first element is of such and such form). Neither does it say anything about the semiotic properties of the expressions so conceived and about the ways of assigning to them these properties. Hence, the need arises for some complementary assumptions to be made. These, it seems, should be approximately of the following form (when applied, for example, to the expression 'I'):

(1) x, y, Z (if x is a token of the form 'I', and y is the person who pronounces x, then Z is an expression of the type 'I' if and only if $Z = x, y$)

(2) x, y, Z (if x is a token of the form 'I' and y is the person who pronounces x, and $Z = x, y$, then Z denotes y).

It can easily be seen that such definitions are constructed according to the same principle as the definitional sentences to which I referred before. Then, the whole issue becomes much more complicated than it actually is when approached in the way advocated in my own analysis. It does not seem, moreover, that any remarkable theoretical advantage can be obtained at the cost of complicating the problem and leading to a greater inconformity with the established terminological habits. For these reasons I prefer to choose another way of solving the problems under discussion.

We are coming towards the end of our considerations. To sum up: this study was intended chiefly to find out those properties of occasional expressions which account for their specific particularities. The main stress has been laid, on the one hand, on the relative character of the semiotic functions of these expressions in regard to the circumstances of use of a certain distinguished kind, as well as on the univocal dependence of the denotations of these expressions on such circumstances; and, on the other hand, on the kind of sentences which can well serve as definitions of occasional expressions. As we have seen, such definitions are conditional sentences, formulated in metalanguage and, moreover, singled out by having the form of semantic rules and by containing, each of them, in the antecedent and in the consequent, the same – let us say – 'circumstantial' variable.

It seems that this analysis brings out just these properties of occasional expressions which are their most significant features and result in their being distinguished as a separate group.

Occasionality is one of the essential properties of natural language. If for no other reasons, the problem of occasionality should be included in the theory of the natural language as one of the main issues.

The University of Warsaw

NOTE

* Printed here by kind permission of Nauwelaerts, Belgium, publishers of *Logique et Analyse*.

H. A. LEWIS

COMBINATORS AND DEEP STRUCTURE
Syntactic and Semantic Functions

1. INTRODUCTION

Combinatory logic is of interest for linguistics in several ways. Its usefulness in syntactic studies was urged before the development of transformational grammar, and some recent authors have renewed its claims [1]. Work in the spirit of Ajdukiewicz has points of contact with combinatory logic [2].

Let me recall some features of combinatory logic that suggest applications to natural languages. 'Uninterpreted' combinatory logic expresses the structure of a domain (unspecified) closed under the 'application' of one object (or, in the noncommittal language of Curry and Feys, 'ob') to another. It is this pure theory of combinators that analyses the role of the variable. Since most logical and mathematical notations make use of variables, an understanding of the role of the variable is useful to anyone who would find in these symbolic notations a representation of 'deep structures' of sentences of natural languages. In the usual interpretation of combinatory logic, 'application' corresponds to the composition of functions [3]. This is why the combinatory theory of functionality is interpretable by an Ajdukiewicz-type categorial grammar: categories other than the basic categories are syntactic functions, and all but those represented by fractional indices whose denominators are basic category symbols are (syntactic) functions of functions – all syntactic categories with fractional indices being defined by means of an associated syntactic function. In pure combinatory logic, where all expressions of the notation denote 'obs', the semantics of the language echoes the syntax in a particularly simple fashion, as Quine has demonstrated [4]. This has at least an indirect interest for linguists, because of the close analogy between the combinatory theory of functionality and categorial grammars.

In several papers, Quine has studied a combinatory version of the first order predicate calculus that he has christened 'predicate-functor logic' [5]. It provides an alternative notation for predicate logic in which variables are 'explained away' and should be of interest to anyone who is attracted by the idea that predicate logic exhibits deep structures of natural languages.

In this paper I shall discuss chiefly the *semantics* of combinatory and predicate-functor logic: but a brief defence of the claim that deep structures could be represented by any logical notation at all may be required, or excused.

2

Why should deep structures be representable by a logical notation? *If* we aim at a grammar in which a semantic account of whole sentences depends only on their deep structures, the answer is immediately forthcoming: a semantic 'component' worthy of the name must reveal entailment-relations among sentences. Any symbolism, and so any representation of deep structures, that reveals entailment-relations among such structures is *ipso facto* a logical notation. But need it be one of the familiar logical notations? My answer is that it need not, but it must 'contain' familiar notations in the sense that it expresses at least as much as they do of logical form. And of course the sequents (logical consequence-relations) of natural language that are most certain are just those that our familiar logical notations are designed to express.

These remarks are framed for a grammar that permits a contrast (defined in the grammar) between deep and surface structures. Chomsky has argued that (some) semantic information is represented only in surface structure[6]. His arguments presuppose a particular grammar, however, and it is open to others to devise a generative grammar on other lines. Our intuitions about what follows from what, especially when the consequence depends on other aspects of a sentence than lexical choice, seem to me a more secure criterion of deep structure than syntactical intuitions. Henceforward I shall be concerned with the nature of the semantic account rather than with the source (in the syntax) of the semantically important information about sentence-structure.

3

Even if a grammar purports to use a logical notation for its 'deep structures', its semantics may be problematic. Linguists hope to capture in their semantic accounts more than predicate logic captures, at least given our usual rough and ready techniques of 'translation'. If we restrict our attention to the aspects of natural languages that appear to correspond to our logical notions, it is not the choice alone of a logical notation to represent deep structures that determines the semantic account, but the choice of notation *together with its interpretation*. Formal presentations of logical systems include a 'semantic component' in the form of a recursive definition of truth, classically via a definition of satisfaction (of a formula by a sequence of objects). By choosing a logical notation for his deep structure, a grammarian puts himself under the obligation to provide such a semantic account. He thus enters the same business as philosophically motivated students of logic who have discussed the correct form of semantics for our basic predicate logic, and for its extensions to capture further concepts – necessity, tenses and the like.

Combinatory and predicate-functor logic provide alternatives to predicate logic in its familiar form. Are they, from the point of view of grammar, genuine alternatives? This question, obscure as it stands, invites the following subordinate questions: Are the syntactic differences between combinatory and standard notations important? Is the generation of natural language sentences from their intended combinatory surrogates achievable in a smoothly-running grammar? In the light of the last paragraph we must also ask: does combinatory or predicate-functor logic invite a semantics different from the usual semantics of predicate logic?

In pure combinatory logic the only syntactic operation used in forming names of 'obs' are prefixing and bracketing. The closure properties of the domain are mirrored in closure properties for these names, so that the syntax is extremely simple. In the study of combinatory completeness, the predicate of identify forms sentences. The syntax of predicate-functor logic is not so simple, because of restrictions on contexts in which symbols may occur. If its syntax is compared with standard (infixing) notations, two differences stand out: connectives occur as prefixes in Polish style, and the role of variables in keeping track of argument-places in predicates is taken over by functors, also prefixed. A recursive definition of 'sentence' for predicate-functor logic can easily be given via a definition of 'predicate of degree n' (with sentences as predicates of degree 0). Of these notations, pure combinatory logic alone is describable by a context-free phrase-structure grammar. Predicate-functor logic and predicate logic in standard form (but without vacuous quantifiers) require a context-dependent grammar [7].

I had hoped to be able to show, by the time of this conference, evidence that the production of sentences of at least one language from predicate-functor logic was likely to be possible and even convenient. Such evidence would be provided if we could detect further combinatory devices in (say) English along the lines of Geach's example [8]. (Ordering transformations are combinators, to be sure, but such devices can be expected in any grammar.) All I can tell you now is that some colleagues of mine in Leeds are working along these lines [9]. I am confident that predicate-functor logic will yield further syntactic insights.

The usual semantic account of combinatory logic has already been mentioned: the objects include functions, and all strings of combinators name objects. This account contrasts with the standard interpretation of predicate logic, according to which not all symbols are names [10]. Combinatory logic provides our first example of an interpretation in which functions are assigned to symbols in such a way that semantic functions mirror the syntactic functions governing the construction of sentences.

The interpretation of predicate-functor logic proves a less straightforward matter than might be expected, given that it is but a notational variant of standard predicate logic. Since predicate-functor logic is indeed intertranslatable with standard predicate logic, there is an uninteresting way of interpreting a sentence of predicate-functor logic – transform it (syntactically) into a sentence of standard predicate logic and then read off a given interpretation of the transform. A more interesting way of defining truth for predicate-functor logic is given by Wallace [11]. He indicates how the satisfaction of a predicate-functor sentence by a sequence may be defined directly by allowing the combinatory adjustments wrought by Quine's functors to be mirrored by appropriate adjustments in the ordering in the sequence. In this construction, variables are needed in the metalanguage (the ordering of the sequence expresses ordering features of occurrences of variables in the standard form of predicate logic). In his paper 'Truth and Disquotation' Quine has indicated a third path to the definition of truth for predicate-functor logic [12]. He provides a recursive definition of a one-place predicate 'sat' of expressions that fulfils the requirements of Tarski's Convention T in the strong sense that in

(1)   sat '...' ↔ ...

the blanks are always filled by the *same* expressions. The role of variables in the metalanguage is implicit in the '↔', which is formal equivalence of degree fixed by the flanking predicates. Given this use of '↔', it might seem possible to render *any* predicate-forming operator on predicates thus:

(2)   sat ('0' x) ↔ 0 sat x

A strictly disquotational definition of satisfaction could thus be given for modal operators... but this would be at the cost of false clauses in the recursion.

The functors of predicate-functor logic are revealed by this definition in their role of predicate-forming operators on predicates. But '↔' is not the '↔' of the otherwise similar construction on combinatory logic: there is no role yet for reference to functions of functions in the definition of satisfaction. Moreover intensional operators of all kinds resist this disquotational definition of satisfaction for predicate-functor logic just just as they resist the more usual construction for standard predicate logic.

Let us now turn to higher things. Modal operators, and other intensional operators, are normally written as prefixes and so predicate-functor logic may be extended to admit such operators – perhaps with more syntactic elegance than such extensions of standard notations possess. The convenience of the category 'operator' or 'functor', understood as including connectives, quantifiers, combinators and modal operators is merely superficial, however, at the level of syntax, since the contextual restrictions on occurrences of the various members of the category hold as usual. The proposal to regard some devices of natural language as (formal) 'operators' that some recent writers have put forward derives its chief interest from the accompanying semantics[13]. Here we may seem to move away from the discussion of combinatory logic and towards the discussion of the semantics of modal logic, but there is an interesting overlap of these topics. One writer, indeed, has used predicate-functor logic with modal operators to express deep structures for English[14]. It is not the choice of the predicate-functor notation that links the topics most closely, however, but the offering of a semantic account according to which semantic functions mirror syntactic.

From the family of such semantic accounts I pick for discussion one that is clearly expounded in print and also has the merit of being built on an Ajdukiewicz-type categorial grammar, and so of displaying clearly the functional character of the deep-structure representations: the account of David Lewis, in 'General Semantics'[15]. Lewis uses a categorial grammar with three primitive categories, S (sentence), N (name) and C (common noun). His interpretation is given in general terms thus: the syntactic category of an item determines its 'appropriate intension'; the appropriate intension for an item of category S is any function from indices (series of coordinates) to truth-values; for an item of category N, any function from indices to things; and for an item of category C, any function from indices to sets. He mentions as coordinates in the indices a possible-world coordinate, some contextual coordinates, and an assignment coordinate. Appropriate intensions for derived categories are of this form: where c, $c_1, ...,$ and c_n are any categories, basic or derived, an appropriate intension for an item of category $c/c_1 ... c_n$ is any n-place function from c_1-intensions, ..., and c_n-intensions to c-intensions. Thus, for example, items of the derived category C/C are interpreted by functions from common-noun intensions to common-noun intensions.

For any language (or set of deep structures) for which a categorial grammar can be given, this method gives a way of specifying the type of the intension of lexical items. If, on the other hand, Ajdukiewicz categories are assignable to all the lexical items of a language, does it follow that such a semantic account is the right one? The account has an appearance of weakness and generality that could lead one to think that it does follow. I think, however, that there are reasons for doubt.

Two reasons I mention to put to one side. One might take exception to the suggested coordinates in the intensions for the basic categories, for example to the possible-world coordinate. To do so would not be to undermine the whole conception, however, since there might be other more acceptable coordinates: moreover, the relation between basic and derived categories might hold if there were no coordinates at all, or quite different ones. It could also be objected that an account of the meaning of expressions that assigns functions of the same type to lexical items of the same syntactic category is mistaken because of well-known semantic differences between items of the same category. Such difficulties as there are here are in the *application* (via transformations) of the categorial grammar to natural language, not in the semantic theory itself.

The interpretation is specified by means of reference to functions (in the metalanguage). If the interpretation of a sentence is spelled out, its translation (according to the interpretation) involves reference to objects (functions) about which the original did not appear to be. (Remember the difference between the interpretations of combinatory logic and of predicate-functor logic.) A simplified illustration must suffice:

(3) the sequence $\langle a...\rangle$ satisfies 'x_1 is red' if and only if a is red

contrasts with

(4) the sequence $\langle a...\rangle$ satisfies 'x_1 is red' if and only if a is a member of the set of red things

The latter clause, but not the former, represents 'a is red' as talking about the set of red things. Lewis follows a style of setting out his interpretations that involves free reference to functions as (4) refers to a set. It is odd to claim that our ordinary talk is about such functions[16].

What I have said so far does not touch the central idea of the proposed interpretation – the idea that the pattern of interpretation is in terms of functions mirroring syntactic functions. This idea has formal elegance but seems to me to be defective in articulating the way in which we come to understand compound sentences. The defect that I see is in the manner of interpreting the derived categories, and I shall try to exhibit it for adverbs of category $(S/N)/(S/N)$[17].

The syntactic category S/N requires as intensions functions that map name-intensions (functions from indices to individuals) onto sentence-intensions (functions from indices to truth-values). Let us call such S/N intensions *VP-intensions*. Thus the derived category of adverbs that form verb phrases from verb phrases requires intensions that map *VP*-intensions onto *VP*-intensions. For example, the interpretation of 'in London' maps the intension of 'sleeps' onto the intension of 'sleeps in London' and also the intension of 'lives' onto the intension of 'lives in London'. (The intension of 'sleeps in London' is a function that maps name-intensions onto the intensions of sentences of the form 'x sleeps in London'.) Thus, to be sure, 'in London' is given a single interpretation, and it always plays the same (semantic) role in a sentence – it introduces the single function mentioned above. In an adequate semantic account of English we indeed want to reveal the single role of 'in London' in both the verb-phrases – surely it makes the same difference to the meaning of both? But it is an illusion that the assignment of a single function to 'in London' does this, on its own. The *way* the value-range of the function is determined is the important thing, but specification of the values for all arguments leaves this part out. Provided that Lewis has an extensional notion of function, nothing in his framework allows us to assign a single interpretation to 'in London' that fixes in advance the interpretation of 'new' verb-phrases containing it, in such a way that its single meaning is revealed. The account allows that 'in London' means what we mean by 'in Paris' in some of its contexts but not others. My objection is that this account is too weak to bring us near to articulating our way of understanding sentences including words of derived categories. I do not wish to deny that such functions could be associated with items of derived categories, at least subject to haggling over indices.

A different example illustrates the objection. Conjunction can take two verb-phrases and produce one verb-phrase, thus: 'is tall and bald'. Such a compound predicate has a logical form in virtue of which anyone who satisfies it is both tall and bald. But there is no room in Lewis's account for a general statement of this role of 'and'[18]. To say only that 'is tall and bald' is a verb-phrase, and so takes functions from name-intensions to sentence intensions as its own intensions, is not yet to give a way of showing how the complex verb-phrase means what it means.

The mode of interpreting adverbs gives no account of the ability of a language-user to understand new verb-phrases compounded of known adverbs and known intransitive verbs. My understanding of 'in London' and 'plays hockey' would enable me to understand 'plays hockey in London' if I had never put the two together before. This understanding is not made comprehensible by pointing out that we can uncover here two functions[19].

I conclude that there are reasons for saying that a semantic account of natural languages along the lines of the standard semantics for combinatory logic (according to which the syntactic functions are interpreted by corresponding semantic functions) would be, not formally defective, but unrevealing.

On the wider issue, I do not claim to have shown that a base component of a grammar using predicate-functor logic would be superior to all competitors, but I hope that the view of Quine and Geach that combinatory logic is a source of syntactic insights into natural languages will soon receive further justification.

University of Leeds

NOTES

[1] See H. B. Curry and R. Feys, *Combinatory Logic*, North-Holland Publishing Company, Amsterdam 1958, pp. 274–275; H. B. Curry, 'Some Logical Aspects of Grammatical Structure' in *Proceedings of Symposia in Applied Mathematics*, Vol. 12: *Structure of Language and its Mathematical Aspects* (ed. by R. Jakobson), Providence, R. I. 1961, pp. 56–68; W. V. O. Quine, 'Logic as a Source of Syntactical Insights' in *The Ways of Paradox*, Random House, New York 1966, p. 47; P. T. Geach, 'A Program for Syntax', *Synthese* 22 (1970) 3–17; T. Parsons, *A Semantics for English* (duplicated), 1968.

[2] See in particular Curry and Feys, and Geach (Note 1).

[3] For example (taken from H. B. Curry, 'Logic, Combinatory' in *Encyclopedia of Philosophy* Vol. 4, (ed. by Paul Edwards) pp. 504–509): 'If X means the addition function of natural numbers and Y means the number 1, then XY will mean a form of the successor function, and if Z means the number 2, then XYZ will mean the number 3.'

[4] W. V. O. Quine, 'Truth and Disquotation' (known to me at present only in a typescript kindly sent to me by Professor Quine), §§ 2, 3.

[5] W. V. O. Quine, 'Variables Explained Away' in *Selected Logic Papers*, Random House, New York, 1966, pp. 227–235; 'Truth and Disquotation' (see previous note); 'Algebraic Logic and Predicate Functors' in *Logic and Art: Essays in Honor of Nelson Goodman* (ed. by Richard S. Rudner and Israel Scheffler), Bobbs-Merrill, (in press), §§ 2–4; 'Predicate-Functor Logic' in the *Proceedings of the Second Scandinavian Logic Symposium* (ed. by J. E. Fenstad), North-Holland, Amsterdam 1971.

[6] N. Chomsky, 'Deep Structure, Surface Structure and Semantic Interpretation', reproduced by the Indiana University Linguistics Club, January 1969.

[7] The *exclusion* of quantifiers that fail to bind a free variable in the open sentence that follows is clearly a matter of context. It is interesting to note that a categorial grammar strengthened by Geach's recursive rule (see 'A Program for Syntax', Note 1 above) also appears to go beyond the resources of a context-free phrase-structure grammar.

[8] Geach shows how certain operators of Quine's type will render

          Anybody who hurts anybody who hurts him hurts himself
as   Univ Imp Univ Imp H Cnv H Ref H

[9] In the Department of Computational Science at Leeds, work is in progress on a NLQA system that involves the generation of English sentences from predicate-functor logic. My colleague Dr T. C. Potts is working on a grammar (announced

in his paper to this Section of the Congress) that incorporates combinatory features and a categorial grammar (this volume, pp. 245–285).

[10] The interpretation of all symbols as names is not forced on us by the notation, of course. Conversely, logic in standard notations can be understood this way, as does F. B. Fitch (see his *Symbolic Logic*, Ronald Press, New York, 1952, on for example the universal quantifier, pp. 132–133).

[11] John Wallace, 'On the Frame of Reference', *Synthese* **22** (1970), 117–150, Section V.

[12] Quine, 'Truth and Disquotation' (cf. Note 4).

[13] Cf. Parsons *op. cit.* (Note 1), and Terence Parsons, 'The Logic of Grammatical Modifiers', *Synthese* **21** (1970) 320–334, especially Note 12, p. 333.

[14] Parsons, *A Semantics for English* (cf. Note 1)

[15] David Lewis, 'General Semantics', *Synthese* **22** (1970) 18–67.

[16] Cf. Wallace's remarks at the end of his Section III, and his Section V (Modal Logic).

[17] Curry, Geach and D. Lewis have all assigned some adverbs to this category.

[18] The weakness is in Lewis's categorial grammar, lacking, as it does, the recursive rule.

[19] I am indebted here to Donald Davidson, 'Theories of Meaning and Learnable Languages', *Proceedings of the 1964 International Congress for Logic, Methodology and Philosophy of Science* (ed. by Yehoshua Bar-Hillel) Amsterdam, North-Holland, 1965, pp. 383–394.

I. A. MEL'ČUK

LINGUISTIC THEORY AND 'MEANING ⇔ TEXT' TYPE MODELS

1. Our basic assumption is as follows: natural language should be treated by theoretical linguistics mainly as a device ensuring the transmission of information, i.e. from the viewpoint of its *communicative*, or *referential*, function. Then language could be described as a transformer 'meanings ⇔ ⇔ texts'; and, if so, one of the most important tools of linguistic analysis and description should be – for a pure linguist, in contradistinction to a psycholinguist, brain psychologist or learning theorist – models of the 'Meaning ⇔ Text' type.

2. A 'Meaning ⇔ Text' model (**MTM**) is not a generator of texts (in the sense of Chomsky), but rather a translator. Ideally, it can be thought of as a logical device capable of converting a given meaning into all texts (of the language under consideration) that convey this meaning and a given text – into its meaning (or meanings, if the text is ambiguous). Clearly, meaning should be recorded in a special formalized language which we could call 'semantic language'.

3. Meaning is for us a construct – a conventional representation of an invariant of synonymic transformations, which hold, by definition, between equisignificant texts; equisignificance is taken as a primitive (i.e., an intuitive, or empirical) concept. If our invariant representation of any equisignificant texts is called their meaning, then a given meaning may contain a contradiction, an (analytically) false assertion or even be 'meaningless' (in the current sense of the word). The analysis of meaning seems to go beyond the frames of **MTM** whose task is to transform any given meaning ('meaningless' meaning included) into optimal text(s), or to extract meaning – whatever it is – from a given text.

4. Formally, meaning is a constructive object: a graph, where apices (nodes) are *semes* – elementary meanings from a finite pre-established set, and arcs (branches) are relations of a predicate to its arguments. (Semes can be predicates, like 'begin', 'cause', 'good', etc.)

5. Between meanings and texts there exists a very complicated many-to-many correspondence (due to synonymy and homonymy so widespread in natural languages). Therefore, it seems advisable to break up the transition from meanings to texts and vice versa into a number of stages, or levels. In the **MTM** under consideration (this author participates in its development jointly with A. K. Žolkovsky), the following levels are distinguished:

(1) *Semantic level* – meaning representation, or semantic graph (see Section 4 above).

(2) *Syntactic level*:

(2a) Deep syntactic representation (**DSR**). Its main component is deep syntactic structure: dependency tree whose nodes are labelled with so-called generalized lexemes (including symbols of full words, idioms, and lexical functions), and branches – with symbols of 6–8 universal deep-syntax relations[1]. Besides this tree, **DSR** contains also information about topic – comment division, anaphoric relations, and the like.

(2b) Surface syntactic representation (**SSR**). Like **DSR**, the main component here is surface syntactic structure, or a dependency tree too, but having in its nodes symbols of all lexemes only, and on its branches – symbols of syntagmas, or constructions, specific for the language in question. Similarly, the **SSR** includes data about topicalization, etc. (cf. above).

(3) *Morphological level*:

(3a) Deep morphological representation, or ordered string of completely characterized lexemes (lexeme symbols provided with all grammatical features). The linear ordering of lexemes takes place just during the transition from (2b) to (3a).

(3b) Surface morphological representation, or string of morphemes.

(4) *The text itself*: either in phonemic or phonetic transcription, or in conventional spelling[2].

6. An essential part of the **MTM** is a special kind of dictionary which describes in detail the meaning of every item as well as its compatibility with other items[3].

7. This approach is clearly based on the principles put forth and developed by Chomsky (1965); the nearest approaches this author knows of

seem to be those of Sgall (1966) and Lamb (1966), while similar ideas have been also suggested by M. Kay (1970) and W. Hutchins (1971).

Institute of Linguistics,
Academy of Sciences of U.S.S.R., Moscow

NOTES

[1] Neither deep nor surface syntactic structures have linear order imposed on their nodes.
[2] For more details on **MTM**, see *Linguistics* **57** (1970), 10–47.
[3] See *Studies in Syntax and Semantics* (ed. by F. Kiefer), D. Reidel, Dordrecht, 1969, pp. 1–33.

BIBLIOGRAPHY

Chomsky, N., 1965, *Aspects of Theory of Syntax*, The MIT Press, Cambridge, 251 pp.
Hutchins, W., 1971, *The Generation of Syntactic Structures from a Semantic Base*, North-Holland, Amsterdam-London, 197 pp.
Sgall, P., 1966, 'Ein mehrstufiges generatives System', *Kybernika* **2**, 181–190.
Lamb, S., 1966, *Outline of Stratificational Grammar*, Washington, 109 pp.
Kay, M., 1970, 'From Semantics to Syntax', in *Progress in Linguistics* (ed. by M. Bierwisch and K. E. Heidolph), Mouton, The Hague pp. 114–126.

GABRIEL ORMAN

PROPERTIES OF THE DERIVATIONS ACCORDING TO A CONTEXT-FREE GRAMMAR

1. In this paper some problems concerning the derivations according to a context-free grammar will be investigated. First, we shall present the basic definitions and notations regarding the context-free grammars and languages as well as those relating to graphs, as they are given in Berge (1967), Ginsburg (1966), and Marcus (1964).

Let Σ be an *alphabet* and Σ^* the *free semigroup* (with identity) generated by Σ.

DEFINITION 1. A context-free grammar is a quadruple $G = (V, \Sigma, P, \sigma)$ where

(a) V is a finite alphabet;
(b) $\Sigma \subseteq V$ is a finite alphabet (the terminal alphabet);
(c) P is a finite set of ordered pairs (μ, v) called *productions*, with μ in $V - \Sigma$ and v in V^* (it follows that P is a subset of $(V - \Sigma) \times V^*$);
(d) σ is in $V - \Sigma$ (the initial symbol).

The elements of the set $V - \Sigma$ are called *variables* (or nonterminals). The elements of Σ are often called *terminals*.

Let $G = (V, \Sigma, P, \sigma)$ denote a context-free grammar. For the words u and w in V^* write $u \Rightarrow w$ if there exist z_1, z_2, μ and v such that $u = z_1 \mu z_2$, $w = z_1 v z_2$ and $\mu \rightarrow v$ is in P. Write, also, $u \stackrel{*}{\Rightarrow} w$ if either $u = w$ or there exist words $u_0, u_1, \ldots, u_k$ such that $u_0 = u$, $u_k = w$ and $u_i \Rightarrow u_{i+1}$ for each i.

DEFINITION 2. The sequence $u_0, u_1, \ldots, u_k$ above is called a *derivation* of length k and it is denoted by $u_0 \Rightarrow u_1 \Rightarrow \ldots \Rightarrow u_k$, $(u_0, u_1, \ldots, u_k$ are called the *steps* of the derivation).

Given a context-free grammar $G = (V, \Sigma, P, \sigma)$, the set

$$L(G) = \{u \text{ in } \Sigma^* \mid \sigma \stackrel{*}{\Rightarrow} u\}$$

is called a *context-free language* (it is the language generated by G). Let us define now the notions of multigraph and semi-degree of a vertex of it.

DEFINITION 3. Let Ω be a set of points and Γ a mapping of Ω into 2^Ω. The pair (Ω, Γ) is said to be a *graph*, and will be denoted $\mathfrak{G} = (\Omega, \Gamma)$.

The elements of the set Ω are called *vertices* of the graph. A pair of vertices (x, y), such that $y \in \Gamma_x$, (Γ_x being the set of vertices corresponding to the vertex x by the mapping Γ), is said to be an *arc* of the graph. For the arc (x, y), x is said to be *initial extremity* and y *terminal extremity*.

Therefore, if we shall denote by U the set of arcs of the graph $\mathfrak{G} = (\Omega, \Gamma)$, we may write this graph so: $\mathfrak{G} = (\Omega, U)$. From now on, we shall use the last form for denoting a graph.

A *path* in a graph $\mathfrak{G} = (\Omega, U)$ is a non-empty finite sequence of arcs $(x_1, y_1), (x_2, y_2), \ldots, (x_k, y_k)$ such that the terminal extremity of each arc coincides with the initial extremity of the next arc.

We can associate to the graph $\mathfrak{G} = (\Omega, U)$ a relation of order in the following way: given two vertices x_i, x_j of the graph $\mathfrak{G}$, we write $x_i \leqslant x_j$ if either $x_i = x_j$ or there is a path from x_i to x_j.

DEFINITION 4. If more distinct arcs connecting the same pairs of vertices exist one gets a *multigraph*.

DEFINITION 5. Let $\mathfrak{G} = (\Omega, U)$ be a multigraph with the vertices $x_1, x_2, \ldots, x_n$. One calls *external semi-degree* of a vertex x_i, $(1 \leqslant i \leqslant n)$, the number $d^+(x_i)$ of arcs which have the initial extremity in x_i. The number $d^-(x_i)$, $(1 \leqslant i \leqslant n)$ of arcs which have the terminal extremity in x_i is called *internal semi-degree* of the vertex x_i.

If $d^+(x_j) \neq 0$ and $d^-(x_j) = 0$, $(1 \leqslant j \leqslant n)$ we say that the vertex x_j is an *input*, while if $d^+(x_j) = 0$ and $d^-(x_j) \neq 0$, $(1 \leqslant j \leqslant n)$ we say that x_j is an *output*.

2. Now, the notion of partial derivation of a given derivation will be introduced. To this end, let $G = (V, \Sigma, P, \sigma)$ be a context-free grammar generating the language $L = L(G)$, and

$$\sigma = \omega_0 \Rightarrow \omega_1 \Rightarrow \omega_2 \Rightarrow \ldots \Rightarrow \omega_{k-2} \Rightarrow \omega_{k-1} \Rightarrow \omega_k = \omega$$

a derivation D, of length k, according to G.

DEFINITION 6. Any derivation $\omega_i \stackrel{*}{\Rightarrow} \omega_j$, ($\omega_i$ and ω_j in V^*) such that $\sigma \stackrel{*}{\Rightarrow} \omega_i$ and $\omega_j \stackrel{*}{\Rightarrow} \omega$ is said to be a *partial derivation* of D.

Evidently, one can obtain several partial derivations (having or not the same length) of the derivation D. These lead to some functions which are able to characterize D well enough. They will be examined in another paper.

We agree to represent each step of the derivation D by a point, two points ω_i and ω_{i+1} being connected by an arc if and only if $\omega_i \Rightarrow \omega_{i+1}$. In this way, we define a graph $\mathfrak{G} = (\Omega, M)$ where Ω is the set of vertices (respectively the steps of D), and M is the set of its arcs.

Let us associate to the graph $\mathfrak{G}$ a relation of order as it follows: for two vertices ω_i and ω_j of it we write $\omega_i \leqslant \omega_j$, $(i, j = 0, 1, 2, \ldots, k)$ if either $\omega_i = \omega_j$ or there exists a path from ω_i to ω_j (that is to say: if $\omega_i \overset{*}{\Rightarrow} \omega_j$).

We now consider all the partial derivations of D and identify each partial derivation with a path into the graph $\mathfrak{G}$. Evidently, two vertices will be connected by more distinct arcs. Thus, we obtain a multigraph which will be denoted $\mathfrak{G}^{(D)} = (\Omega, U)$, where Ω is the set of vertices of $\mathfrak{G}^{(D)}$, (the number of elements of Ω is equal to the number of the steps of D) and U is the set of all its arcs.

We shall call $\mathfrak{G}^{(D)} = (\Omega, U)$ the *associated multigraph to the derivation D*. The number of vertices of the multigraph will be denoted by $|\Omega|$ and the number of its arcs by $|U|$.

Example 1. Consider the derivation D

$$\sigma_0 = \omega_0 \Rightarrow \omega_1 \Rightarrow \omega_2 \Rightarrow \omega_3 \Rightarrow \omega_4 \Rightarrow \omega_5 = \omega$$

of length $k = 5$ according to a context-free grammar $G = (V, \Sigma, P, \sigma_0)$. Then $\mathfrak{G}^{(D)} = (\Omega, U)$ is the following:

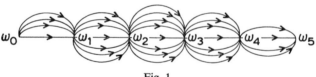

Fig. 1.

As can be easily seen, the associated multigraph of a derivation according to a context-free grammar characterizes this derivation. Hence, we are going to give some results obtained from their study. Conventionally, we shall consider that a derivation of length 1 has not partial derivations (in the opposite case, we must consider that the partial derivations have the length zero, that is just the steps of the given derivation). Therefore in the following only derivations having the length $k \geqslant 2$ will be considered.

THEOREM 1. *Let $G = (V, \Sigma, P, \sigma)$ be a context-free grammar and $\sigma \overset{*}{\Rightarrow} \omega$ a derivation D, of length k, according to G. Denote by $\mathfrak{G}^{(D)} = (\Omega, U)$ the associated multigraph to the derivation D. (1) If k is odd there is exactly a* vertex of the multigraph having the external semi-degree higher than the external semi-degrees of the other vertices. (2) If k is even there are two vertices of the multigraph having the same external semi-degree and satisfying this condition.

Proof. Compute effectively the semi-degree of each vertex, denoting the elements of the set Ω by $\omega_0, \omega_1, \ldots, \omega_{k-1}, \omega_k$.

(1) Suppose that k is odd. Then, the derivation $\sigma = \omega_0 \overset{*}{\Rightarrow} \omega_k = \omega$ has $k+1$ steps, namely an even number. Hence it follows that the associated multigraph of this derivation has an even number of vertices (i.e. $|\Omega| = k+1$ is even).

– The external semi-degree of the vertex ω_0 is equal to $k-1$. These arcs are initial arcs into the paths going from ω_0;

– in ω_1 have the initial extremity the following arcs: $k-2$ arcs which belong to the paths going from ω_0 and $k-1$ arcs which belong to the paths going from ω_1. Hence, in all $2k-3$ arcs. In other words, $d^+(\omega_1) = 2k-3$;

– in ω_2 have the initial extremity the following arcs: $k-3$ of those which belong to the paths going from ω_0, $k-2$ of those which belong to the paths going from ω_1 and $k-2$ of the arcs which belong to the paths going from ω_2. Totalizing, we get $d^+(\omega_2) = 3k-7$; and so on.

– in $\omega_{(k-1)/2}$ have the initial extremity $\frac{1}{4}(k^2 + 2k - 3)$ arcs. These are: $(k-1)/2$ arcs which belong to the paths going from ω_0, $(k+1)/2$ arcs which belong to the paths going from ω_1, $(k+1)/2$ arcs which belong to the paths going from ω_2. Hence, we have $d^+(\omega_{(k-1)/2}) = (k-1)/2 + (k-1)/2 \cdot (k+1)/2 = \frac{1}{4}(k^2 + 2k - 3)$.

Analogously, we obtain

$$d^+(\omega_{(k+1)/2}) = \tfrac{1}{4}(k^2 + 2k - 7);$$
$$d^+(\omega_{(k+3)/2}) = \tfrac{1}{4}(k^2 + 2k - 19);$$
$$d^+(\omega_{k-3}) = 3k - 7;$$
$$d^+(\omega_{k-2}) = 2k - 3;$$
$$d^+(\omega_{k-1}) = k - 1.$$

It is easy to see that the vertex $\omega_{(k-1)/2}$ has the highest external semi-degree. This proves the first part of the theorem.

(2) Suppose now that k is even. In this case the derivation $\sigma = \omega_0 \overset{*}{\Rightarrow} \overset{*}{\Rightarrow} \omega_k = \omega$ has an odd number of steps. Hence, it follows that the associated multigraph to this derivation has an odd number of vertices, that is

$|\Omega|=k+1$ is odd. By a similar argument, of that for k odd, we now get:

$d^+(\omega_0) = k-1$;
$d^+(\omega_1) = k-3$;
$d^+(\omega_2) = 3k-7$;

$d^+(\omega_{(k-2)/2}) = \frac{1}{4}(k^2 + 2k - 4)$;
$d^+(\omega_{k/2}) = \frac{1}{4}(k^2 + 2k - 4)$.
$d^+(\omega_{(k+2)/2}) = \frac{1}{4}(k^2 + 2k - 12)$;

$d^+(\omega_{k-3}) = 3k-7$;
$d^+(\omega_{k-2}) = 2k-3$;
$d^+(\omega_{k-1}) = k-1$.

(Observe that the vertex $\omega_{k/2}$ is in the middle of the ordered string of vertices of $\mathfrak{G}^{(D)}$).

Therefore, there exist two vertices $\omega_{(k-2)/2}$ and $\omega_{k/2}$ of which external semi-degree has the highest value. That is to say $d^+(\omega_{(k-2)/2}) = d^+(\omega_{k/2}) = \frac{1}{4}(k^2 + 2k - 4)$. Hence, the second case is proved. In addition, we know the maximal number of arcs having the initial extremity in any vertex of $\mathfrak{G}^{(D)}$, both for k odd and even.

From Theorem 1 two corollaries follow:

COROLLARY 1. Let D be a derivation of length k. For k even, there is a vertex of the associated multigraph to D having the external semi-degree equal to the internal semi-degree. For k odd such a vertex does not exist.

Indeed, if k is even, from Theorem 1 follows that $d^-(\omega_{k/2}) = d^+(\omega_{k/2}) = \frac{1}{4}(k^2 + 2k - 4)$. Hence, $\omega_{k/2}$ is the vertex having this property. Evidently for k odd no vertex of the associated multigraph to D, having this property, exists.

COROLLARY 2. The external semi-degree of the input ω_0 of the associated multigraph to D is equal to the internal semi-degree of its output.

This result follows immediately from Theorem 1 by observing that $d^+(\omega_0) = d^-(\omega_k) = k-1$, both for k odd and even.

Since in the multigraph $\mathfrak{G}^{(D)}$ the set of vertices is ordered, it follows that for two consecutive vertices ω_i and ω_{i+1} we have $d^+(\omega_i) = d^-(\omega_{i+1})$. But this fact proves the following theorem:

PROPERTIES OF THE DERIVATIONS

THEOREM 2. *Let $\mathfrak{G}^{(D)}$ be the associated multigraph to a derivation D, of length k, according to a context-free grammar. If k is odd, there is just a vertex of $\mathfrak{G}^{(D)}$ having the internal semi-degree higher than the internal semi-degree of the other vertices. If k is even there are two vertices of $\mathfrak{G}^{(D)}$ having the same internal semi-degree and satisfying this condition.*

From the proof of Theorem 1 it results that, for k odd, the vertex satisfying the condition of Theorem 2 is $\omega_{(k+1)/2}$, while for k even the vertices satisfying the same condition are $\omega_{k/2}$ and $\omega_{(k+2)/2}$. This means that for k odd there is just a pair of consecutive vertices connected by a maximal number of arcs, while for k even there are two such pairs of consecutive vertices.

In Example 1 an illustration of this property for k odd is given. For k even, see Example 2.

Example 2. Let D be a derivation of length $k = 6$, according to a context-free grammar $G = (V, \Sigma, P, \sigma_0)$

$$\sigma_0 = \omega_0 \Rightarrow \omega_1 \Rightarrow \omega_2 \Rightarrow \omega_3 \Rightarrow \omega_4 \Rightarrow \omega_5 \Rightarrow \omega_6 = \omega.$$

Then, the associated multigraph to this derivation is the following

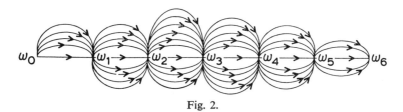

Fig. 2.

Now, a new notion which will permit us to characterize a given derivation according to a contex-free grammar will be introduced.

DEFINITION 7. Denote by $\omega_0, \omega_1, \ldots, \omega_k$, ($\omega_i \leqslant \omega_{i+1}$ for any $0 \leqslant i \leqslant k - 1$) the vertices of the associated multigraph $\mathfrak{G}^{(D)}$ of the derivation D, of length k, according to a context-free grammar. We call *the depth* of a consecutive pair of vertices (ω_i, ω_{i+1}), of $\mathfrak{G}^{(D)}$, the number of arcs connecting the vertices ω_i and ω_{i+1}.

The depth of the pair of vertices (ω_i, ω_{i+1}) will be denoted by Dep (ω_i, ω_{i+1}).

The following lemma is obvious.

LEMMA 1. Let D be a derivation of length k according to a context-free grammar and $\mathfrak{G}^{(D)}$ its associated multigraph. If $\omega_0, \omega_1, \ldots, \omega_k$ are the vertices of $\mathfrak{G}^{(D)}$, then the difference between the external and internal semi-degree of a vertex ω_i, $(1 \leqslant i \leqslant k-1)$ is equal to the difference between the depth of the pair of vertices containing ω_i as an initial element and the depth of the pair of vertices containing it as a second element.

DEFINITION 8. The maximal depth of a pair of consecutive vertices of the multigraph $\mathfrak{G}^{(D)}$ is said to be *the depth of the derivation D*.

If D is a derivation of length k, the depth of D will be denoted by $Dep(D)$. Then from Definition 8 it follows that

$$Dep(D) = \max\{Dep(\omega_i, \omega_{i+1}) \mid \omega_i \in \Omega, \omega_{i+1} \in \Omega \text{ and}$$
$$\omega_i \leqslant \omega_{i+1}, \text{ for any } 0 \leqslant i \leqslant k-1\},$$

where Ω is the set of vertices of $\mathfrak{G}^{(D)}$.

We shall denote, also, the minimal depth of a pair of vertices of $\mathfrak{G}^{(D)}$ by $Dep_m(D)$. Therefore

$$Dep_m(D) = \min\{Dep(\omega_j, \omega_{j+1}) \mid \omega_j \in \Omega, \omega_{j+1} \in \Omega$$
$$\text{and } \omega_j \leqslant \omega_{j+1}, \text{ for any } 0 \leqslant j \leqslant k-1\}.$$

By Theorem 1 the following lemma is obvious.

LEMMA 2. Let $\mathfrak{G}^{(D)}$ be the associated multigraph to the derivation D, of length k, according to a context-free grammar $G = (V, \Sigma, P, \sigma)$,

(a) If k is odd then

(1) $\quad Dep(D) = Dep(\omega_{(k-1)/2}, \omega_{(k+1)/2}) = \frac{1}{4}(k^2 + 2k - 3)$

and

$$Dep_m(D) = Dep(\omega_0, \omega_1) = Dep(\omega_{k-1}, \omega_k) = k - 1;$$

(b) If k is even then

(2) $\quad Dep(D) = Dep(\omega_{(k-2)/2}, \omega_{k/2}) = Dep(\omega_{k/2}, \omega_{(k+2)/2}) =$
$\qquad = \frac{1}{4}(k^2 + 2k - 4)$

and

$$Dep_m(D) = Dep(\omega_0, \omega_1) = Dep(\omega_{k-1}, \omega_k) = k - 1.$$

From here we come to the conclusion that the depth of a derivation D of odd length k is $Dep(D) = d^+(\omega_{(k-1)/2}) = d^-(\omega_{(k+1)/2})$, while the depth of a derivation D' of even length k' is $Dep(D') = d^+(\omega_{(k'-2)/2}) = d^-(\omega_{k'/2}) = d^+(\omega_{k'/2}) = d^-(\omega_{(k'+2)/2})$.

COROLLARY 3. The depth of a derivation D is always a multiple of 4 divided by 4, that is

$$Dep(D) = M_4/4.$$

Indeed, if k is odd putting $2n-1$ instead of k we obtain from (1) $Dep(D) = \frac{1}{4}(4n^2 - 4) = M_4/4$. Similarly, if k is even putting $2n$ instead of k we obtain from (2), $Dep(D) = \frac{1}{4}(4n^2 + 4n - 4) = M_4/4$.

LEMMA 3. Let D be a derivation of length k according to a context-free grammar and $\mathfrak{G}^{(D)}$ its associated multigraph,

(a) If k is odd then

(3) $\quad Dep(D) = \frac{1}{4}(k-1)(k+3)$, for any $k \geq 3$;

(b) If k is even then

(4) $\quad Dep(D) = \frac{1}{4}[(k+1)^2 - 5]$, for any $k \geq 2$.

Proof (a) Suppose k is odd. Then $Dep(D) = \frac{1}{4}(k^2 + 2k - 3)$ (cf. Lemma 1). But the polynomial $k^2 + 2k - 3$ has the integral solutions 1 and -3. Hence, $k^2 + 2k - 3 = (k-1)(k+3)$ and the first case is proved. (b). Suppose now, k is even. By Lemma 2 it follows that $Dep(D) = \frac{1}{4}(k^2 + 2k - 4)$. Since the polynomial $k^2 + 2k - 4$ has the irrational solutions $k + 1 - \sqrt{5}$ and $k + 1 + \sqrt{5}$ we get $k^2 + 2k - 4 = (k+1)^2 - 5$, completing the proof.

Observation. If D is a derivation of length 1 (i.e. $k = 1$), we obtain from (3) that $Dep(D) = 0$. This result corresponds to the fact that a derivation of length 1 has no partial derivations (as we had specified in the beginning of this paragraph).

Conventionally, we shall denote a derivation D of length k by D_k. Thus, D_{2n-1} will signify a derivation of odd length $k = 2n - 1$, while D_{2n} will signify a derivation of even length $k = 2$. In this way, the relations (3) and (4) may be written as:

(5) $\quad Dep(D_{2n+1}) = n(n+2)$, for $n = 1, 2, 3, \ldots$

and

(6) $\quad Dep(D_{2n}) = n^2 + n - 1$, for $n = 1, 2, 3, \ldots$

It is easy to prove the following lemma:

LEMMA 4. Let $\omega_0, \omega_1, \ldots, \omega_k$ be the vertices of the associated multigraph $\mathfrak{G}^{(D)}$ to a derivation D of length k. (a) If k is odd the values of the depths of the pairs of vertices into the string $(\omega_0, \omega_1), (\omega_1, \omega_2), ..., (\omega_{k-1}, \omega_k)$ are symmetrically disposed to the depth of the pair of vertices $(\omega_{(k-1)/2}, \omega_{(k+1)/2})$ which represents the depth of D; (b) If k is even the values of the depths of the pairs of vertices, into the string above, are symmetrically disposed to the depths of the pairs of vertices $(\omega_{(k-2)/2}, \omega_{k/2})$ and $(\omega_{k/2}, \omega_{(k+2)/2})$ which are equal and represent the depth of D.

Finally we shall attempt to express the depth of a pair of consecutive vertices of the associated multigraph to a given derivation by means of the depth of this derivation.

THEOREM 3. Let $\mathfrak{G}^{(D)}$ be the associated multigraph to a derivation D of odd length k and $\omega_0, \omega_1, ..., \omega_k$ its vertices. The depth of the first pair of vertices (ω_0, ω_1) can be computed with the formula

(7) $$A_1 = Dep(\omega_0, \omega_1) = Dep(D) - \left(\sum_{j=1}^{n} a_j\right)_{a_j \in I'}$$

where $I' = \{1, 3, 5, ..., (2n-1)\}$.

Proof. In order to prove the theorem let us consider the first $k + 1/2$ values in the sequence of values obtained by Theorem 1 for k odd. We denote them so:

$A_1 = Dep(\omega_0, \omega_1) = k - 1$
$A_2 = Dep(\omega_1, \omega_2) = 2k - 1$

$A_{(k-1)/2} = Dep(\omega_{(k-3)/2}, \omega_{(k-1)/2}) = \frac{1}{4}(k^2 + 2k - 7)$
$A_{(k+1)/2} = Dep(D) = Dep(\omega_{(k-1)/2}, \omega_{(k+1)/2}) = \frac{1}{4}(k^2 + 2k - 3)$

Going from $A_{(k+1)/2} = Dep(D)$ let us compute the difference between it and the depth of the previous pair of vertices; then, we shall compute the difference between the depth of the last but one pair of vertices and that of the pair preceding it, and so on. Finally, we get:

$A_{(k+1)/2} - A_{(k-1)/2} = 1$
$A_{(k-1)/2} - A_{(k-3)/2} = 3$
$A_{(k-3)/2} - A_{(k-5)/2} = 5$

$A_4 - A_3 = k - 6$
$A_3 - A_2 = k - 4$
$A_2 - A_1 = k - 2$

Therefore
$$A_{(k-1)/2} = Dep(D) - 1$$
$$A_{(k-3)/2} = (Dep(D) - 1) - 3 = Dep(D) - (1 + 3)$$
$$A_{(k-5)/2} = Dep(D) - (1 + 3) - 5 = Dep(D) - (1 + 3 + 5)$$

and so on. Finally we shall obtain

$$A_1 = Dep(D) - (1 + 3 + 5 + \ldots)$$

where the sum of parentheses has $(k-1)/2$ terms. Since k is odd, by hypothesis ($k \geqslant 3$) let us put $k = 2n + 1$. Then $(k-1)/2 = n$ and denoting by I' the set $\{1, 3, 5, \ldots, (2n-1)\}$, it follows that

$$A_1 = Dep(D) - \underbrace{(1 + 3 + 5 + \cdots)}_{n \text{ terms}} = Dep(D) - \left(\sum_{j=1}^{n} a_j\right)_{a_j \in I'}$$

Thus the theorem is proved.

From Theorem 3 results the following rule for computing the depths of any pairs of consecutive vertices of the associated multigraph to a derivation D of odd length k: the depth of the first pair of vertices (ω_0, ω_1) is given by (7); the depth of the next pair of vertices (ω_1, ω_2) is obtained from (7) by taking into parenthesis $n-1$ terms only; the depth of the pair of vertices (ω_2, ω_3) is obtained from (7) by taking into parenthesis $n-2$ terms, and so on. The depths of the pairs of vertices following to the pair of vertices having the maximal depth, can be written by symmetry (cf. Lemma 4).

By a similar argument the following theorem will be proved:

THEOREM 4. Let $\omega_0, \omega_1, \ldots, \omega_k$ be the vertices of the associated multigraph to a derivation D of even length k. The depth of the first pair of vertices (ω_0, ω_1) is given by the formula

(8) $$A_1 = Dep(\omega_0, \omega_1) = Dep(D) - \left(\sum_{j=1}^{n} b_j\right)_{b_j \in I''}$$

where $I'' = \{2, 4, 6, 8, \ldots, 2n\}$.

The rule for computing the depths of the several pairs of consecutive vertices of the associated multigraph to a derivation, resulted from Theorem 3, remains available in the case of Theorem 4 too.

University of Braşov,
Braşov, Romania

BIBLIOGRAPHY

[1] Berge, C., *Théorie des Graphes et ses Applications*, Dunod, Paris, 1967.
[2] Chomsky, N., 'On Certain Formal Properties of Grammars', *Inform. and Control* **2** (1959) 137–167.
[3] Cudia, D. F., 'General Problems of Formal Grammars', *Journal of the ACM* **17** (1970) 31–43.
[4] Ginsburg, S., *The Mathematical Theory of Context-Free Languages*, McGraw-Hill, New York, 1966.
[5] Marcus, S., 'Teorija grafov, lingvističeskie oppozicii i invariantnaja struktura', in *Problemy strukturnoi lingvistiki* **1** (1962) 22–30.
[6] Marcus, S., *Gramatici si automate finite* (Ed. Academiei), București 1964.
[7] Marcus, S., *Introduction Mathématique à la Linguistique structurale*, Dunod, Paris, 1967.
[8] Ore, O., 'Theory of Graphs', *American Math. Soc.* (Colloquim publications) **38** (1962).
[9] Orman, G., 'A Characterization of Left-Linear Context-Free Languages', *The VIth Yugoslav International Symposium on Information Processing*, B-4, Bled, September 23–26, 1970.

BOHUMIL PALEK

THE TREATMENT OF REFERENCE IN LINGUISTIC DESCRIPTION*

In the description of natural languages no great use has hitherto been made of reference. There are even cases where any such possibility is denied, one example of which is Katz' approach to semantics (Katz, 1966). I have earlier attempted to show (Palek, 1968) that reference does have a place in language description, especially in the description of those features which are conditioned by context. A quite different matter, however, is to find a fitting conception of reference in language description, and another question is whether the accounts of reference, or the very theory of reference, as presented in works based on logical foundations, or formalised languages as the case may be, are adequate enough to be applied in the description of natural language.

This paper presents an attempt at a certain modification of the theory of reference in the light of what has just been said.

Roughly speaking, reference is a term which we use for the expression of the relation between a name and the object designated by that name. In works which give a more specific treatment to these problems, e.g. Quine (1951, 1960), the distinction is drawn between the characters of individual names (singular and general terms). These characters determine the respective content of the relation, namely as the extension of the term in the case of general terms, and the actual assignment of name to object, i.e. reference in a narrower sense of the word, e.g. Föllesdal (1966), in the case of singular terms. At no cost to precision we might say that the term 'reference' primarily means that by the use of certain expressions of language (names) the idea of certain extralinguistic qualities or properties of a given object arises in the language-user's mind. In what follows we shall designate this role of names as *naming*.

Strictly speaking, we cannot here talk of the relation between name and object but only of that between a name and a certain selection of the properties of the object explicitly denoted by that name. The selection may be motivated by various circumstances, such as the speaker's intention, the situation, contextual restrictions, the speaker's impression of his auditor's attitude to himself, the character of the respective object, and so on. Therefore, it is not possible to express *all* the properties of the relevant object by the use of an expression of natural language which is a token in the given text (leaving aside here expressions which are proper names or indexical signs). Some of the properties are omitted, whether we will or no, which leads to a certain distortion of the relation between name and object. For this idea of 'object' arising out of the selective and distortional character of names we use the term *denotate*. The distinction between object and denotate, which we consider to be essential to the description of certain phenomena of language, is amplified in Palek (1968) and we shall continue to use the term denotate in the present paper.

The item expressing a certain denotate may of course be more complex than the kind of expression usually to be found in classical examples used to demonstrate denotation in logical works. The relation between such an expression and its denotate is also often more complex, i.e. the expression tells the language-user more than simply the quality determined by what we have called naming. If the theory of reference confined itself to dealing with naming, it would probably not exhaust all the relations between name and denotate which an analysis of material would bring to light.

As soon as we begin to work with referential relations in the description of natural language, the question of what is over and above naming becomes quite an essential one. This evidently stems from differences in the character of the terms used for individual denotates and from the requirement of analysing continuous texts for the description of natural language.

We shall begin the demonstration of what we have in mind in a primitive way, using isolated sentences like those employed in works of logic.

Let us take three sentence contexts:

(a) --- is a feline beast of prey
(b) --- is rare
(c) --- is loved by children.

From the point of view of meaning, these contexts may be characterized as 'placing in a higher category', 'determination of a quality', and 'description of an attitude'. All three contexts are so formed as to give general sentences when the blanks are filled. The expressions filling the blanks are in an extensional position which is even an extensional pure position (Föllesdal, 1966).

Let us now take the following names for insertion into the contexts:

(a) lion
(b) bachelor
(c) kitten

In all three examples there is a choice of ways of forming a general sentence from the given context by the insertion of the respective expression. (Only in (b7) was it necessary to adapt --- *is rare* to --- *is a rare thing* for purely grammatical reasons; likewise we observe congruence of number between subject and verb.) It is essential to note that the extension of the inserted expression and the truth function of the sentences created remain without change. The differences between the variants are not differences in naming. (Unacceptable eventualities are asterisked.)

Examples:

(a) (1) a lion is a feline beast of prey
 (2) the lion is a feline beast of prey
 (3) every lion is a feline beast of prey
 (4) any lion is a feline beast of prey
 (5) lions are feline beasts of prey
 (6) all lions are feline beasts of prey
(b) (7) a bachelor is a rare thing
 (8)*the bachelor is (a) rare (thing)
 (9)*every bachelor is (a) rare (thing)
 (10)*any bachelor is (a) rare (thing)
 (11) bachelors are rare
 (12) all bachelors are rare
(c) (13) a kitten is loved by children
 (14)*the kitten is loved by children
 (15) every kitten is loved by children
 (16) any kitten is loved by children
 (17) kittens are loved by children
 (18) all kittens are loved by children

In every case we have given all the theoretical means of forming general sentences. However, some of the sentences created are not acceptable to the language of the examples. A glance is sufficient to see that the range of possibilities for different names and different contexts varies considerably.

Taking examples (a), let us now substitute for *lion* the expression *king of the animals*:

(19)   *a king of the animals is a feline beast of prey
(20)   the king of the animals is a feline beast of prey
(21)   *every king of the animals is a feline beast of prey
(22)   *any king of the animals is a feline beast of prey
(23)   *kings of the animals are feline beasts of prey
(24)   *all kings of the animals are feline beasts of prey.

The new sentences, except for (20), are unacceptable, which in itself is an indication that, with respect to the given context, there exists no relation of synonymy between the expressions *lion* and *king of the animals*. What is important from our point of view, however, is that the acceptability of the two sets of sentences varies. Since the naming in all the sentences of either set remains unchanged, the difference in acceptability must once more be caused by the presence of some other function which also affects the naming expression.

From the point of view of the language-user's intuition, the variants within examples (a), (b), and (c) are not identical as to meaning, and they are far from being as freely interchangeable as might seem, in spite of the fact that it is difficult to give explicit expression to the differences between them. Let us take example (c). Except for (14), all the sentences are acceptable as general sentences. That the communication --- *is loved by children* applies to all kittens is common to all the sentences. They differ however in the way in which this allness is presented.

(13) and (17) differ from the remaining three in that while implying allness, this is in no way emphasized. The difference between (13) and (17) seems to be a matter of frequency and style.

(15), (16) and (18) all expressly guarantee allness, (15) suggesting allness irrespective of time and place, (18) being the emphasized version of (17), and (16) allowing for the most various physical qualitative differences among kittens.

(I do not wish to claim that this is the precise state of affairs in view of the acknowledged difficulties of explicitly formulating intuitively felt differences.)

If a variant *some kittens* were used, the naming would indeed remain unchanged but there would be an important change in the scope of the denotate so designated and thus a change in extension within the limits of the sentence.

If we take an example of a singular sentence (and this is the commonest type of sentence in natural language), we encounter even more complex possibilities and even greater differences between the individual variants. For example, in the sentence:

(25)   Respecting language, I willingly hold communion in that spoken by my grandmother...

it is theoretically possible to replace *that* by any of the following expressions: *one, the one, the ones, some, all (those), every one, each one, any, any one*, etc., some of which are themselves open to more than one interpretation.

From the point of view of the intuitive feeling of the language-user, it would be difficult to describe all the possible versions of sentence (25) as mere variants (just as in the case of *some kitten* as opposed to *all kittens*) of a single underlying sentence. The resulting senses are too far apart. However, confining reference to what we have called naming, we can treat them as variants.

Even from these few examples it is also clear that the differences between these variants cannot be recorded by the use of logical quantifiers. The stock of these is not large enough to capture the full wealth of variety in natural language, not even in the isolated general sentences adduced above. Moreover the quantifiers do not have any function at all in contextual cohesion as becomes apparent when we come to interpret relations (e.g. those of cross-reference) which go beyond the framework of a single sentence.

Having tried to show that in addition to naming there is something else carried by nominal expressions in a referential position, I wish here to introduce the term *instauration* to cover this function, and the term *instaurators* for the language means used to express it.

The raison d'être of this approach lies in my conviction that a minimal expression of the type under consideration transmits not only information as to the quality and properties of each denotate but also, and separately (by specific means), as to the grouping (external and internal) of the denotates the communication is concerned with, i.e. information as to their number, scope, identity, previous occurrence, interrelations, etc. Thus an instaurator may be described as a grammatical means which serves to create at a given place in the text which is occupied either by a significant expression (i.e. expression which can denote) together with some such grammatical means, or by such grammatical means alone, the idea of a denotate or group of denotates, such a denotate having then the disposition of being an actant in the flow of the text.

Instauration as such manifests itself in a variety of ways. To give a more concrete idea of them we might mention some of the types of instaurators which it seems reasonable to distinguish in the description of natural language: Pronouns, numerals and certain adjectives (e.g. *different, individual,* etc.) and adverbs are the most frequent instaurators.[1] We also consider the category of number in names in referential position as an instaurator. Thus from the structural point of view, in languages like English or Czech, there is a broad scale of complexity among the cases where instauration is expressed by the category of number alone, or by an instaurator alone, or by instaurator (as a separate item) in combination with a noun.

From the point of view of their role we might then distinguish the following types of instaurators:

(1) The addition to a name affords information as to the number of units in the framework of the denotate, or as to its wholeness (e.g. *each, few, some, whole, total, overall*).

(2) The use of an instaurator indicates that the relevant denotate is in a certain particular relation to another foregoing denotate (e.g. *another, different, further, last, next, twenty-seventh*); see 'alterators' in (Palek, 1968).

(3) The use of an instaurator expresses identity between a given denotate and a foregoing denotate (e.g. *the same, that, this, he, it, the former*); see 'indicators' (*ibid.*).

(4) The role of the instaurators is to point directly to an object, i.e. deixis and ostension (e.g. *this, that, yonder*).

(5) An instaurator is used to indicate inherent appurtenance of one denotate to another (e.g. *his, their, someone's*).

(6) An instaurator is used to indicate our want of knowledge of a denotate (e.g. *who?, which?*).

(7) An instaurator furnishes information as to the non-existence of a denotate (e.g. *nothing, nobody, no*).

It is quite evident that the problem of instauration is a complex one, and from the few types given it is clear that in some cases instaurators have a definitely contextual function (i.e. 2, 3 and 5), while in others they do not, remaining a textual matter (1), and in yet other cases their description must take the extralinguistic situation into account (4), (6), (7).

It goes without saying that there is more than one way of investigating instaurators. One way is to proceed like, for example, Döhman (1962) using logical means (i.e. operators) for their interpretation. This however may lead to distortion of the final conception of the instaurators, as we have already indicated: different instaurators in natural language may often be synonymous from the point of view of a logical operator, and some instaurators cannot be recorded at all.

For this reason I think it is expedient in the interpretation of instaurators to proceed from the language-user's evaluation and therefore to treat the question at least in part as an empirical matter.

From the methodological point of view I adopt the standpoint that it is more expedient in such an approach to formulate a certain intermediate range of problems independent of any overall conception of language description[2] (which we do here and have done in (Palek, 1968) and with as much precision as possible. Results arrived at in this way are thus not influenced by any methodological principles of the given conception. It is to be hoped that an overall description of natural language which would have at its disposal such independent intermediate descriptions would present a more adequate reflection of natural language than when the reverse procedure is applied.

Charles University, Prague

NOTES

* This paper is a slightly modified version of a paper published in *Teorie a metoda* 3 (1971) 53–60. It represents a small part of a work being prepared in cooperation with G. Fischer and D. Short on instauration in German, English and Czech. The aim is to find the instaurators with which these languages work and to formulate the theoretical consequences of such an investigation.

[1] Textual contrasts of the type *large table: small table* do not involve instauration. Here we have a distinction in naming expressed by the adjectives *large* and *small*. A similar but purely instaurational distinction would be, for example, *this table: another table*.

[2] A very definite set of problems would emerge if we were to undertake the task of strictly defining the entire set of instaurators in any given natural language from the point of view of the requirements of the language of science.

BIBLIOGRAPHY

Döhman, K., 'Die sprachliche Darstellung der Quantifikatoren', *Logique et Analyse* 5 (1962) 33.

Föllesdal, D., 'Referential Opacity and Modal Logic', *Filosofiske Problemer* 32 (1966), Oslo.

Katz, J. J., 'M. Pfeifer on Questions of Reference', *Foundations of Language* 2 (1966) 241.

Palek, B., 'Cross-Reference. A Study from Hyper-Syntax', *AUC Monographia* XXI (1968), Prague.

Quine, W. O., *Mathematical Logic*, Cambridge, Mass., 1951.

Quine, W. O., *Word and Object*, Cambridge, Mass., MIT Press., 1960.

TIMOTHY C. POTTS

FREGEAN CATEGORIAL GRAMMAR

1. Categories

In *Begriffsschrift* (1879) and *Grundgesetze der Arithmethik* (1893), Frege expounds a finite type theory which can be presented as a categorial grammar. The theory rests upon two distinctions: between complete and incomplete expressions and between different levels of incomplete expression. I call incomplete expressions *functors*, belonging to functorial categories; complete expressions belong to basic categories.

Frege recognized only one basic category, proper name (*Eigenname*). It included most of the proper nouns of grammarians, but also many sentences, together with expressions beginning with the definite article. He also held that (declarative) sentences are proper names of truth-values and that functors designate functions. None of these assumptions is essential to a categorial grammar based upon his analytical principles and they will not be made here. The etymological associations of 'functor' should therefore be disregarded, nor need it be supposed that a functor designates anything at all. For purposes of exposition, however, I follow Frege for the most part in using only one basic category, in order to bring out as clearly as possible, without any extraneous distractions, the characteristics of Fregean grammar. The letter 'N' will be used for this category; both proper nouns and sentences are to be expressions of category N.

Functors are obtained by removing one or more occurrences of an expression of known category from another expression of known category; the category of the resulting functor is computed from the two previously known categories. The procedure begins by removing one or more occurrences of the same expression of category N from another expression of category N, e.g. removing a proper noun from a sentence. The result is a *one-place* functor, since it contains one empty place which can be filled by an expression of category N. Should this functor still contain any further expressions of that category, the procedure may be applied recursively until none remains. At any stage we shall have an i-place functor for i applications of the procedure. Members of the series of functors thus obtained may therefore be distinguished by the number of empty places which each contains.

This distinction may be exhibited by *category-labels*, adapted from Ajdukiewicz (1935, 1967). We begin with 'N' as a basic category- label. Then, given one application of the procedure just described, the resulting functor is assigned to category :NN, given two applications, to category ::NNN, and so on. Functors of category :NN thus turn one expression of category N into another, functors of category ::NNN turn two expressions of category N into an expression of category N, etc. The number of initial colons in the category-label shows the number of empty places in the functor, the letter which immediately follows them shows the category of the expression obtained when all of the empty places in the functor are correctly filled (I call this expression the *valuor* of the functor) and the subsequent letters show, in order, the categories of the expressions which are to fill the empty places in the functor (I call these expressions the *argumentors* of the functor and the places, accordingly, its argumentor-places).

A recursive *definition of 'category'* may be built upon the foregoing notation for category-labels. Every functorial category-label will consist of three groups of signs: first, the initial colons; second, a basic category-label giving the valuor of the functor (its *valuor-label*); and, third, a series of category-labels giving the categories of its argumentors (its *argumentor labels*). The general form of a category-label will therefore be $:_m \alpha \beta_1 \ldots \beta_m$, where '$:_m$' abbreviates m colons and the Greek minuscules are schematic letters for which category-labels may be substituted (a different label may be substituted for each β_i); α will be the valuor-label and $\beta_1, \ldots, \beta_m$ the argumentor-labels. Basic category-labels may be regarded as special cases in which $m = 0$. Then, leaving it open how many basic categories there are to be:

(1) every basic category is a category;
(2) if α is a basic category and $\beta_1, \ldots, \beta_m$ are categories, then $:_m \alpha \beta_1 \ldots \beta_m$ is a category;
(3) (closure condition).

The information contained in category-labels is used to construct corresponding *category-symbols*, which are schematic signs for which any expression of that category may be substituted and which are themselves made up of category-labels. The necessity for having category-symbols in addition to category-labels will be explained in Section II. For basic categories, label and symbol are the same; for functorial categories, the category-symbol consists of the original label, which I call the *main* label of the symbol, followed by category-labels marking the argumentor-places of the functor. Thus we can show not only how many argumentor-places each category of functor has, but also what category of expression may fill each of them. In order to make it clear that these labels mark empty places, they will be written in italicized minuscules. I also adopt the convention that the argumentor-places are to be shown in reverse order to that of the argumentor-labels. The rule according to which category-symbols are constructed from their main labels is therefore:

$$:_m \alpha \beta_1 \ldots \beta_m \rightarrow :_m \alpha \beta_1 \ldots \beta_m \, \beta_m \ldots \beta_1.$$

So if we had a category-label ': : ABC', the corresponding symbol would be ': : ABC $c\ b$'.

Functors which take expressions of *basic* categories as their argumentors are functors of *first*-level; but given an expression which has been assigned to a first-level functorial category, one or more occurrences of *this* may also be removed from an expression of category N. The result will be a functor which takes a first-level *functor* as its argumentor: this is a second-level functor. Second-level functors, like first-level ones, can have any number of argumentor-places, for the procedure just described can again be applied recursively. In addition, the first-level functors which they take as argumentors may themselves have any number of argumentor-places. These distinctions will be reflected in their category-labels. Thus the label for a functor which turns a one-place first-level functor into an expression of category N is : N : NN; for a functor which turns two one-place first-level functors into an expression of category N, : : N : NN : NN; but for a functor which turns one two-place first-level functor into an expression of category N, : N : : NNN.

We can also have functors of *mixed* levels. For example, if we are able to remove *both* an expression of category N *and*, subsequently, a one-place first-level functor from an expression of category N, the resulting functor will be of category : : NN : NN. It is often simpler, however, to reckon the level of a functor as one level higher than that of its highest-level argumentor, thus counting ::NN:NN as a second-level functor. In any case, we cannot always describe a functor exactly just by giving its level and the number of its empty places, for such a description would not distinguish : N : NN from :N: :NNN; we also need to know how many empty places each of its argumentors has. Hence the need for category-labels, which encode all of this information unambiguously.

In constructing category-symbols for higher-level functors (those of second-level and above), the rule stated above is to be applied recursively, starting with β_m and working through to β_1. That is to say, if any of the βs is a functorial category-label, it must be taken as $:_m\alpha\beta_1 \ldots \beta_m$ in a further application of the rule. The reason for this will be explained in Section II. Where the rule has only been applied once to the main label, I call the result a *partial* category-symbol; for basic and first-level functorial categories, there will be no difference between category-symbols and partial category-symbols.

Where the rule has to be applied a second time to one of the β's of the first application, there will certainly be some labels to the left of the β concerned and there may also be some to its right. To allow for this, using Γ and Δ for any sets of labels, the rule must be amended to:

(S1) $\Gamma :_m\alpha\beta_1 \ldots \beta_m \Delta \rightarrow \Gamma :_m\alpha\beta_1 \ldots \beta_m \beta_m \ldots \beta_1 \Delta$

with the following typographical stipulations:

> **if** Γ is empty, **then** the βs are to be written in italicized minuscules; **else if** $:_m\alpha\beta_1 \ldots \beta_m$ is written in unitalicized minuscules **then** the βs are to be written in italicized minuscules;
> **else** the βs are to be written in unitalicized minuscules.

Applying (S1) to the second-level category-labels ':N:NN', ':N::NNN' and '::N:NN:NN', we then obtain:

:N:NN :*nn*   : N: :NNN : :*nnn*   : :N:NN:NN :*nn* :*nn*
:N:NN :*nn* n   :N: :NNN ::*nnn* n n   ::N:NN:NN :*nn* n: *nn* n

For every application of (S1) in which Γ is not empty, a second rule is now to be applied immediately after each application of (S1). In order to state this rule, I use

$:_m\alpha\beta_1 \ldots (:_n\gamma\delta_1 \ldots \delta_n)_i \ldots \beta_m$

to indicate that $:_n\gamma\delta_1\ldots\delta_n$ is the ith argumentor-label of $:_m\alpha\beta_1\ldots\beta_m$, i.e. it is β_i. The aim of the rule is to introduce $\delta_1\ldots\delta_n$ as sub-scripts of $:_m\alpha\beta_1\ldots\beta_m$, but the latter may already have sub-scripts as a result of previous applications of the rule, so let these be Σ. The new sub-scripts are to be added to the *left* of Σ, so we have;

(S2) $\quad \Gamma :_m\alpha\beta_1\ldots(:_n\gamma\delta_1\ldots\delta_n)_i\ldots\beta_{m\,\Sigma}$
$\qquad\qquad\qquad \beta_m\ldots(:_n\gamma\delta_1\ldots\delta_n\;\delta_n\ldots\delta_1)_i\ldots\beta_1\,\Delta\;\rightarrow$
$\quad \Gamma :_m\alpha\beta_1\ldots(:_n\gamma\delta_1\ldots\delta_n)_i\ldots\beta_{m\;\delta_1\ldots\delta_n\,\Sigma}$
$\qquad\qquad\qquad \beta_m\ldots(:_n\gamma\delta_1\ldots\delta_n\;\delta_n\ldots\delta_1)_i\ldots\beta_1\,\Delta$

We shall now have *two* occurrences of $\delta_1\ldots\delta_n$ in the category-symbol; I call the δs *index-labels*, the first occurrence of each index-label being the sub-scripted occurrence introduced by (S2), the *second* occurrence that introduced by (S1). The δs introduced by (S2) are to be written in the same typography in the consequent of the rule as in the antecedent.

Applying (S2) to the examples above, we shall get:

$:N:NN_n\;:nn$ n $\quad :N::NNN_{n\;n}::nnn$ n n $\quad :N:NN:NN_{n\;n}\;:nn$ n $:nn$ n

In the last two examples, we have two index-labels of the same category, with the result that it is impossible to tell which of the second occurrences corresponds to which of the first occurrences. To eliminate the ambiguity, numerical sub-scripts must be added to the index-labels. I shall follow the convention of adding numerical sub-scripts to the *first* occurrences of index labels working from left to right; but so far as the *second* occurrences are concerned, we may follow whatever order we wish. Thus for each of these will be two possibilities:

$\quad:N::NNN_{n_1 n_2}::nnn$ n_2 $n_1 \qquad ::N:NN:NN_{n_1 n_2}:nn$ n_2 $:nn$ n_1
$\quad:N::NNN_{n_1 n_2}::nnn$ n_1 $n_2 \qquad ::N:NN:NN_{n_1 n_2}:nn$ n_1 $:nn$ n_2

In adding numerical sub-scripts to index-labels, account must be taken of any previously introduced *unitalicized* index-labels whose first occurrences are in Σ or in Γ, another number being taken for new ones; since δ's are introduced to the left of Σ, our convention may involve some re-numbering of previous sub-scripts as we go along.

From second-level functors we can ascend to third-level functors, from third to fourth, and so on indefinitely. Thus, by removing a second-level functor from an expression of category N, we shall obtain a third-level functor e.g. one whose category is :N:N:NN. The rules which have just been given will produce the correct category-symbol for functors of any level from the main label, provided that we continue to apply them to any new functorial category-labels which they introduce, the first occurrences of index-labels excepted. Thus the argumentor-places of third-level functors will be given index-labels and with fourth-level functors, the second occurrences of index-labels of the main label will, in addition, receive index-labels of their own.

Examples of a third-level functor and of a mixed second-third-level functor are:

$$:N:N:NN \quad :n:nn \quad (S1)$$
$$:N:N:NN \quad :n:nn \ :nn \quad (S1)$$
$$:N:N:NN_{:nn} \quad :n:nn \ :nn \quad (S2)$$
$$:N:N:NN_{:nn} \quad :n:nn \ :nn \ n \quad (S1)$$
$$:N:N:NN_{:nn} \quad :n:nn_n \ :nn \ _n \quad (S2)$$

$$:N::NN:NN \quad ::nn:nn \quad (S1)$$
$$:N::NN:NN \quad ::nn:nn \ :nn \ n \quad (S1)$$
$$:N::NN:NN_{n:nn} \quad ::nn:nn \ :nn \ n \quad (S2)$$
$$:N::NN:NN_{n:nn} \quad ::nn:nn \ :nn \ n \ n \quad (S1)$$
$$:N::NN:NN_{n:nn} \quad ::nn:nn_n \ :nn \ _n \ n \quad (S2)$$

Numerical sub-scripts are not required in the second example, for although we have two index-labels of category N, one is italicized.

One example of a fourth-level functor will suffice:

$$:N:N:N:NN \quad :n:n:nn \quad (S1)$$
$$:N:N:N:NN \quad :n:n:nn \ :n:nn \quad (S1)$$
$$:N:N:N:NN_{:n:nn} \quad :n:n:nn \ :n:nn \quad (S2)$$
$$:N:N:N:NN_{:n:nn} \quad :n:n:nn \ :n : nn : nn \quad (S1)$$
$$:N:N:N:NN_{:n:nn} \quad :n:n:nn_{:nn} \ :n:nn \ :nn \quad (S2)$$
$$:N:N:N:NN_{:n:nn} \quad :n:n:nn_{:nn} \ :n:nn \ :nn \ n \quad (S1)$$
$$:N:N:N:NN_{:n:nn} \quad :n:n:nn_{:nn} \ :n:nn_n \ :nn \ n \quad (S2)$$

It will be seen from these examples of third and fourth-level functors that while the last application of (S1) and (S2) always produces a label of a basic category, it is unitalicized for even-levelled functors but italicized for odd-levelled functors.

Alternative category-labels are available for higher-level functors. These could be constructed directly from main category-labels, but for a reason which will appear in Section II, I show instead how to obtain them from the category-symbols already specified, which I now call the *standard* symbols. The antecedent of the rule can be the same as the consequent of (S2), but as we operate only upon the main label, Γ and Δ will always be empty and so can be omitted. The basic idea is that we delete the main label and its sub-scripted index-labels, replacing the second occurrences of the latter by majuscules of the same category. For example,

instead of :N:NN$_n$:nn n we may have :nn N
and instead of :N:N:NN$_{:nn}$:n:nn$_n$:nn n we may have :n:nn$_n$:NN n

These alternative symbols exhibit the kinship between even-levelled category-symbols, on the one hand, and odd-levelled ones, on the other.

The situation is complicated, however, by many-place functors: we may then delete one argumentor-label of the main label at a time, so that there will be several alternative symbols. In this case, the main label will not be deleted, but one of its argumentor-labels will be. The rule must therefore cover two cases:

(S3) if $m > 1$
then $:_m\alpha\beta_1\ldots(:_n\gamma\delta_1\ldots\delta_n)_i\ldots\beta_{m\,\Pi\,\delta_1\ldots\delta_n}\,\Sigma$
$\beta_m\ldots(:_n\gamma\delta_1\ldots\delta_n\;\delta_n\ldots\delta_1)_i\ldots\beta_1 \to\; :_{m-1}\alpha\beta_1\ldots\beta_{i-1}\beta_{i+1}\ldots\beta_{m\,\Pi\Sigma}$
$\beta_m\ldots(:_n\gamma\delta_1\ldots\delta_n\,\delta_n\ldots\delta_1)_i\ldots\beta_1$
if $m = 1$
then $:\alpha:_n\gamma\delta_1\ldots\delta_{n\delta_1}\ldots\delta_n\,:_n\gamma\delta_1\ldots\delta_n\,\delta_n\ldots\delta_1 \to$
$\to\; :_n\gamma\delta_1\ldots\delta_n\,\delta_n\ldots\delta_1$
where $\delta_1\ldots\delta_n$ in the consequents are to be written in majuscules.

This rule is optional; if all of the argumentor-labels of the main label are to be deleted, then the first clause is applied to each β in turn until only one remains; the second clause is then applied to effect the final deletion. It is to be observed that the other β's in the first clause, apart from β_i, may be followed by index-labels.

Mixed first-second-level functors provide examples in which only partial deletion of the main category-label is possible, e.g.

           $::NN:NN_n$ $:nn$ n n
           becomes $:NN$ $:nn$ N n
But $::N:NN:N:NN_{n:nn}$ $:n:nn_n$ $:nn$ n $:nn$ n
           becomes $:N:N:NN_{:nn}$ $:n:nn_n$ $:nn$ n $:nn$ N
           and then $:n:nn_n$ $:NN$ n $:nn$ N
while $:N::NN:NN_{n:nn}$ $::nn:nn_n$ $:nn$ n n
           becomes $:N:NN_n$ $::nn:nn_n$ $:NN$ n n
           and then $::nn:nn_n$ $:NN$ n N

2. Derivations

Frege's procedure for obtaining functors is equally a procedure for analyzing an expression into two constituents, the first a functor and the second one of its argumentors. The latter is then not removed, but fills the appropriate argumentor-place of the functor. It should therefore be possible to devise a parallel procedure for replacing a category-symbol by two new ones, the first a functorial symbol and the second so related to it that any expression substituted for the second will be an argumentor of any expression substituted for the first. In order to show this, we must allow italicized minuscules in a category-symbol to be replaced by a category-symbol; thus, given the category-symbols ': NN n' and 'N', we may replace 'n' by 'N' to yield ': NN N'. This is parallel to filling the argumentor-place in a functor of category : NN with an expression of category N; I therefore call category-symbols or partial category-symbols combined in this way *structures*, a single category-symbol or partial category-symbol being allowed as a limiting case.

It will be convenient in what follows to be able to refer to a functorial category-symbol as a functor and to category-symbols which, in a structure, fill its argumentor-places as its argumentors. This is to be understood merely as an abbreviation to avoid clumsiness, *not* as allowing the substitution of category-symbols, along with linguistic expressions, for category-symbols. Since we shall be concerned, in this section and the next, only with structures, it should occasion no confusion.

Our aim is then to formulate a production rule such that, given any category-symbol, we may replace it by a structure consisting of two category-symbols of which the first is a functor and the second one of its argumentors. If the rule is to be applicable to its own results, it must not be restricted to lone category-symbols, but must allow us to replace any category-symbol occurring in a structure. Starting with a single category-symbol, we shall then be able to derive, by successive applications of the rule, infinitely many structures each of which is, as a whole, of the same category as the original symbol. I call a series of such structures, in which each is obtained from the preceding structure by one application of a production rule, *derivations*; whereas logical proofs preserve truth, derivations preserve categoriality.

Each line of a derivation will consist of a structure; the latter must be expressed in category-symbols and not just main labels in order to show unambiguously the internal structure of an expression which has been analyzed into constituents. An illustration: in

::NNN N,

using only main labels, we cannot tell whether 'N' fills the first or the second argumentor-place in '::NNN'; but, given category-symbols, we can distinguish

::NNN *n* N   from   ::NNN N *n*

A convention (cf. combinatorial logic) would also remove the ambiguity; the price, deprivation of one or other of the alternatives, is unacceptable.

Structures also provide the justification for including index-labels in the category-symbols of higher-level functors. For suppose that we have a structure consisting of a one-place second-level functor and its argumentor; the latter will then also be a functor, so that its argumentor-place will be shown. Hence if the category-symbol for ':N:NN' were merely

:N:NN :*nn*

the result of filling its argumentor-place with a category-symbol ':NN *n*' would be

:N:NN :NN *n*

which still contains an empty place, in contradiction with the explanation of category :N:NN given in Section I. The index-label introduced by rule (S1) shows that the second-level functor fills the empty place, e.g.

:N:NN :NN n

and is therefore a necessary consequence of the convention that argumentor-places are always shown *after* the main label in a category-symbol.

The other occurrence of the index-label, introduced by rule (S2), serves to link the main label of the higher-level functor with the index-label introduced by rule (S1), so we could regard the two occurrences of the index-label in

$$:N:NN_n:NN\ n$$

as like the tail and head of an arrow pointing from the main label of the higher-level functor to the empty places in its argumentor in order to show that it fills the latter (cf. Bourbaki, pp. 10–13, 35). It will be possible to give a further and stronger justification for the sub-scripted index-labels in Section III; meanwhile, it may be said that the function of index-labels is solely to show how the higher-level functor and its argumentor are combined. Index-labels have no meaning in isolation, nor are they, by themselves, schematic signs for which linguistic expressions may be substituted. They are integral and unseparable parts of the category-symbols for higher-level functors.

If our production rule is to be applicable to any category-symbol occurring in a structure, we must have a means of identifying the argumentors of that symbol, should it be functorial; for the structure which replaces it will have the same number and type of argumentor-places as the original symbol and these must be filled by the same argumentors as in the previous structure. Similarly, if the category-symbol selected is itself an argumentor of some other functor, the argumentor-place which it fills must subsequently be filled by the structure which replaces it. Hence we must be able to identify any functor of which it is an argumentor and which argumentor-place in that functor it fills.

Given such a method, the production rule may be applied to the category-symbol in isolation; i.e. we can first remove it from the original structure, then use the production rule to replace it by a structure, and finally replace the new structure in the correct position in the original structure. In that case we need not, in formulating the production rule itself, take any account of the context in which the category-symbol occurs. Nor will the production rule ever be applied again to the *structure* resulting from its first application. The first application will yield a structure consisting of two category-symbols; the production rule will, of course, be applicable to either of these, but if we wish so to apply it, the method will be to remove the symbol concerned from that structure and to apply the rule to the symbol alone, subsequently replacing the symbol in the result of the first production by the structure resulting from the second.

A further simplification is possible if we design the production rule to operate upon *partial* category-symbols, for it will then be unnecessary for the rule to take account of index-labels. In that case, however, index-labels must be added where required immediately after the production rule has been applied, before the new structure is substituted, in any preceding structure, for the category-symbol from which it has been derived. Partial category-symbols will therefore be accounted as structures and admitted to derivations, but it will never be allowed to end a derivation with a structure containing a partial category-symbol.

If α and β are category-labels, let α^{β_i} be the category-label obtained by inserting β between the i-1th and the ith argumentor-labels of α and adding an extra intial colon. The definition of 'category' ensures that α^{β_i} will indeed be a category-label. Frege's method of analysis can then be captured in the following production rule:

(R1) $\quad \alpha \, \varphi_m \ldots \varphi_1 \to \alpha^{\beta_i} \, \varphi_m \ldots \varphi_i \, \beta \, \varphi_{i-1} \ldots \varphi_1$

where $\varphi_m \ldots \varphi_1$ are the argumentor-places of α, being written, therefore, in italicized minuscules, while β is to be written in majuscules, since it will be a main label.

Derivations are to consist of a series of successively numbered lines, beginning with 0, a single structure appearing on each line. The first line of any derivation is to consist of a single partial category-symbol, obtained by selecting a main category-label and applying (S1) to it once, unless it is a basic category-label. Each subsequent line is to be derived from the previous line by a rule, which is to be cited at the end of the line, including the value of i for (R1). So far, four rules are available, (R1), (S1), (S2) and (S3). If (R1) is used, (S1) and (S2) must be used immediately afterwards as many times as they are applicable; in order to save space I shall omit intermediate steps in applying (S1) and (S2), citing both rules at the end of the line against the final structure.

The two structures distinguished earlier may now be derived by taking

$\alpha =\, :\mathrm{NN}$ and $\beta = \mathrm{N}$; the difference between them depends upon the value given to i:

0	:NN n	(S1)
1	::NNN n N	(R1)/$i = 1$
0	:NN n	(S1)
1	::NNN N n	(R1)/$i = 2$

Index-labels must be supplied in the next two examples:

0	N	
1	:N:NN :NN	(R1)/$i = 1$
2	:N:NN$_n$:NN n	(S1), (S2)

0	:N:NN :nn	(S1)
1	::N:NN:NN :nn :NN	(R1)/$i = 1$
2	::N:NN:NN$_{n_1 n_2}$:nn n$_2$:NN n$_1$	(S1), (S2)

In any derivation using (R1), derived structures will have the same number of argumentor-places, of the same type and in the same order, as the partial symbol from which they are derived. This can be expressed formally by using the archaic Greek majuscule F (di-gamma) as a schematic letter for which any *structure* may be substituted and writing $F(\varphi_m, \ldots, \varphi_1)$ to indicate that the structure F contains the argumentor-places $\varphi_m, \ldots, \varphi_1$, occurring in that order. The general form of a derivation is then:

$$\alpha\ \varphi_m \ldots \varphi_1$$
$$\vdots$$
$$F(\varphi_m, \ldots, \varphi_1)$$

Since we wish to be able to apply (R1) not merely to isolated partial category-symbols, but to any category-symbol occurring in a structure, it must be completed by a second rule which allows us to transform one *derivation* into another: the kind of rule which Stoic logicians called a *thema*, by contrast with (R1), which transforms one *structure* into another and is a *schema*. The derivation to be transformed must be of the general form given above and, after the fashion of subordinate proofs in a number of well-known natural deduction systems of logic, it will be shown as a subordinate derivation by indenting it against a vertical line (cf. e.g. Fitch, 1952).

We can say that α must be a main category-label occurring in the previous line. Other category-labels may occur to the left of α; let these be Γ. Suppose, first, that α does not fall within the scope of any higher-level functor in Γ, i.e. that no second occurrence of an index-label whose first occurrence is in Γ comes to the right of α. If α is a functorial category-label there will be further labels to the right of α which must be correlated with the argumentor-labels of α, namely $\varphi_1 \ldots \varphi_m$. It will be necessary to specify a procedure later for making this correlation, but in simpler cases the labels will coincide with $\varphi_m \ldots \varphi_1$, apart from typographical differences. Having made the correlations, we may find that there remain yet other labels to the right of φ_1; let these be Δ. Then if we use $\lambda_m \ldots \lambda_1$ for the labels between α and Δ such that each label or group of labels λ_i is correlated with φ_i, the structure upon which we operate will have the form:

$$\Gamma \; \alpha \; \lambda_m \ldots \lambda_1$$

The *thema* will then allow us, given the subordinate derivation, to substitute the structure $\text{F}(\lambda_m, \ldots, \lambda_1)$ for $\alpha \, \varphi_m \ldots \varphi_1$, where $\text{F}(\lambda_m, \ldots, \lambda_1)$ is obtained from $\text{F}(\varphi_m, \ldots, \varphi_1)$ by replacing each φ_i with the λ_i correlated with it. The rule is therefore:

(Q)
$$\Gamma \; \alpha \; \lambda_m \ldots \lambda_1 \; \Delta$$
$$\left| \begin{array}{l} \alpha \; \varphi_m \ldots \varphi_1 \\ \vdots \\ \text{F}(\varphi_m, \ldots, \varphi_1) \end{array} \right.$$
$$\Gamma \; \text{F}(\lambda_m, \ldots, \lambda_1) \; \Delta$$

where α is any main category-label occurring in the last line of the derivation before (Q) is applied. To mark what has been chosen as α in that structure, it will be italicized. The first line of the subordinate derivation is then obtained by one application of (S1) to α, unless α is a basic category-label. This will ensure that the φ's are in italicized minuscules; the typography of Γ, Δ and the λ's must remain unchanged throughout the rule.

Some examples may now be given in which the only differences between each λ_i and φ_i are typographical. The final structure in the first example illustrates the main problem which can arise in determining the λ's:

```
0    N
1    :NN  N                    (R1)/i = 1   λ₁ = N
2    |   :NN  n                (S1)         φ₁ = n
3    |   ::NNN  n  N           (R1)/i = 1
4    ::NNN  N  N               (Q)
5    |   N
6    |   :NN  N                (R1)/i = 1
7    ::NNN  :NN  N  N          (Q)
```

If we had taken $\alpha = \text{::NNN}$ in line 4, λ_2 would have been N; the actual derivation, however, replaces that occurrence of N with :NN N in line 7. Consequently, if we took $\alpha = \text{::NNN}$ in line 7 for a further derivation, φ_2 must be correlated with the whole sub-*structure* :NN N as λ_2. The procedure for determining the λs must reckon with this possibility.

Applying (R1) to partial category-symbols circumvents problems which would otherwise arise if α is a higher-level category-label. Consider, for example, the following derivation:

```
0    N
1    :N:NN  :NN                              (R1)/i = 1
2    :N:NNₙ  :NN  n                          (S1), (S2)
3    |   :N:NN  :nn                          (S1)
4    |   ::N:NN:NN  :nn  :NN                 (R1)/i = 1
5    |   ::N:NN:NNₙ₁ₙ₂  :nn  n₂ :NN  n₁      (S1), (S2)
6    ::N:NN:NNₙ₁ₙ₂  :NN  n₂ :NN  n₁          (Q)
```

In determining the λs in line 2, the index-label is ignored, the relevant structure being that of line 1. The second use of (S1) and (S2) in line 4 introduces a corresponding index-label, so that it only remains to replace :nn by :NN in line 5.

If we now drop the assumption that α in rule (Q) does not fall within the scope of a higher-level functor in Γ, index-labels whose first occurrence is in Γ but whose second occurrence is to the right of α cannot be ignored, since they fill argumentor-places in α. Thus if the first :NN in line 6 of the derivation above be taken as α, $\lambda_1 = n_2$. The derivation might then continue:

```
6    ::N:NN:NNₙ₁ₙ₂  :NN  n₂  :NN  n₁          (Q)    λ₁ = n₂
7    |   :NN  n                                (S1)
8    |   ::NNN  N  n                           (R1)/i = 2
9    ::N:NN:NNₙ₁ₙ₂  ::NNN  N  n₂ :NN  n₁       (Q)
```

I now formulate, in quasi-Algol, the procedure for determining the λs. It is assumed that the argumentor-places of α have already been read, so that we have the values of m and of $\varphi_m \ldots \varphi_1$. If α has any index-letters, we ignore their second occurrences, letting the remaining labels to the right of α be $a_1 \ldots a_k$; I assume that these, too, have been read. One departure from Algol which is unavoidable is the notation 'c := [c⌣ a_j]'; this means that the value of c is to be set to a *string* consisting of the previous value of c (which may, therefore, itself be a string), followed by a space, followed by the value of a_j. The values of c will therefore be strings of category-labels, except at the first assignment in each cycle, when the value will be a single category-label.

λ-procedure: **begin**

> **integer** φ, λ, a_j, b, c; **integer** i, j, m; **comment** Values of the first group of integer-variables are category-labels, of the second group numbers;
> **integer procedure** cat(b, a_j); **integer** b, a_j; **comment** This procedure, not formally specified here, is used to compute the category of an expression consisting of two category-symbols by means of their main labels and the rule:
> A1: $\alpha^\beta \beta \to \alpha$
> We put $b := \alpha^\beta$ and $a_j := \beta$. The substitution made for α is then assigned to cat(b, a_j). The rule A1 is slightly adapted from Ajdukiewicz (1935, p. 10: 1967, p. 643: cf. Geach, 1970, p. 4: 1972, p. 484);
> A: **begin** i := m; j := 1 **end**
> B: **begin** b := a_j; c := a_j **end**
> C: *if* b = φ_1 **then begin** print "λ"; print i; print "="; print c **end**
>   **else begin** j := j+1; c := [c⌣ a_j]; b := cat(b,a_j); **goto** C **end**
> D: **if** i = 1 **then goto** E
>   **else begin** i := i−1; j := j+1; **goto** B **end**
> E: **end**

As an example, take the structure on line 8 of the last derivation, with $\alpha = ::N:NN:NN$. Then $m = 2$, $\varphi_2 = :NN$ and $\varphi_1 = :NN$; α has two index-labels, n_1 and n_2, so we ignore their second occurrences. Then $a_1 = ::NNN$ $a_2 = N, a_3 = :NN$. We begin by comparing φ_2 (:NN) with a_1 (::NNN); they are not the same, so we set $c = [a_1 \leftharpoonup a_2]$, i.e. ::NNN N. Rule A1 tells us that the category of this sub-structure is :NN, so now $b = \phi_2$. Hence λ_2 is ::NNN N. We now go to compare φ_1(:NN) with a_3(:NN). These are the same, so $\lambda_1 = $:NN. The value of i is now 1, so the procedure terminates.

No provision has yet been made for the derivation of structures containing alternative symbols for higher-level functors. Since main category-labels play a central role in the method of derivation, it is evident that if alternative symbols are allowed at intermediate stages of a derivation, then the rules will have to be amended. This is likely to be a complicated business. The method of generating a structure containing alternative symbols will therefore be to generate a corresponding structure with the standard symbol first, using (S3) to replace it with an alternative symbol at the end of the derivation.

If more than one standard symbol in a structure is to be replaced by an alternative symbol, we begin with the right-most symbol and work from right to left. The restriction is, however, to be imposed that the rule may not be applied to any symbol in a structure unless it falls within the scope of a higher-level functor. A couple of simple examples will show the point of this. First, applying (S3) to

:N:NN$_n$:NN n   we get   :NN N,

a structure which can already be derived, more easily, without (S3). Second, applying (S3) to

::N:NN:NN$_{n_1 n_2}$:NN n$_2$:NN n$_1$ we get   :NN N :NN N

The derivation would then no longer preserve categoriality, for this consists of two sub-structures, each of the form :NN N and thus of category N (by A1). Nothing is left to bind them together into a single structure and we should thus have produced two separate structures of category N from a single one of that category. Generative though the grammar may be, it is not intended to produce a linguistic population explosion!

3. THE COMBINATION PROBLEM

The first use to which Frege put his *Begriffsschrift* was the formulation of first-order predicate logic. In his version, only three functors are used:

negation, which he assigned to category :NN; implication, assigned to category ::NNN; and the universal quantifier, assigned to category :N:NN. However, he also tells us how conjunction and (inclusive) disjunction may be expressed with the aid of negation and implication and also how existential quantification may be expressed with the aid of negation and universal quantification. This is tantamount to defining conjunction, disjunction and the existential quantifier in terms of the three primitive notions, so we may reasonably infer from his elucidations that conjunction and disjunction should be assigned with implication to category ::NNN, while the existential quantifier is to be assigned with the universal to category :N:NN.

Paradoxically, the very *Begriffsschrift* which was the fruit of Frege's analytical method led him to results which that method could not justify. Two key examples will bring this out: first, the use of negation within the scope of a quantifier, as in the formula $\bigwedge_x \neg Fx$; and, second, the definition of $\bigvee_x Fx$ as $\neg \bigwedge_x \neg Fx$ already mentioned. Given only the rules of the previous section, of which (R1) was intended to summarize Frege's method, we cannot generate a structure of which either of these formulas would be an instance. For $\bigwedge_x \neg Fx$ we should require

(A) :N:NN_n :NN:NN n

and for $\neg \bigwedge_x \neg Fx$ the structure

(B) :NN :N:NN_n :NN:NN n

Frege thus provided no justification for his own practice; nor could he allow, without offending against his own prohibition of piecemeal definition, that the category-symbols should be endowed, as an afterthought, with new powers of combination:

Ist ein Grundbegriff eingeführt, so muss er in allen Verbindungen eingeführt sein, worin er überhaupt vorkommt. Man kann ihn also nicht zuerst für *eine* Verbindung, dann noch einmal für eine andere einführen. ... Wir dürfen sie nicht erst für die eine Klasse von Fällen, dann für die andere einführen, denn es bliebe dann zweifelhaft, ob ihre Bedeutung in beiden Fällen die gleiche wäre und es wäre kein Grund vorhanden, in beiden Fällen dieselbe Art der Zeichenverbindung zu benützen (Wittgenstein, 1922, 5.451).

If the meanings of the categories are to be determinate, all of their possible combinations must then be introduced at a single blow. It is to the solution of this *Verbindungsproblem* that this paper is primarily directed.

Frege's practice, or his notation, may be justified by strengthening his method of analysis with two new production rules. These are both recursive and hence themas, so their use will again involve subordinate derivations; but they can both be used repeatedly in nested derivations. The gap between Frege's theory and his practice is caused, as structures (A) and (B) illustrate, by the presence of higher-level functors in the grammar, with the consequence that functors of a lower level may lie within their scope. The new rules will therefore normally be used within a subordinate derivation introduced by rule (Q); otherwise they will only yield alternative derivations of structures which are already derivable in another way without them. This means that both rules can be framed, like (R1), to operate upon a single, isolated partial category-symbol, which can be withdrawn from its context for the purpose with the aid of (Q). Since the rules are to be recursive, it must be possible to take a derivation licensed by one of them as a subordinate derivation in a further application, thus giving rise to nested derivations.

The basic idea behind the first recursive rule is that, given a subordinate derivation in which the main label of the partial category-symbol in the first line is α and the two main labels in the last line are β and γ (in that order), we may derive from a partial category-symbol whose main label is α^δ, a structure whose two main labels are β and γ^δ.

In order to work out how the rule must be formulated, suppose first that no higher-level functors occur in any of the structures involved, so that we do not have to consider index-labels. The simplest case will then be one in which the subordinate derivation consists only of an application of (R1); δ must then be the ith argumentor-place of α^δ, the value of i being given by (R1), and we shall have:

$$\alpha^\delta\ \varphi_m \ldots\ \varphi_i\ \delta\ \varphi_{i-1} \ldots\ \varphi_1$$
$$\begin{vmatrix} \alpha\ \varphi_m \ldots\ \varphi_1 \\ \vdots \\ \beta\ \varphi_m \ldots\ \varphi_i\ \gamma\ \varphi_{i-1} \ldots\ \varphi_1 \end{vmatrix}$$
$$\beta\ \varphi_m \ldots\ \varphi_i\ \gamma^\delta\ \delta\ \varphi_{i-1} \ldots\ \varphi_1$$

where α, β, γ, α^δ and γ^δ are written in majuscules and all other labels in italicized minuscules

This rule is already enough to allow structure (A) to be derived, the φ's being empty:

0	N		
1	:N:NN :NN		(R1)/i = 1
2	:N:NN$_n$:NN n		(S1), (S2)
3	$\|$:NN n		(S1)
4	$\|$ $\|$ N		
5	$\|$ $\|$:NN N		(R1)/i = 1
6	$\|$:NN :NN n		
7	:N:NN$_n$:NN :NN n		(Q)

The derivation of an analogue of structure (A) provides an example in which the φs are not empty, φ_1 occurring first as the first argumentor-place of α^δ and then as the second:

0	N	
1	:N::NNN ::NNN	(R1)/i = 1
2	:N::NNN$_{n_1 n_2}$ ::NNN n$_2$ n$_1$	(S1), (S2)
3	$\|$ ::NNN n n	(S1)
4	$\|$ $\|$:NN n	(S1)
5	$\|$ $\|$ ::NNN N n	(R1)/i = 2
6	$\|$ ::NNN :NN n n	
7	:N::NNN$_{n_1 n_2}$ ::NNN :NN n$_2$ n$_1$	(Q)
8	$\|$ ::NNN n n	(S1)
9	$\|$ $\|$:NN n	(S1)
10	$\|$ $\|$ ::NNN n N	(R1)/i = 1
11	$\|$ ::NNN n :NN n	
12	:N::NNN$_{n_1 n_2}$ ::NNN :NN n$_2$:NN n$_1$	(Q)

Now consider the position if we wish to use the result of one application of the rule as the subordinate derivation of a second application. In the final line of the rule given above, δ is an argumentor-place of γ^δ, not of β; in the first line, however, it is an argumentor-place of α^δ along with the φs. Consequently, if the rule is applied a second time, we must know where to insert δ among the argumentor-places of γ. Let the latter be $\psi_n \ldots \psi_1$. The subordinate derivation must preserve the number, type and order of argumentor-places from its first line through to its last. So if we show the last line as

$$\beta \; \varphi_m \ldots \varphi_1 \; \gamma \; \psi_n \ldots \psi_1 \; \varphi_{i-1} \ldots \varphi_1$$

the first must be shown as

$$\alpha \; \varphi_m \ldots \varphi_1 \; \psi_n \ldots \psi_1 \; \varphi_{i-1} \ldots \varphi_1$$

For the first line of the main derivation, we shall be free to insert δ at any point among the ψ's. It must then occupy a corresponding position among the ψ's in the derived structure.

This gives us

$$\alpha^{\delta_j} \; \varphi_m \ldots \varphi_i \; \psi_n \ldots \delta \ldots \psi_1 \; \varphi_{i-1} \ldots \varphi_1$$
$$|\;\alpha \; \varphi_m \ldots \varphi_i \; \psi_n \ldots \psi_1 \; \varphi_{i-1} \ldots \varphi_1$$
$$|\;\vdots$$
$$|\;\beta \; \varphi_m \ldots \varphi_1 \; \gamma \; \psi_n \ldots \psi_1 \; \varphi_{i-1} \ldots \varphi_1$$
$$\;\;\;\beta \; \varphi_m \ldots \varphi_i \; \gamma^{\delta_j-(i-1)} \; \psi_n \ldots \delta \ldots \psi_1 \; \varphi_{i-1} \ldots \varphi_1$$

where α, β, γ, α^δ and γ^δ are written in majuscules and all other labels in italicized minuscules

This reduces to the simplest case when $n=0$. In applying the rule, we are free to pick what argumentors of α^δ we please as the φs, the ψs and δ, provided that the arrangement given by the rule is preserved. Our choice will be guided by what we want to derive.

As an example in which the rule must be used twice, with $\psi_1 = n$ in the second application, I give the derivation of another analogue of structure (A):

```
0        N
1        :N ::NNN  ::NNN                          (R1)/i = 1
2        :N ::NNN_{n₁n₂}  ::NNN  n₂  n₁           (S1), (S2)
3        |   ::NNN  n  n                          (S1)
4        |   |   :NN  n                           (S1)
5        |   |   |   N
6        |   |   |   :NN  N                       (R1)/i = 1
7        |   |   |   :NN  :NN  n                  j = 1
8        |   |   :NN  ::NNN  n  n                 j = 2
9        :N ::NNN_{n₁n₂}  :NN  ::NNN  n₂  n₁      (Q)
```

It has so far been assumed that no higher-level functors occur in uses of the new rule. I now suppose that β is a higher-level functor but that $n=0$. In that case, the final structure of the subordinate derivation will contain labels which have been introduced by (S1) and (S2) after (R1)

Again I begin with the simplest case, in which the subordinate derivation consists only of an application of (R1) followed by as many applications of (S1) and (S2) as are necessary. If the index-labels are $\rho_1 \ldots \rho_r$, the subordinate derivation will then have the form:

$$\alpha\ \varphi_m \ldots \varphi_1$$
$$\vdots$$
$$\beta_{\rho_1\ldots\rho_r}\ \varphi_m \ldots \varphi_1\ \gamma\ \rho_r \ldots \rho_1\ \varphi_{i-1} \ldots \varphi_1$$

The ρs will then fill the argumentor-places of γ, like the ψs in the previous version of the rule, but unlike the latter they will not occur in the first line of the derivation, having been introduced by (S1) and (S2); further, they will be written in unitalicized minuscules.

In the first line of the derivation, δ must again occur between φ_i and φ_{i-1}; but in the derived structure, we may introduce it into any position that we please among the ρs. This gives us, as an initial version of the rule:

$$\alpha^\delta\ \varphi_m \ldots \varphi_i\ \delta\ \varphi_{i-1} \ldots \varphi_1$$
$$\left|\ \begin{array}{l} \alpha\ \varphi_m \ldots \varphi_1 \\ \vdots \\ \beta_{\rho_1\ldots\rho_r}\ \varphi_m \ldots \varphi_i\ \gamma\ \rho_r \ldots \rho_1\ \varphi_{i-1} \ldots \varphi_1 \\ \beta_{\rho_1\ldots\rho_r}\ \varphi_m \ldots \varphi_i\ \gamma^{\delta_k}\rho_r \ldots \delta \ldots \rho_1\ \varphi_{i-1} \ldots \varphi_1 \end{array}\right.$$

Frege tells us that if we combine a quantifier with a *two*-place first-level functor, the result will be a *one*-place first-level functor, which can thus itself serve as argumentor of another quantifier (1893, p. 36). The rule just stated allows us to derive a structure which will justify this contention:

```
0        :NN n                (S1)
1        | N
2        | :N:NN  :NN          (R1)/i = 1
3        | :N:NNₙ :NN n        (S1), (S2)
4        :N:NNₙ ::NNN n n      k = 1
```

Alternatively, we could have taken j = 2 in line 4, obtaining

```
4        :N:NNₙ ::NNN n n      k = 2
```

and thus showing that the quantifier can be used to fill *either* of the argumentor-places of ::NNN $n\ n$. This example provides the further justification, promised in Section II, for sub-scripted index-labels; for if we use the derivation above as a subordinate derivation under the aegis of rule (Q), then from $:N:NN_n :NN\ n$ we can derive

$$:N:NN_{n_1} :N:NN_{n_2} ::NNN\ n_2\ n_1$$

and here it would be impossible, without the first occurrences of each index-label, to tell which of the two second-level functors fills which of the argumentor-places in the first-level functor.

The rule will not, however, allow us to derive

$$:N:NN_n :::NNNN\ n\ n\ n$$

from $::NNN\ n\ n$; for this, we should have to be able to apply it again to the derivation above. Now since the δ which occurs on its final line marks an argumentor-place of γ^δ, in the next application that δ will become ψ_1. Consequently this hurdle can only be jumped by combining the two versions of the rule. This means that the ψs and the ρs must be treated together as argumentors of γ in the subordinate derivation, the position into which δ is inserted being relative to *both* series. We can, however, always distinguish the ψs from the ρs since the former will be written in italicized but the latter in unitalicized minuscules.

If, then, we write $|\psi_n \ldots \psi_1, \rho_r \ldots \rho_1|$ to signify a series in which the ψ's and the ρ's may be mixed up together, but in such a way that the ordering of each series is preserved, the combined rule can be stated as

(R2)
$$\alpha^{\delta j}\ \varphi_m \ldots \varphi_i\ \psi_n \ldots \delta \ldots \psi_1\ \varphi_{i-1} \ldots \varphi_1$$
$$\begin{vmatrix} \alpha\ \varphi_m \ldots \varphi_i\ \psi_n \ldots \psi_1\ \varphi_{i-1} \ldots \varphi_1 \\ \vdots \\ \beta_{\rho_1 \ldots \rho_r}\ \varphi_m \ldots \varphi_i\ \gamma\ |\psi_n \ldots \psi_1, \rho_r \ldots \rho_1|\ \varphi_{i-1} \ldots \varphi_1 \end{vmatrix}$$
$$\beta_{\rho_1 \ldots \rho_r}\ \varphi_m \ldots \varphi_i\ \gamma^{\delta j+k-1}\ |\psi_n \ldots \psi_1, \delta, \rho_r \ldots \rho_1|\ \varphi_{i-1} \ldots \varphi_1$$

where $\alpha, \beta, \gamma, \alpha^\delta$ and γ^δ are written in majuscules, the ρs in unitalicized minuscules and all other labels in italicized minuscules. The value of k will be fixed by the first application of (R2), when $n = 0$; δ is then inserted between ρ_k and ρ_{k-1}, with a free choice for the value of k. No further ρs can be inserted thereafter, so k remains constant in every subsequent application of the rule within which that application is nested. We can now use an expression of category $:N:NN$ to fill one argumentor-place in a first-level functor which itself has any number of argumentor-places.

It has been assumed in the statement of (R2) that the only index-labels which occur in the subordinate derivation are index-labels of β. However, if γ is also a higher-level functor (as might be the case, for instance, if β were a third-level functor) then it too would attract index-labels. We cannot provide for these in the rule, though, for not only might γ require index-labels, but also the ρs, the index-labels of γ themselves, and so on. This complication can be avoided by specifying that (S1) and (S2) shall only be applied once each to β of (R1) before (R2) is applied; the conclusion of the subordinate derivation will then *only* show index-labels which fill argumentor-places of γ.

It will always be safe to add any further index-labels which are necessary *after* (R2) has been used, for they will always follow immediately after the relevant ρ and cannot be separated from it by any other of the ρs or by any of the ψs. Similarly, if the partial category-symbol which stands on the first line of the subordinate derivation requires index-labels, these too can be added to the structure resulting from the use of (R2). In neither case will these index-labels be added if (R2) is to be used again, but they must all be supplied before any use of (Q). With this provision, all restrictive assumptions concerning (R2) may now be dropped.

The following derivation illustrates the manner of applying (S1) and (S2) just prescribed:

0	$:N:NN\ :nn$	(S1)
1	N	
2	$:N:N:NN\ :N:NN$	(R1)/$i=1$
3	$:N:N:NN_{:nn}\ :N:NN\ :nn$	(S1), (S2)
4	$:N:N:NN_{:nn}\ ::N:NN:NN\ :nn\ :nn$	(R2)/$k=1$
5	$:N:N:NN_{:nn}\ ::N:NN:NN_{n_1n_2}\ :nn\ n\ :nn\ n$	(S1), (S2)

With the aid of (R2), structure (B) can now also be derived. We need only continue the derivation of structure (A) as follows:

7	$:N:NN_n\ :NN\ :NN\ n$	(Q)
8	$:N:NN\ :nn$	(S1)
9	N	
10	$:NN\ N$	(R1)/$i=1$
11	$:NN\ :N:NN\ :nn$	(R2)/$j=1$
12	$:NN\ :N:NN_n\ :nn\ n$	(S1), (S2)
13	$:NN\ :N:NN_n\ :NN\ :NN\ n$	(Q)

But this does not justify the definition of $\bigvee_x Fx$ as $\neg \bigwedge_x \neg Fx$, for it does not derive

$$:NN \quad :N:NN_n \quad :NN \quad :nn \text{ n} \quad \text{from} \quad :N:NN_n \quad :nn \text{ n}$$

In the second part of the derivation, $:N:NN$ is taken as α^δ of (R2) to yield $:NN$ as β and $:N:NN$ as γ^δ; but in the first part, $:NN$, which fills the argumentor-place in the structures just cited, is taken as α^δ, with $:NN$ as β and $:NN$ again as γ^δ.

Thus, given only the preceding rules, we must agree with Quine that

The parts of '$\neg \bigwedge_x \neg$' do not, of course, hang together as a unit;...

Yet, surely, they *ought* to do so, and Quine in effect postulates that they do:

But the configuration of prefixes '$\neg \bigwedge_x \neg$' figures so prominently in future developments that it is convenient to adopt a condensed notation for it (1951, p. 102).

As Russell once rather tartly observed,

The method of "postulating" what we want has many advantages; they are the same as the advantages of theft over honest toil. Let us leave them to others and proceed with our honest toil (1919, p. 71).

In the present instance, the 'honest toil' must consist in formulating a syntax which will *justify* Quine and other logicians in treating '$\neg \bigwedge_x \neg$' as an unit. This is, in part, why a second recursive rule is necessary, though the rule also has the further motivation that it enables us to derive new structures which are underivable by any means without it.

I begin, as before, by expounding the basic idea of the rule. In order to do so, the earlier notation must be extended, so that we can use $\alpha^{\beta_i \gamma_j}$ to represent a higher-level functor whose ith argumentor-label is β^{γ_j}, so that this in turn has γ as its jth argumentor-label. Then, given a subordinate derivation in which the main label of the partial category-symbol in the first line is α^δ and the main labels in the last line are β^ζ and γ (in that order), we may derive, from a partial category-symbol whose main label is $\alpha^{\delta_\varepsilon}$, a structure whose two main labels are $\beta^{\zeta_\varepsilon}$ and γ.

Availing ourselves of the reasoning behind the formulation of (R2), we can start with the following tentative statement of the new rule:

$$\begin{array}{l} \alpha^{\delta_j^{\varepsilon_h}} \varphi_m \ldots \varphi_i \psi_n \ldots \delta^{\varepsilon_h} \ldots \psi_1 \varphi_{i-1} \ldots \varphi_1 \\ \quad \alpha^{\delta_j} \varphi_m \ldots \varphi_i \psi_n \ldots \delta \ldots \psi_1 \varphi_{i-1} \ldots \varphi_1 \\ \quad \vdots \\ \quad \beta^{\zeta_i}{}_{\rho_1 \ldots \rho_r} \varphi_m \ldots \varphi_i \gamma \, |\psi_n \ldots \psi_1, \delta, \rho_r \ldots \rho_1| \, \varphi_{i-1} \ldots \varphi_1 \\ \quad \beta^{\zeta_i^{\varepsilon}}{}_{\rho_1 \ldots \rho_r} \varphi_m \ldots \varphi_i \gamma \, |\psi_n \ldots \psi_1, \delta^{\varepsilon_h}, \rho_r \ldots \rho_1| \, \varphi_{i-1} \ldots \varphi_1 \end{array}$$

This assumes that the subordinate derivation has been effected with the aid of (R2); it is, in fact, unnecessary to provide for the new rule to transform a subordinate derivation which does not use (R2), since what could then be derived with it can already be derived with (R1) (and the (S)-rules) alone. A more important assumption is that the (S)-rules have not yet been applied after (R2), for γ might then have sub-scripted index-labels whose second occurrences would have to be shown after the ψs or ρs.

The formulation given above also omits one essential piece of information: it does not tell us the position of ε among the argumentor-labels of ζ. Now if ζ already has other argumentor-labels, they must be $\rho_1 \ldots \rho_r$; so the position of ε among them must correspond to the position of δ^{ε_h} among the second occurrences of the ρs. This will be the value of k in the applications of (R2) in the subordinate derivation, which, as we have already seen, remains constant from one use of (R2) to the next. A slight modification to the notation for the symbols occupying the argumentor-places of γ will enable us to represent this: I shall now write

$$\gamma \, |\psi_n \ldots \delta_x \ldots \psi_1, \rho_r \ldots \delta_z \ldots \rho_1|$$

to signify that δ, though occurring only *once* in the combined series of the ψs and the ρs, comes at the xth place among the ψs and at the zth place among the ρs. In the new rule, we then have that $x = j - (i - 1)$ and that $z = k$; ε, consequently, must be the kth argumentor-label of ζ^{ε}.

When index-labels are added after the use of (R3), ε must be added to the sub-scripted ρs in order to correspond to its occurrence as an argumentor-label of ζ^{ε}. But whereas (S1) can be left to introduce the second occurrence of ε in the correct position, (S2) would add ε as a sub-script of γ, not of $\beta^{\zeta^{\varepsilon}}$. This difficulty can be overcome by allowing the new rule to introduce the sub-scripted ε, subsequently applying only (S1) to δ^{ε}; ε must then take the kth position among the sub-scripted ρ's. As the final version of the new rule, we then have:

(R3) $\alpha^{\delta_j \varepsilon_h} \varphi_m \ldots \varphi_i \psi_n \ldots \delta^{\varepsilon_h} \ldots \psi_1 \varphi_{i-1} \ldots \varphi_1$
$\quad\quad | \; \alpha^{\delta_j} \varphi_m \ldots \varphi_i \psi_n \ldots \delta \ldots \psi_1 \varphi_{i-1} \ldots \varphi_1$
$\quad\quad | \;\; \vdots$
$\quad\quad | \; \beta^{\zeta_i}{}_{\rho_1 \ldots \rho_r} \varphi_m \ldots \varphi_i \gamma \,|\, \psi_n \ldots \delta_{j-(i-1)} \ldots \psi_1, \rho_r \ldots \delta_k \ldots \rho_1 |$
$\quad\quad\quad\quad\quad\quad\quad\quad\quad\quad\quad\quad\quad\quad\quad\quad\quad\quad \varphi_{i-1} \ldots \varphi_1$
$\quad\quad \beta^{\zeta_i \varepsilon_k}{}_{\rho_1 \ldots \varepsilon \ldots \rho_r} \varphi_m \ldots \varphi_i \gamma \; |\psi_n \ldots \delta^{\varepsilon_h}_{j-(i-1)} \ldots \psi_1, \rho_r \ldots \delta^{\varepsilon_h}_k \ldots \rho_1 |$
$\quad\quad\quad\quad\quad\quad\quad\quad\quad\quad\quad\quad\quad\quad\quad\quad\quad\quad \varphi_{i-1} \ldots \varphi_1$

We can now derive :NN :N:NN$_n$:NN :*nn* n from :N:NN$_n$:*nn* n, thus justifying the definition of the existential quantifier:

0	:N:NN :*nn*	(S1)
1	$\mid$:NN n	
2	$\mid\quad$ N	
3	$\mid\quad$:NN N	(R1)/i = 1
4	$\mid\quad$:NN :NN n	(R2)/j = 1
5	:N:NN$_n$:NN :*nn*	(R3)/h = 1
6	:*N:NN*$_n$:NN :*nn* n	(S1)
7	$\mid$:N:NN :*nn*	(S1)
8	$\mid\quad$ N	
9	$\mid\quad$:NN N	(R1)/i = 1
10	$\mid$:NN :N:NN :*nn*	(R2)/j = 1
11	$\mid$:NN :N:NN$_n$:*nn* n	(S1), (S2)
12	:NN :N:NN$_n$:NN :*nn* n	(Q)

Similarly, we can give an alternative derivation of the two-quantifier structure which shows that the two quantifiers form a sub-structure:

0	N	
1	:N::NNN ::NNN	(R1)/i = 1
2	:*N*::*NNN*$_{n_1 n_2}$ ::NNN n$_2$ n$_1$	(S1), (S2)
3	$\mid$:N::NNN ::*nn* n	(S1)
4	$\mid\quad$:N:NN :*nn*	
5	$\mid\quad\quad$ N	
6	$\mid\quad\quad$:NN N	(R1)/i = 1
7	$\mid\quad$:NN :N:NN :*nn*	(R2)/j = 1
8	$\mid$:N:NN$_n$:N:NN ::*nnn*	(R3)/h = 1
9	$\mid$:N:NN$_{n_1}$:N:NN$_{n_2}$ ::*nnn* n$_2$ n$_1$	(S1), (S2)
10	:N:NN$_{n_1}$:N:NN$_{n_2}$ ::NNN n$_2$ n$_1$	(Q)

FREGEAN CATEGORIAL GRAMMAR 271

To exemplify a structure underivable without (R3), I give the following:

```
0        ::N:NN:NN    :nn  :nn                    (S1)
1        | ::N:NNN    n   :nn
2        | | :N:NN    :nn
3        | | | :NN    n
4        | | | | N
5        | | | :NN    N                           (R1)/i = 1
6        | | | :NN   :NN   n                      (R2)/j = 1
7        | | :N:NNₙ  :NN  :nn                     (R3)/h = 1
8        | :N:NNₙ    ::NNN  n   :nn               (R2)/j = 2
9        :N::NNNₙ₁ₙ₂ ::NNN   :nn   :nn            (R3)/h = 1
10       :N::NNNₙ₁ₙ₂ ::NNN   :nn  n₂   :nn  n₁    (S1)
```

If the two recursive rules introduced in this section are to be used in subordinate derivations under the aegis of rule (Q), two modifications must be made to the λ-procedure. Consider line 6 of the first derivation given above to illustrate (R3). Given the existing λ-procedure, $\lambda_1 = $:NN and :*nn* is relegated to Δ; yet this cannot be correct, for the second occurrence of the index-label 'n' shows that ':*nn*' lies within the scope of :N:NN. Again, consider line 10 of the next derivation; $\varphi_1 = $:NN, but $a_1 = $:N:NN, so we set $b = $:N:NN ::NNN. The category of this substructure is not computable by A1, so λ_1 cannot be determined.

The first modification is therefore to supplement the category-procedure with two further A-rules, corresponding respectively to (R2) and (R3), namely:

A2   if $\beta \gamma \to \alpha$, then $\beta \gamma^\delta \to \alpha^\delta$ (cf. Geach, 1970. p. 5; 1972, p. 485)

A3   if $\beta^\zeta \gamma \to \alpha^\delta$, then $\beta^{\zeta\varepsilon} \gamma \to \alpha^{\delta\varepsilon}$

A2 will then be used in the category-procedure for the second example cited above, for since :N:NN :NN $\to$ N by A1, :N:NN ::NNN $\to$:NN by A2. This is not enough to deal with the first example, however, for the procedure does not allow us to consider :NN :*nn*; so even if $b = \varphi_1$ in clause C of the procedure, we must now look at the next a before concluding that $\lambda_1 = b$.

Here we encounter another difficulty: if, as before, we ignore second occurrences of index-labels of α, two separate argumentors of α could be ac-

counted as a single sub-structure. For example, if we take
$$\alpha = ::N:NN:NN$$
in the structure
$$::N:NN:NN_{n_1 n_2} :NN\ n_2 :NN\ n_1,$$
the original λ-procedure gives the correct result, that $\lambda_2 = :NN$ and $\lambda_1 = :NN$. But if we now have to consider the next label after :NN before determining λ_2, we find that we have :NN :NN, which by A2 will be a sub-structure of category :NN, giving the incorrect result that $\lambda_2 = :NN$:NN.

In order to overcome this difficulty, it is necessary to take account of the *second* occurrences of index-letters. However, we do need to know that they are index-letters and not main labels, so the following procedure must be incorporated into the λ-procedure:

> **Boolean procedure** index (a_j); **integer** a_j;
>   **if** a_j is written in unitalicized minuscules **then** index: = **true**
>   **else** index: = **false**

Further, in the earlier version of the λ-procedure, we could always guarantee a result for applications of the 'cat' procedure; but this will no longer be so, and consequently a terminating criterion must be included in the algorithm for that procedure. If no result is obtainable, the null element will be assigned to 'cat'; this will mean that the two functors are not combinable by our rules, or, to use Ajdukiewicz's terminology, that they do not form a *syntactically coherent* expression. In order to use this as a criterion in the λ-procedure, we shall need a third procedure, as follows:

> **Boolean procedure** syncoh (b, a_j); **integer** b, a_j;
>   **if** cat $(b, a_j) = \emptyset$ **then** syncoh: = **false else** syncoh: = **true**

It remains to amend clauses C and D; we split up C into C and D, replacing the old clause D by a new clause E, as follows:

> C: **if** $b = \varphi_i$ **then**
>   **begin** j: = j + 1;
>   **if** index (a_j) **then**
>   **begin** print "λ"; print i; print " = "; print c;
>   j: = j − 1 **end**
>   **else if** syncoh $(b, a_j) \wedge$ cat $(b, a) = \varphi_i$ **then**

>   begin c: = [c⌴a_j]; b: = cat (b, a_j); **goto C end**
>   else begin print "λ"; print i; print " = "; print
>   c; j: = j − 1; **goto E end**
> **end**
> D: **else begin** j: = j + 1;
>   **if** index (a_j) **then goto D**
>   **else begin** c: = [c⌴a_j]; b: = cat (b, a_j); **goto C**
>   **end**
>   **end**
> E: **if** i = 1 **then goto G**
>   **else begin** i: = i − 1; F: j: = j + 1;
>   **if** index (a_j) **then goto F else goto B**
> G: **end**

Despite the complexity of the seven rules formulated in this paper, the actual derivation of structures required in grammatical analysis is not unduly difficult, for it is seldom that every contingency envisaged by one of the rules arises in a single application of that rule. In order to avoid mistakes, it is a good method to use several different basic category-symbols in sketching out a derivation; the principle that every derivation must preserve the number, type and order of argumentor-places, together with the convention that argumentor-places must occur in inverse order to the corresponding argumentor-labels, then usually make it clear where a newly-introduced label must be placed and whether or not there is a choice as to its position. The different basic category-symbols can finally be converted into a lesser number or, as here, into a single one. This may, indeed, offer a means of simplifying the rules, but, if it does so, I do not see at present how the simplification is to be accomplished.

It should not occasion any surprise that the rules are complex, for the problem has been essentially that of generalizing the rules of substitution in first order predicate logic. It is well-known that among these the rule of substitution for first-level functors is particularly difficult to formulate correctly:

An inadequate statement of this rule for the pure functional calculus of first order appears in the first edition of Hilbert and Ackermann (1928). There are better statements of the rule in Carnap's *Logische Syntax der Sprache* and Quine's *A System of Logistic* (1934), but neither of these is fully correct. In the first volume of Hilbert and Bernays's *Grundlagen der Mathematik* (1934) the error of Hilbert and Ackermann is noted, and a correct statement of a rule of substitution for functional variables is given for the first time. ... In the case of logistic systems which involve operators other than quantifiers, such as the abstraction operator λ or the description operator $\imath$ ..., correct statement of the rule of substitution for functional variables becomes still more troublesome and lengthy (Church, 1956, pp. 289–290. and n. 461).

Small wonder, then, that the rules of substitution in a system allowing for basic categories and functors of any level or mixture of levels, both in what is substituted and in that for which it is substituted, should turn out to be complex and involved; indeed, the key to the problem has been to replace *rules* of substitution by a *procedure* for making substitutions, namely derivation, itself conducted in accordance with rules.

4. Natural languages

It should now be clear that the complexity of Fregean categorial grammar arises primarily from the distinction between different levels of functor; if it were not for the higher-level functors with their index-labels, the rules could be greatly simplified. It is therefore natural to ask whether such a sophisticated apparatus is really necessary for the analysis of natural languages.

Our answer will largely depend upon what we conceive to be the aims of grammar. One well-known formulation of the latter is that grammar seeks to explain how thoughts may be expressed in a particular natural language, how the meaning of an expression is related to its form, phonetic or graphic. On this account, we must be able, at the very minimum, to assign a structure to each expression of the language in terms of which its meaning can be explained; I call this a *semantic* structure. Since it is usually possible to express the same thought in different natural languages, but often only by using a different phraseology as well as different words, it seems that we must also be able to assign a second type of structure to each expression, which is proper to a given natural language and which provides the medium in which *that* thought may be expressed in *that* language; I call this a *phonographic* structure. We then have to show how the two types of structure are related.

Fregean categorial grammar is a theory of semantic, not of phonographic structures: natural languages do not, in general, observe the convention that functors precede their argumentors, not do they contain anything corresponding to the index-labels of higher-level functors. A theory of semantic structures may be judged by two standards. The first is whether it can tell us, from the structure which it assigns to each complex expression, whether or not the latter has a meaning; the second, whether, given a complex expression which has a meaning, that meaning may be elucidated by recourse to the structure assigned to the expression. We cannot, of course, expect a structural description to tell us the meanings of *simple* constituent expressions, only to identify them. Given their meanings, however, it should enable us to explain the meaning of the whole expression by showing us the manner in which they are combined.

A theory which met the first standard would be one in which it is impossible to write down nonsense, but there are two obstacles in the way of applying this test to the Fregean grammar. First, it cannot be applied to a theory of semantic structures in isolation. In order to know what expressions of a given natural language a certain semantic structure will yield, we must be able to say what phonographic structure or structures are correlated with it and, in order to know what expressions may be substituted for each of the category-symbols which it contains, we must know which are the simple expressions of the language and how each is to be categorized.

The second obstacle is that the intuitions even of native-born speakers of a natural language are a poor guide to the borderline between sense and nonsense. Intuition has its place in guiding our first steps in the formulation of a theory: there are many cases in which there would either be unanimous agreement that a certain expression had a meaning or that it did not. This could rightly lead us to rule out some theories from the start. But there will inevitably also be many borderline cases, in which our theory will have to guide our intuitions and not our intuitions our theory. Furthermore, while it is relatively easy to distinguish nonsense from what is *contingently* false, it is much more difficult to distinguish nonsense from what is *logically* false; yet, so far as this first standard is concerned, the theory must only exclude the former while leaving room for the latter.

Nevertheless, the experience which has been gained since 1879 in 'translating' formulas of predicate logic into sentences of natural languages gives us an informal idea of what may be expected from the Fregean categorial grammar, since any complex expression whose structure can be represented in predicate logic can also be represented in the grammar. If, then, we follow Frege in classifying a proper noun like 'John' as belonging to category N and 'not', as a sign of negation, to category : NN, it should be open to us to make the substitution 'not John' in the structure ': NN N', which is generable in the grammar. Further, by rule A1 this is a structure of category N, so we ought to be able to substitute 'not John', *salva congruitate*, for 'John' in the sentence 'Mary gave a present to John'. Again, suppose that the verb 'yawn' is categorized as : NN. Then (ignoring tense), we should be able to substitute it for both occurrences of : NN in the structure ': NN : NN N', obtaining e.g. 'yawned yawned John'; but whatever assumptions we care to make about the phonographic structure for English corresponding to ': NN : NN N', it will be impossible to turn 'yawned yawned John' into an acceptable English expression.

If nonsensical expressions can be obtained so easily from structures generated by the grammar, it seems that we can confidently conclude that it will not meet the first standard which has been proposed. But Frege was quite aware of these and similar difficulties. His solution was to lay down, in addition to the categorization of each expression of the language, provisions respecting its use which made examples like 'Mary gave a present to not John' and 'John yawned yawned' automatically *false*, while yet not excluding them as nonsense. He thus took his stand on the very ground where intuition, as a weapon which might be deployed against him, is of least efficacy.

Another way in which the Fregean grammar can be made more selective is to increase the number of basic categories. Suppose, for example, that we follow Ajdukiewicz in distinguishing two basic categories, S (sentence) and N (name), and while still categorizing 'John' under N, put 'not' under : SS and 'yawn' under : SN. Then we can no longer obtain 'not John', since ': SS N' is not a derivable structure, nor 'yawned yawned John', for neither is ': SN : SN N'. However, if negation belongs to category : SS, conjunction must belong to category : : SSS, and then, as ': : SSS N N' is not derivable, expressions like 'John and Mary' must be rejected as nonsense too. Many people to whose intuitions the first result is agreeable will find the second repugnant.

It is not my purpose to settle here the question how many basic categories are needed for a grammar applicable to natural languages. The immediate moral to be drawn is rather that an independent justification is required for our choice of basic categories, which does not rely upon an appeal to its effects with regard to the borderline between sense and nonsense. Frege, of course, did offer a justification for restricting himself to a single basic category and the nature of his argument shows that the issue is properly a philosophical one, to be decided *a priori*.

The first standard sets a necessary, but not a sufficient, condition for meeting the second. Here, too, a theory of semantic structures cannot be judged in isolation; we must first have a method of determining the meaning of the simple expressions and, in addition, an account relating their meanings to those of complex structures composed from them. Yet, even in the absence of these complements, there is indirect confirmation that the Fregean grammar marks a significant advance upon previous theories towards attaining the second standard.

The source of this assurance is logic. Logic may advertize itself as interested in the validity of arguments rather than in the meanings of propositions, but arguments are composed of propositions and the logical consequences which may be validly drawn from a proposition depend ultimately upon its meaning, not upon whether it happens to be true or false. Though they may not depend upon *every* aspect of its meaning, in talking of the 'logical force' of a proposition we are appealing to an aspect of its meaning which is of such fundamental importance that it is doubtful whether any other aspect could survive were not this one presupposed.

In the first instance, logicians classify *patterns* of argument as valid or invalid, a particular valid argument being one which exhibits a valid pattern. In order to do so, they must form hypotheses about the structures of the propositions from which particular arguments are composed. If, then, these structures successfully distinguish valid from invalid argument-patterns, such structures must tell us something about the logical force of propositions which exhibit them and, hence. also about their meanings.

It is to Frege's grammar that we owe the two most definitive advances in logic since the end of the middle ages. First, it allows us to analyze propositions in more than one way. This is reflected in the possibility of alternative derivations of the same structure, to which attention has been drawn several times in Sections II and III. Logic demands this facility, because many instances of a given argument-pattern will have a more complex structure than the argument-pattern can itself show, for if it did, any simpler structures would thereby be excluded as instances. Second, by distinguishing between functors of second-level and expressions of basic categories, Frege was able to solve at a single blow the problems which had so greatly exercised medieval logicians concerning arguments from premisses involving, as we should now say, multiple quantification. The details have been chronicled by historians of logic and need not be repeated here (cf. Geach, 1962, Ch. 4; Barth, 1971). While we may not be able to claim confidently that Fregean categorial grammar can be supplemented so as to meet the two standards for a theory of semantic structures, therefore, we can be quite sure that no grammar will meet them unless it both allows for alternative derivations of the same structures and distinguishes between the levels of functors.

It is evident that the distinction between levels of functors presupposes that between complete and incomplete expressions, but it is not so obvious that we cannot consistently espouse the latter distinction without also being committed to the former. It would indeed be possible to have a categorial grammar limited to basic categories and first-level functors; we need only to amend clause 2 of the definition of 'category' in Section I by restricting $\beta_1 \ldots \beta_m$ to basic categories. In a similar way, an upper level to the functors allowed in the grammar can be set at any point.

What *is* impossible is to allow Frege's method of analysis, i.e. that an expression of any known category may be removed from another of known category to yield an expression of a new category computable from those already known, and yet reject the distinction of levels. The point must be laboured, for whereas the difference between complete and incomplete expressions has been almost universally adopted by logicians and increasingly finds favour among linguists, I know of no logic book (save Bourbaki), nor any work on linguistics, which correctly explains the distinction between first and second-level functors.

Ajdukiewicz's classification of the quantifiers as belonging to category : SS is a typical example (1935, pp. 15–18); although logicians do not usually deploy a categorial apparatus, the same implicit classification is commonplace amongst them. Church, for example, says that if Γ is well-formed and a is an individual variable, then $\wedge_a \Gamma$ is well-formed (1956, p. 159). As Geach has pointed out,

> This ... entirely leaves out of account the most important feature of such operators – the fact that they "bind" variables. Ajdukiewicz's theory would show no difference e.g., between the syntax of "For some x, John loves x" and of "For some x, John loves z." In my opinion, the trouble arises from trying to assign categories to "for some x"

and the bound variable "x" *separately*. As the vernacular equivalent "John loves *somebody*" might suggest, "for some x" and "x" are just two pieces of a logically unitary expression, like a German separable verb; "for some x" and "z" form no expression at all, and so "For some x, John loves z" is ill-formed (Ajdukiewicz, 1967, p. 635, n.**).

It is scarcely better to stipulate, as Hilbert and Ackermann do, that if $\Gamma(x)$ is any formula in which the variable x occurs as a free variable, then $\bigwedge_x \Gamma(x)$ and $\bigvee_x \Gamma(x)$ are also formulas (1950, p. 66), for this merely introduces a categorial ambiguity into the notion of a formula. If formulas are expressions of a basic category, then the quantifiers must be one-place first-level functors, and the stipulation that they must contain a certain constituent is purely arbitrary, nor could we understand how the two xs in $\bigwedge_x \Gamma(x)$ were related. If, on the other hand, free variables are not constituents in a formula but, like our italicized minuscules, mark argumentor-places in a functor, then the '$\Gamma(x)$' which is called a formula is a functor, while the '$\bigwedge_x \Gamma(x)$' and '$\bigvee_x \Gamma(x)$' which are called formulas are complete expressions. Furthermore, the 'x' in the first formula is not the same sign as the 'x' in the second, first 'free', then 'bound', but rather the sign of an argumentor-place in the first and an index-label in the second.

Frege once described this assimilation of second to first-level functors as 'the grossest confusion possible'. This is especially interesting, because his solution to the problem of inferences involving multiple quantification turned upon distinguishing second-level functors from expressions of *basic* categories, not from first-level functors, so that it would have been very understandable if, elated by his discovery, he reserved his harshest criticism for the mistake which had for so long blocked a solution to this problem. Yet I believe that his prophetical reprimand was rightly directed: the kinship between second-level functors and expressions of basic categories, exhibited by the alternative category-symbols for the former, shows that their assimilation, while a serious mistake, is not an entirely stupid one. The persisting confusion between first and second-level functors, however, shows that the very distinction between complete and incomplete expressions has still not been fully grasped. It is a remarkable tribute to the perspicuity of Frege's notation that in spite of this almost total misunderstanding, it has carried its many users through the dangerous shoals of logic without serious formal errors. In the application of Fregean grammar to natural languages and in philosophical logic, however, the initial mistake has begotten many more.

A final objection which may be raised against Fregean grammar is that it is far *too* powerful for the analysis of natural languages. The latter, it may be urged, are more limited in two respects: first, in that simple expressions are rarely, if ever, functors of more than three argumentor-places; and, second, that they probably contain no functors of higher then third level. Even if these suppositions are correct, they give no great cause for concern. Functors with an unlimited number of argumentor-places will still be needed in order to preserve the full range of alternative derivations for sentences containing connectives, of which a sentence may contain any number; for although the final line of the derivation may contain no functor of more than three argumentor-places, some of the intermediate lines may demand functors with a much larger number. On the second point, if no functors of higher than third-level are required, we can easily curtail the grammar by amending the definition of 'category' so as to exclude them.

The whole objection is in any case premature, for in the absence of any extensive functorial analysis of natural languages we simply do not know, at present, how largely the resources of Fregean grammar must be tapped in order to describe them. Through failure to understand the distinction of levels, these resources still await exploitation. Apart from quantifiers, even the second level remains almost totally unexplored, while the discerning reader of current linguistic literature will find many examples of a confusion between third and first-level functors which affords an instructive parallel to that between second-level functors and expressions of basic categories.

APPENDIX: SUMMARY OF RULES

(S)-Rules

(S1) $\Gamma :_m \alpha \beta_1 \ldots \beta_m \Delta \to \Gamma :_m \alpha \beta_1 \ldots \beta_m \beta_m \ldots \beta_1 \Delta$
where **if** Γ is empty **then** the βs are to be written in italicized minuscules **else if** $:_m \alpha \beta_1 \ldots \beta_m$ is written in unitalicized minuscules **then** the βs are to be written in italicized minuscules **else** the βs are to be written in italicized minuscules

(S2) $\Gamma :_m \alpha \beta_1 \ldots (:_n \gamma \delta_1 \ldots \delta_n)_i \ldots \beta_{m \Sigma} \beta_m \ldots (:_n \gamma \delta_1 \ldots \delta_n \delta_n \ldots \delta_1)_i \ldots$
$\qquad \qquad \qquad \qquad \qquad \qquad \qquad \qquad \qquad \ldots \beta_1 \Delta \to$
$\Gamma :_m \alpha \beta_1 \ldots (:_n \gamma \delta_1 \ldots \delta_n)_i \ldots \beta_{m \, \delta_1 \ldots \delta_n \Sigma}$
$\qquad \qquad \beta_m \ldots (:_n \gamma \delta_1 \ldots \delta_n \delta_n \ldots \delta_1)_i \ldots \beta_1 \Delta$

FREGEAN CATEGORIAL GRAMMAR 281

where $\delta_1 \ldots \delta_n$ are to be written in the same typography in the consequent as in the antecedent.

(S3) if $m > 1$
then $\Gamma :_m \alpha \beta_1 \ldots (:_n \gamma \delta_1 \ldots \delta_n)_i \ldots \beta_{m \Pi \delta_1 \ldots \delta_n \Sigma} \beta_m \ldots (:_n \gamma \delta_1 \ldots$
$\ldots \delta_n \delta_n \ldots \delta_1)_i \ldots \beta_1 \Delta \to \Gamma :_{m-1} \alpha \beta_1 \ldots \beta_{i-1} \beta_{i+1} \ldots \beta_{m \Pi \Sigma}$
$\beta_m \ldots (:_n \gamma \delta_1 \ldots \delta_n \delta_n \ldots \delta_1)_i \ldots \beta_1 \Delta$
if $m = 1$
then $\Gamma : \alpha :_n \gamma \delta_1 \ldots \delta_n {}_{\delta_1 \ldots \delta_n} :_n \gamma \delta_1 \ldots \delta_n \delta_n \ldots \delta_1 \Delta \to$
$\to \Gamma :_n \gamma \delta_1 \ldots \delta_n \delta_n \ldots \delta_1 \Delta$

provided that, if (S3) is used in a derivation, Γ contains a higher-level functor within whose scope the functor, to which the rule is applied, falls; and where $\delta_1 \ldots \delta_n$ in the consequents are to be written in majuscules.

(R)-Rules

(R1) $\alpha \, \varphi_m \ldots \varphi_1 \to \alpha^{\beta_i} \varphi_m \ldots \varphi_i \, \beta \, \varphi_{i-1} \ldots \varphi_1$
where α, β and α^β are to be written in majuscules and all other labels in italicized minuscules

(R2) $\alpha^{\delta_j} \, \varphi_m \ldots \varphi_i \, \psi_n \ldots \delta \ldots \psi_1 \, \varphi_{i-1} \ldots \varphi_1$
$\quad \alpha \, \varphi_m \ldots \varphi_i \, \psi_n \ldots \psi_1 \, \varphi_{i-1} \ldots \varphi_1$
$\quad \vdots$
$\quad \beta_{\rho_1 \ldots \rho_r} \, \varphi_m \ldots \varphi_i \, \gamma \, |\psi_n \ldots \psi_1, \rho_r \ldots \rho_1| \, \varphi_{i-1} \ldots \varphi_1$
$\beta_{\rho_1 \ldots \rho_r} \, \varphi_m \ldots \varphi_i \, \gamma^{\delta_{j+k-1}} |\psi_n \ldots \delta_{j-(i-1)} \ldots \psi_1, \rho_r \ldots \delta_k \ldots \rho_1|$
$\hfill \varphi_{i-1} \ldots \varphi_1$

where α, β, γ, α^δ and γ^δ are written in majuscules, the ρs in unitalicized minuscules and all other labels in italicized minuscules.

(R3) $\alpha^{\delta_j{}^{\varepsilon_h}} \varphi_m \ldots \varphi_i \, \psi_n \ldots \delta^{\varepsilon_h} \ldots \psi_1 \, \varphi_{i-1} \ldots \varphi_1$
$\quad \alpha^{\delta_j} \varphi_m \ldots \varphi_i \, \psi_n \ldots \delta \ldots \psi_1 \, \varphi_{i-1} \ldots \varphi_1$
$\quad \vdots$
$\quad \beta^{\zeta_1}{}_{\rho_1 \ldots \rho_r} \varphi_m \ldots \varphi_i \, \gamma \, |\psi_n \ldots \delta_{j-(i-1)} \ldots \psi_1, \rho_r \ldots \delta_k \ldots \rho_1|$
$\hfill \varphi_{i-1} \ldots \varphi_1$
$\beta^{\zeta_1 \varepsilon_k}{}_{\rho_1 \ldots \varepsilon \ldots \rho_1} \varphi_m \ldots \varphi_i \, \gamma \, |\psi_n \ldots \delta^{\varepsilon_h}_{j-(i-1)} \ldots \psi_1, \rho_r \ldots \delta^{\varepsilon_h}_k \ldots \rho_1|$
$\hfill \varphi_{i-1} \ldots \varphi_1$

where α^δ, β^ζ, γ, $\alpha^{\delta\varepsilon}$ and $\beta^{\zeta\varepsilon}$ are written in majuscules, ε and the ρs in unitalicized minuscules and all other labels in italicized minuscules

Rule (Q)

(Q) $\quad \Gamma\, \alpha\, \lambda_m \ldots \lambda_1\, \Delta$
$\quad\quad\quad |\ \alpha\, \varphi_m \ldots \varphi_1$
$\quad\quad\quad |\ \vdots$
$\quad\quad\quad |\ F(\varphi_m, \ldots, \varphi_1)$
$\quad\quad \Gamma\, F(\lambda_m, \ldots, \lambda_1)\, \Delta$

where: 1. α is any majuscule label;

2. $\alpha\, \varphi_m \ldots \varphi_1$ is derived from α by one application of (S1);

3. $F(\varphi_m, \ldots, \varphi_1)$ is any structure to which (S2) and (S2) are inapplicable;

4. $\lambda_m, \ldots, \lambda_1$ are determined by the following

λ-procedure: **begin**
integer φ, λ, a, b, c; **integer** i, j, m;
Boolean procedure index (a_j); **integer** a_j;
    if a_j is written in unitalicized minuscules **then** index: = **true**
    else index: = **false**
    integer procedure cat (b, a_j); **integer** b, a_j; **comment**
    procedure not specified here, but uses the rules
      A1: $\alpha^\beta\, \beta \to \alpha$
      A2: if $\beta\, \gamma \to \alpha$ then $\beta\, \gamma^\delta \to \alpha^\delta$
      A3: if $\beta^\zeta\, \gamma \to \alpha^\delta$ then $\beta^{\zeta\varepsilon}\, \gamma \to \alpha^{\delta\varepsilon}$
where b is set to the first of each pair before the arrow, a_j to the second and 'cat' to the third. If the procedure does not yield a result for 'cat', cat: = $\emptyset$;
Boolean procedure syncoh (b, a_j); **integer** b, a_j;
    if cat (b, a_j) = $\emptyset$ **then** syncoh: = **false else** syncoh: = **true**
A: **begin** i: = m; j: = 1 **end**
B: **begin** b: = a_j; c: = a_j **end**
C: **if** b = φ_i **then**
    begin j: = j + 1;
      if index (a_j) **then**
        begin print "λ"; print i; print " = "; print c;

$$j := j - 1 \text{ end}$$
$$\text{else if syncoh } (b, a_j) \wedge \text{cat } (b, a_j) = \varphi_i \text{ then}$$
$$\text{begin } c := [c \sqcup a_j]; \ b := \text{cat } (b, a_j); \text{ goto } C$$
$$\text{end}$$
$$\text{else begin print "}\lambda\text{"; print i; print " = "; print}$$
$$c; j := j - 1; \text{ goto } E \text{ end}$$
end
D: else begin $j := j + 1;$
   if index (a_j) *then goto* D
   else begin $c := [c \sqcup a_j]; \ b := \text{cat } (b, a_j); \text{ goto } C$
   end
end
E: if $i = 1$ then goto G
   else begin $i := i - 1; \ F: j := j + 1;$
   if index (a_j) **then goto** F **else goto** B
G: end

ACKNOWLEDGMENTS

To Professor P. T. Geach I am indebted for first setting my feet on the path of categorial grammar and for constant encouragement to persevere in it; to Mr. R. M. White for a better understanding of 'the great works of Frege'; and to Dr. D. H. Sleeman and Mr. B. S. Marshall for invaluable assistance in formulating the λ-procedure.

The University of Leeds

BIBLIOGRAPHY

Ajdukiewicz, Kazimierz, 1935, 'Die syntaktische Konnexität', *Studia Philisophica* **1**, 1-27.
Ajdukiewicz, Kazimierz, 1967, 'On Syntactical Coherence', *The Review of Metaphysics* **20**, 635-647 (part I of (1935), translated by P. T. Geach from the Polish text in Ajdukiewicz, *Jezyk y Poznawie*, Warsaw, 1960).
Barth, Else Margarete, 1971, *De Logica van de Lidwoorden in de traditionelle Filisophie*, Universitaire Pers Leiden, Leiden.
Bourbaki, N., 1954, *Eléments de mathématique*, première partie, *Les structures fondamentales de l'analyse*, livre 1, *Théorie des ensembles*, Hermann et Cie., Paris.
Church, Alonzo, 1956, *Introduction to Mathematical Logic*, vol. 1, Princeton University Press, Princeton, New Jersey.

Fitch, Frederic Benton, 1952, *Symbolic Logic: An Introduction*, The Ronald Press Company, New York.
Frege, Gottlob, 1879, *Begriffsschrift*, L. Nebert, Halle.
Frege, Gottlob, 1893, *Grundgesetze der Arithmetik*, vol. 1, H. Pohle, Jena.
Geach, Peter Thomas, 1962, *Reference and Generality*, Cornell University Press, Ithaca.
Geach, Peter Thomas, 1970, 'A Program for Syntax', *Synthese* **22**, 3–17.
Geach, Peter Thomas, 1972, re-print of (1970) in D. Davidson and G. Harman (editors), *Semantics of Natural Language*, Reidel, Dordrecht.
Hilbert, D. and Ackermann, W., 1950, *Principles of Mathematical Logic*, Chelsea Publishing Co., New York. Translation from the second German edition, 1938.
Quine, Willard Van Orman, 1951, *Mathematical Logic*, Harper and Row, New York.
Russell, Bertrand, 1913, *Principia mathematica*, Cambridge University Press, Cambridge.
Russell, Bertrand, 1919, *An Introduction to Mathematical Philosophy*, Allen and Unwin, London.
Wittgenstein, Ludwig, 1922, *Tractatus Logico-Philosophicus*, Routledge and Kegan Paul, London. Re-printed with a new translation, 1961.

MARIAN PRZEŁECKI

A MODEL-THEORETIC APPROACH TO SOME PROBLEMS IN THE SEMANTICS OF EMPIRICAL LANGUAGES

Model theory may be regarded as a modern form of logical semantics. It deals with relations between languages and what these languages speak about, i.e. with typical semantical relations. What is characteristic of the model-theoretic approach might be briefly put as follows: it is a theory of the relations between formalized languages and their models. All languages studied in model theory are formalized ones, and the fragments of reality they speak about are conceived of as certain set-theoretic entities called *models* (structures, relational systems). The model-theoretic investigations have thus far been restricted to mathematical languages almost exclusively. But this seems to be an unjustifiable limitation. Empirical – that is, non-mathematical – languages also seem to constitute a proper object of the model-theoretic studies. Any empirical language can be formalized – if need be. And that what it speaks about may always be thought of as a certain model; the set-theoretic concept of model is broad enough to cover all fragments of reality.

In what follows I shall give an example of a model-theoretic approach to some problems concerning empirical languages. The object of our analysis will be the procedure of extending a given empirical language by introducing into it new extralogical constants. And our main concern will be with the problem of *interpretation* of the language considered. I shall try to give an account of that interpretation by means of some simple model-theoretic concepts. The main ones are: the concept of *model of a set of sentences*, and the concepts of *extension* and of *elementary extension of models*. The analysis is meant as an example of application of these concepts to the semantics of empirical languages. Our considerations will be confined to empirical languages of the simplest kind: languages which can be formalized within first-order predicate calculus (with identity), and which contain only predicates as their extralogical constants.

Let L_1 be such a language, and $r_1, ..., r_n$ its extralogical predicates. A *model* (semi-model, realization) of language L_1 is any sequence

$$\mathfrak{M}_1 = \langle U, R_1, ..., R_n \rangle$$

where U, the *universe* of model $\mathfrak{M}_1$, is a non-empty set, and $R_1, ..., R_n$ are *relations* on U having the same number of arguments as the corresponding predicates of L_1. Each model of L_1 determines a possible interpretation of this language: it represents a fragment of reality which L_1 can speak about. The fundamental model-theoretic notion is that of a *sentence's being true in a model of L_1*. If all sentences of a set K_1 are true in model $\mathfrak{M}_1$, we say that $\mathfrak{M}_1$ is a model of the set K_1, and write $\mathfrak{M}_1 \in M(K_1)$. Two other model-theoretic concepts which will be made use of in our analysis are those of extension and elementary extension of models. They will be symbolized as follows: $\mathfrak{M}_1 \subset \mathfrak{M}_1'$ ($\mathfrak{M}_1'$ is an extension of $\mathfrak{M}_1$), and $\mathfrak{M}_1 < \mathfrak{M}_1'$ ($\mathfrak{M}_1'$ is an elementary extension of $\mathfrak{M}_1$).

Now, the actual interpretation of language L_1 is usually conceived of as given by a single model of L_1, called its *intended* (or *proper*) model. It represents that fragment of reality which L_1 does *actually* speak about. Thus an interpreted language is usually identified with a couple $\langle L_1, \mathfrak{M}_1 \rangle$. However, with respect to a large class of languages – including all empirical ones – such a concept of interpretation seems to be too restrictive. The different pragmatical factors which decide what an empirical language actually speaks about do not determine its intended model in a unique way. The notorious vagueness of all empirical terms reflects this feature. What is determined by these factors is not a single model $\mathfrak{M}_1$, but rather a certain *class of models* M_1, containing more than one member. If that class fulfils certain conditions – if it is a non-empty and proper subclass of the class of all models of L_1, language L_1 may be regarded as an *interpreted* (though, of course, ambiguously interpreted) language. Thus, in what follows, we shall identify an interpreted empirical language with a couple $\langle L_1, M_1 \rangle$, where M_1 is a class of models which fulfils the conditions mentioned. We shall here also accept the usual assumption that the meaning of predicates $r_1, ..., r_n$ is governed by a *set of postulates*, say P_1. So all the intended models of L_1 are assumed to be models of set P_1: $M_1 \subseteq M(P_1)$. The converse, however, cannot hold. If L_1 is to be an empirical language, the class M_1, which determines its interpretation, must be a proper subclass of the class $M(P_1)$: $M_1 \nsubseteq M(P_1)$. And it must be isolated from the latter by some non-verbal procedures – e.g. by the so-called ostensive definitions of certain predicates of L_1. In consequence, postulates P_1 (and their logical consequences) are the only sentences of L_1 whose truth in models of the class M_1 is guaranteed in advance – by the very way the class M_1 is determined (Przełecki, 1969).

Now, let us suppose that language L_1, interpreted in the way described, is extended to a language L_2 by adding new extralogical predicates q_1, ..., q_m. Let P_2 be the set of postulates for the new predicates. Models of language L_2 are sequences of the kind: $\mathfrak{M}_2 = \langle U, R_1, ..., R_n, Q_1, ..., Q_m \rangle$. By $\mathfrak{M}_{2|1}$ we shall denote the fragment of model $\mathfrak{M}_2$ corresponding to language L_1, i.e. the model of L_1 obtained from $\mathfrak{M}_2$ by eliminating from it the denotations of predicates $q_1, ..., q_m$. How is the interpretation of language L_2 – given by a class of models M_2 – to be accounted for? Our analysis will be restricted to situations in which L_2 is to be a 'conservative' extension of L_1. It seems that in such situations the interpretation of L_2 must satisfy two conditions: (1) it must determine the interpretation of new predicates $q_1, ..., q_m$ in accordance with postulates P_2; (2) it must preserve the existing interpretation of old predicates $r_1, ..., r_n$ as given by class M_1. The first condition simply demands that every model $\mathfrak{M}_2$ of class M_2 be a model of postulates P_2: $\mathfrak{M}_2 \in M(P_2)$. But the second condition might be understood in different – more or less rigorous – ways. They can be accounted for by means of the model-theoretic concepts introduced above. The most rigorous explication requires that the fragment of any model $\mathfrak{M}_2$ of class M_2 corresponding to language L_1 be identical with some model $\mathfrak{M}_1$ of class M_1: $\mathfrak{M}_1 = \mathfrak{M}_{2|1}$; the more liberal one demands that it be an elementary extension of the latter: $\mathfrak{M}_1 < \mathfrak{M}_{2|1}$; and the most liberal explication requires – instead of elementary – a simple extension only: $\mathfrak{M}_1 \subset \mathfrak{M}_{2|1}$. We arrive thus at three different definitions of class M_2:

D1 $M_2 = \{\mathfrak{M}_2 : \mathfrak{M}_2 \in M(P_2)$ and $\mathfrak{M}_1 = \mathfrak{M}_{2|1}$, for some $\mathfrak{M}_1 \in M_1\}$,

D2 $M_2 = \{\mathfrak{M}_2 : \mathfrak{M}_2 \in M(P_2)$ and $\mathfrak{M}_1 < \mathfrak{M}_{2|1}$, for some $\mathfrak{M}_1 \in M_1\}$,

D3 $M_2 = \{\mathfrak{M}_2 : \mathfrak{M}_2 \in M(P_2)$ and $\mathfrak{M}_1 \subset \mathfrak{M}_{2|1}$, for some $\mathfrak{M}_1 \in M_1\}$.

Now, what are the consequences of these definitions? In the case of D1, the former interpretation of language L_1, as given by M_1, is preserved in the strictest sense. The universe of discourse and the denotations of predicates $r_1, ..., r_n$ are under the present interpretation, as given by M_2, exactly the same as before. According to D2 and D3, the universe of discourse may at present contain some new elements. However, within the old part of the universe, the predicates $r_1, ..., r_n$ are interpreted in the same way as before. Only in the new part of the universe they may be interpreted differently. In the case of D2, these are differences which are inexpressible in language L_1, and hence unnoticeable for the user of that language. As $\mathfrak{M}_1 < \mathfrak{M}_{2|1}$, the predicates $r_1, ..., r_n$ must in $\mathfrak{M}_{2|1}$ be interpreted in such a way which does not change the truth-value of any statement of L_1 containing these predicates. All that was true about them in $\mathfrak{M}_1$ (and expressible in language L_1) must remain true in $\mathfrak{M}_{2|1}$. No such restriction is implied by D3.

We may, I think, assume that there are actual scientific procedures of extending empirical languages which conform to the patterns just described. In situations corresponding to D1, the extension of a given language does not involve any extension of its former universe. Such situations seem to be characteristic of all definitional extensions. In situations corresponding to D2 and D3, the extension of language L_1 is accompanied by some extension of its universe. In the case of D2, the new objects must be of the same kind as the old ones – they must possess the same properties (expressible in language L_1) as the elements of the former universe. In the case of D3, the new objects may differ from the old ones. L_1 may here be extended to L_2 by postulating the existence of certain essentially new kinds of objects. (E.g. L_1 may be a language which speaks about macro-objects only, while L_2 may refer to micro-objects as well.) There seem to occur in actual scientific practice situations of all the kinds mentioned.

If the definitions D1, D2 and D3 are to provide L_2 with an adequate interpretation, the postulates P_2 are bound to fulfil certain conditions. What these conditions amount to depends on some general semantical assumptions as to the nature of the language considered. In what follows, we shall assume that L_2 is a meaningful language independently of any empirical findings. Experience may decide only whether a given sentence of L_2 is true or false, but not whether it is meaningful or meaningless. Its meaningfulness must be guaranteed in advance. Now, L_2 may be treated as an interpreted, and hence meaningful, language only if the class M_2 is a non-empty one: $M_2 \neq 0$. We shall thus require that this fact be guaranteed in advance. In the case of D1, the statement: $M_2 \neq 0$ reads, in an explicit formulation, as follows:

(*) For some $\mathfrak{M}_1 \in M_1$ there is some $\mathfrak{M}_2 \in M(P_2)$ such that $\mathfrak{M}_1 = \mathfrak{M}_{2|1}$.

Now, the truth of (*) is independent of experience if, and only if, the following general statement holds as well:

C1   For every $\mathfrak{M}_1 \in M(P_1)$ there is some $\mathfrak{M}_2 \in M(P_2)$ such that $\mathfrak{M}_1 = \mathfrak{M}_{2|1}$.

Postulates P_1 are the only sentences whose truth in models of M_1 is guaranteed in advance; so if we want to know in advance that (*) is true, we must know that C1 is true as well. In the case of D2 and D3, the corresponding conditions read as follows:

C2   For every $\mathfrak{M}_1 \in M(P_1)$ there is some $\mathfrak{M}_2 \in M(P_2)$ such that $\mathfrak{M}_1 < \mathfrak{M}_{2|1}$.

C3   For every $\mathfrak{M}_1 \in M(P_1)$ there is some $\mathfrak{M}_2 \in M(P_2)$ such that $\mathfrak{M}_1 \subset \mathfrak{M}_{2|1}$.

If the definitions D1, D2, and D3 are to characterize the interpretation of L_2 in a proper way, the postulates P_2 must fulfil the conditions C1, C2 or C3 respectively. What do the conditions amount to? All of them are semantical in nature. But in contrast to C1, the conditions C2 and C3 have simple syntactical counterparts:

C2' $Cn(P_2) \cap L_1 \subseteq Cn(P_1)$,
C3' $Cn(P_2) \cap A_1 \subseteq Cn(P_1)$,

where L_1 symbolizes the set of all sentences of the language L_1, and A_1 symbolizes the set of all purely universal sentences of that language. C2' states that all sentences of L_1 which follow from P_2 follow from P_1. C3' asserts that all purely universal sentences of L_1 which follow from P_2 are consequences of P_1. C2' is the standard condition of non-creativeness of the set P_2 (with regard to the set P_1). It is entailed by the condition C1, but it does not entail the latter. (C2' entails C1 only in certain special cases, e.g. if P_2 consists of purely universal sentences of L_2, or if P_1 implies a 'condition of finiteness', limiting to some n the number of all individuals.)

Now, the postulates which are actually accepted in the procedure of extending a given language might not meet the above requirements. The new predicates are often characterized by the whole set of axioms of an empirical theory, and such a set cannot satisfy any of the conditions mentioned above. There arises then a problem of isolating from the actually given set of postulates – let us call it S_2 – a set P_2 satisfying those conditions. One way in which this can be done may be briefly described as follows. The set of postulates P_2 is here characterized by two conditions. On the one hand, P_2 must be a set weak enough to meet the corresponding condition of non-creativeness: C1, C2', or C3'. On the other hand, it must be a set strong enough to include all meaning postulates 'contained' in the set S_2. This is a rather vague requirement, which can be made precise in more than one way. One of the possible explications reads as follows. All the sentences of L_1 which are entailed by S_2, but which cannot be entailed by P_2 (because of the latter's non-creativeness) belong – in the case of D1 and D2 – to the set: $Cn(S_2) \cap L_1 - Cn(P_1)$, and – in the case of D3 – to the set: $Cn(S_2) \cap A_1 - Cn(P_1)$. So, the set P_2 should be strong enough to yield – together with the sets mentioned – a set equivalent to S_2. In the case of D1 and D2, this requirement amounts to the following condition:

A. $\quad Cn((Cn(S_2) \cap L_1 - Cn(P_1)) \cup P_2) = Cn(S_2);$

and in the case of D3, to the condition:

B $\quad Cn((Cn(S_2) \cap A_1 - Cn(P_1)) \cup P_2) = Cn(S_2).$

Thus P_2 may be characterized as a set which – for given P_1 and S_2 – fulfils:
(1) in the case of D1 – conditions C1 and A;
(2) in the case of D2 – conditions C2' and A;
(3) in the case of D3 – conditions C3' and B.

The problem of uniqueness and existence with regard to the characteristics given above has been examined. No pair of the conditions mentioned determines the set P_2 in a unique way. The problem of existence of sets P_2 has a negative answer in the case (1) and (2), and a positive one in the case (3). It may be shown that for some sets P_1 and S_2 there is no set P_2 satisfying conditions C1 and A, and no set P_2 satisfying conditions C2' and A. On the other hand, for every set P_1 and S_2 there can always be constructed a set P_2 which fulfils conditions C3' and B.

University of Warsaw

BIBLIOGRAPHY

M. Przełecki, *The Logic of Empirical Theories*, Routledge & Kegan Paul, London, 1969.

I. I. REVZIN

METHODOLOGICAL RELEVANCE OF LANGUAGE MODELS WITH EXPANDING SETS OF SENTENCES

One of the very fruitful notions introduced in the theory of generative grammars is that of *deep structure*. The deep structure of a sentence, as distinguished from the *surface structure*, has to reflect the relation of a given surface structure to structures which underlie it in the derivation process. The essence of generation consists in a recursive enumeration of the set of all possible sentences of a language. Recently, some followers of Chomsky tend to understand generation in a different way – as a synthesis of a set of sentences departing from the given sense. Therefore, the notion of deep structure was identified with the semantic structure of a sentence. This identification was facilitated by some views expressed by Chomsky himself. Already in 1959, in his review of a book of Skinner (in *Language* 35(1959)), Chomsky criticized the behaviorist conception of the language acquisition in the childhood, and suggested as an alternative the hypothesis of the innate character of language categories. This rebirth of Kantianism has been criticized already by some scholars (for instance, by H. Hiż and Mrs. El. Charney) stating that it is not a necessary implication of the Chomskian theory.

Here we want to produce another argument: to build a generative grammar which is also associated with an idea of deep structure but which seems to be consistent with the learning theory.

We analyse a language $L = \langle V, \alpha, \beta, P, \rho \rangle$, where V is a *vocabulary* (interpreted as consisting of *word-tokens*), α – a finite set of *exemplary* sentences, β – a finite set of *prohibited* strings, P – a partition of V into two subsets D (*deictic* signs) and $V-D$. ρ is a relation defined on α: $f\rho g$ meaning that the sense of the sentence f is *included* in the sense of sentence g.

We denote the set of words belonging to $V-D$ and contained in a string or sentence f through $\{f\}$. A sentence g is called a *subsentence* of f if (1) $g\rho f$, and (2) for each $x \in V-D$ from $x \in \{g\}$ it follows that $x \in \{f\}$. Thus g is a subsentence of f if $g\rho f$ and some words contained in f are omitted or/and replaced by words from D.

We say that x *semidominates* y in f, and write $x \to y(f)$, if for each subsentence g of f such that $y \in \{g\}$ it holds that $x \in \{g\}$. We say that x *dominates* y in f, and write $x \to y(f)$, if (1) $x \neq y$ and $x \to y(f)$ and (2) there exists no z such that $z \neq x$, $z \neq y$, and $x \to z(f)$ and $z \to y(f)$.

A sequence $x_1, \ldots, x_k$, where each $x_i \in \{f\}$, is called a *group* in f, if for each $i, j (1 \leq i, j \leq k)$ it holds $x_i \to x_j(f)$ and $x_j \to x_i(f)$. Let X and Y be groups in f. We say that $X \to Y(f)$, if there exist $x \in \{X\}$ and $y \in \{Y\}$ such that $x \to y(f)$.

We establish $X \Rightarrow Y(f)$, if $X \neq Y$ and there exists no Z ($Z \neq X$ and $Z \neq Y$) such that $X \to Z(f)$ and $Z \to Y(f)$.

We investigate the *graph* $G = \langle \Sigma(f), \Rightarrow \rangle$, where $\Sigma(f)$ is the system of groups in f, and $\Rightarrow$ is the relation of domination defined above.

Proposition 1. The graph G is *connex* if, and only if, there exists at least one word x in f such that for each subsentence g of f it holds $x \in \{g\}$.

A *syntagma* with the nucleus X in f is the sequence of groups dominated by X.

Now we give a simple example. The sentence *The boy plays in the very large room* has the following subsentences:

> *The boy plays in the very large room*; *The boy plays in the large room*; *The boy plays in the room*; *The boy plays there*; *He plays in the very large room*; *He plays in the large room*; *He plays in the room*; *He plays there*; *He plays.*

There is only one non-trivial group: *in the room*. The relation of domination contains the pairs: ⟨*plays, the boy*⟩, ⟨*plays, in the room*⟩, ⟨*room, large*⟩, ⟨*large, very*⟩. A syntagma with the nucleus *room*: *the very large room*.

Under the 'deep structure' of a sentence we understand here a sequence where each group of the sentence is represented together with the deictic word replacing this group in the subsentences. For the sentence cited the 'deep structure' reads as follows:

> *(That boy he) plays (there in the very large – such – room)*.

Since the elements of D admit of two different approaches, the relation ρ has two interpretations:

(a) such words as *he*, *there*, etc., denote variables. Then the relation ρ can be interpreted logically: the corresponding terms fulfill the predicate of a propositional function.

(b) such words as *he, there,* etc., are token-reflexive (Reichenbach) or egocentric (Russell) signs which realize – in the same sense as the categories of person and tense, being the main grammatical vehicles of a sentence – the orientation of the speaker and hearer in time and space with respect to the communication act. Then ρ is a relation of inclusion of 'deep structures'.

Our definition of 'deep structure' may be supported by this second interpretation. It can be corroborated by the following linguistic arguments.

(1) Egocentric signs originate in gesticulation. Most scholars who have investigated the problem of language origins think that human language has developed two different channels of animal communication: the acoustic and the kinetic one, into one complex system. The 'explanatory power' of grammar is not diminished if it reflects some traces of such combination of the two channels.

(2) Investigations of language acquisition in childhood and their farsighted interpretation given by Vygotsky showed that such a combination of the two channels can be detected also here.

(3) This logically reconstructed 'deep structure' can be found in many languages as a regular phenomenon of 'surface structure'. In other languages (as in most European ones) it becomes visible in emphatic speech, word isolation, or simply in moments of hesitation; for instance: *This man here he knows her that wicked girl.*

(4) Deictic meanings can be transfigured into the logical ones. Chomsky's followers would analyse the article in the two sentences (a) **The** *lion is a carnivorous animal,* and (b) **The** *lion is nice* as representing absolutely different deep structures, since in (a) it corresponds to the universal operator (for each x), and in (b) to the iota-descriptor (the only x). We suggest an original deictic meaning for the definite article: 'all these objects in this given communicative situation'. In (b) this meaning is transparent. Since the situation is a specific one, the singular is relevant, and the resulting meaning is 'this only object in this given situation'. In (a) the enunciation is valid for any communicative situation, the singular being irrelevant, and the resulting meaning is 'all objects'. We suggest to analyse this meaning as a secondary one.

Now we turn to the suggested representation of 'deep structure'. It can be interpreted as reconstructing of a 'kernel' which yields the sentence in question. The same representation can be used in order to obtain new sentences. We introduce the following definition.

Let f be a sentence. The notion of *super-sentence* is defined recursively: (a) each sentence f is a super-sentence of itself; (b) let g be a supersentence of f and h be obtained from g by replacing a group X through a syntagma Y with the nucleus X from a sentence g'. Then h is a super-sentence of f.

Proposition 2. If the graph G is a tree and g is a subsentence of f, then f is a super-sentence of g.

It is natural to assume that the inverse is also valid: if g is a super-sentence of f, the graph G is a tree, then f is a sub-sentence of g. It means that when generating new sentences according to the definition of super-sentence, we ascribe a sense to them, i.e. we assume that their sense is constructed out of the senses of underlying sentences.

We denote by $\ulcorner f \& g \urcorner$ a sentence which is obtained from f and g by adding a special element of D. By definition, the operation '&' is possible only if there is $x \in D$ in the 'deep structure' of f, and y in the 'deep structure' of g such that $x = y$ (we consider only the simplest situation). For example, let f be *The man lives in the village* (with the 'deep structure': *This man here he lives there in that village*), and let g be: *I was born there* (with an identical 'deep structure'). The operation '&' is possible, and yields: *The man lives in the village where I was born*. There is much historical and typological evidence that such a way of production is plausible (cf. the work of Benveniste on the origin of relative clauses: *Problèmes de linguistique générale*, Paris 1966).

So far the set of well formed ('grammatical') sentences could be expanded only on the basis of syntactic operations. Now we shall investigate another possibility.

We say that x belongs to the same category as y, if there exist strings g and h such that $gxh \in \alpha$ and $gyh \in \alpha$, and there exist no strings g and h such that $gxh \in \alpha$, and $gyh \in \beta$ or $gxh \in \beta$ and $gyh \in \alpha$. We call a sentence f *semiadmissible* if it is obtained from a well formed sentence by replacing a word with a word belonging to the same category.

We have called such sentences semiadmissible because their proper admissibility must be tested. Our definition admits of an indetermination. A semiadmissible sentence can turn out to be prohibited. We shall consider an abstract example. Let the vocabulary consist of three words $V = \langle x, y, z \rangle$, $\alpha = \langle xyz, zx, yx \rangle$, $\beta = \langle xzy \rangle$. Then z belongs to the same category as y, y belongs to the same category as z, and it follows from the definition that xzy is a semiadmissible sentence, and in the same time it is prohibited.

Here an oracle is assumed who tests semiadmissible sentences for their well-formedness. It is a trying aim to investigate a situation, when the intervention of an oracle leads to a redefinition of the sets α and β. Let us consider the situation when α remains unchanged and β is reduced. It corresponds to the strategy of pure romanticism in the history of literature. The strategy of classicism can be described as follows: α is reduced, β is enlarged. When investigating the process of language acquisition we are interested in situations when both sets α and β are gradually enlarged.

Even if the sets α and β remain unchanged, our model does not exclude that two different users of the same language may reveal different categorisations of one and the same word x, because they part from different words y and z such that $gxh \in \alpha$ and $gxh \in \alpha$ for the first speaker, and $g'xh' \in \alpha$ and $g'zh' \in \alpha$ for the second speaker. Thus this model also provides for an explanation of the fact that two speakers of the same language sometimes fail to understand each other.

An analogous situation arises when new sentences are coined in poetry. Here is an example of lines by Dylan Thomas:

boys
Stabbing, and herons, and shells
That speak seven seas...

The words *seas* and *languages* may belong to one category when the point of departure is sentences like *He knows many languages* and *He knows many seas*. But a presupposition that each speaker tests all pairs of sentences is a too strong one, and it does not surprise anybody if a speaker of English maintains that *Shells speak seven seas* is not a well formed sentence in English. The language competence can be considered neither perfect nor complete and uniquely determined.

We come to the conclusion that both in syntax and in semantics we can work with models which do not suppose anything which is beyond ostensive learning. The only thing which must be taken into account is an ability of orientation in space and time since such an ability is actually presupposed in all operations with deictic words.

Institute of Slavonian and Balkanistic Studies, Moscow

LIANA SCHWARTZ

A NEW TYPE OF SYNTACTIC PROJECTIVITY: SD-PROJECTIVITY

1. Introduction

Dependence and subordination relations occurring in a string were translated from natural language into mathematical language and were subjected to some syntactic restrictions, thus giving rise to various types of *projectivity*.

Projective strings coincide with well-formed strings in a natural language and the last are specific to the scientific language. Thus syntactic dislocations which appear as a deviation from the normal aspect of a string may be studied taking as reference projective property. S. Marcus (see [3], [4]) has proposed a method to establish a hierarchy of syntactic dislocations degrees, taking into account various types of projectivity and some syntactic regularities (such as the existence of a single regent, the small distance between one word of a string and its regent, etc.).

Our aim is to introduce and to study a new type of syntactic projectivity. We shall prove that this projectivity generalizes some of the most important types of projectivity and may be found in several strings which do not satisfy Tesnière's property (see [8]), i.e. in those strings in which there is at least one word having more regents. Such situations are illustrated by the figure of speech named repetition, by the so-called supplementary predicative element in Romanian (see [9]) by homogeneous terms, by repetition of the same grammatical category.

2. SD-projectivity

Let V be the finite vocabulary of a language L and $F = \{f; f = x_1 x_2 \ldots x_n, x_i \in V, 1 \leqslant i \leqslant n\}$ be the set of all strings over V. Let $S = \{f; f = x_1 x_2 \ldots x_n, f$ is a structured string$\} \subseteq F$, be the set of all structured strings over V, that is the set of strings endowed with three types of binary relations: $\leqslant$, $\rightarrow$ and $\Rightarrow$, which represent the natural order relation, the dependence relation and the subordination relation between the words of a string, respectively.

A NEW TYPE OF SYNTACTIC PROJECTIVITY

Let $f = x_1 x_2 \ldots x_n$ be a string belonging to S.

f is said to be *SD-projective* (*SD* from syntactic dislocations) if for any i, $1 \leq i \leq n$, there exist more values of j, and $x_j \to x_i$ (x_i depends upon x_j), such that there exists at least a value k of j, and $x_k \rightrightarrows x_l$ (x_l is subordinate to x_k) for any l, min $(i, k) \leq l \leq$ max (i, k).

Let us give some examples of *SD-projective* strings.
The dependence relations are marked by the arrows.

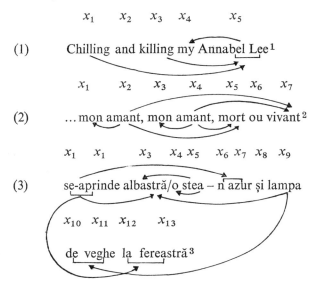

Let us analyse in the first string the word *Annabel Lee*. We may observe that it depends upon the words *chilling* and *killing*, but only *killing* may be taken as an $x_k(x_1 \to x_5, x_3 \to x_5$ hence $i = 5$, $j \in \{1, 3\}$ and $k = 3$) because $x_5 \to x_4$ and $x_3 \to x_5$ implies $x_3 \rightrightarrows x_4$, x_4 being the single word situated between x_3 and x_5. *Chilling* is not an x_k since the words *and* (x_2) and *killing* (x_3) are not subordinate to it, in spite of their places between *chilling* and *Annabel Lee*.

We can show in the same manner that the second word *amant* represent an x_k for the words *mort* and *vivant* if we do not take into account the particle *ou* (example (2)) and the words *se-aprinde* (kindle) and *stea* (star) are both one x_k for the word *albastră* (blue), example (3)).

Proposition 1. The projectivity in the sence of Lecerf and Ihm implies *SD-projectivity*. The convverse is not true.

If we note by ⇒ the logical implication, we obtain the following sequence of implications: monotonical projectivity ⇒ strong projectivity ⇒ projectivity in the restricted sense ⇒ projectivity in the sense of Lecerf and Ihm ⇒ *SD-projectivity* (see [2] and [6]).

We proved in another paper that there is no implication relation between *SD-projectivity* and quasi-projectivity [6].

We shall now introduce a new type of syntactic projectivity generalising *SD-projectivity*.

Let $f = x_1 x_2 \ldots x_n$ be a string belonging to S.

f is said to be *SD-quasi-projective* if for any value of i and m, $1 \leqslant i$, $m \leqslant n$, $i \neq m$, can exist more values of the index j where $x_j \to x_i$ and $x_j \to x_m$ such that there exists at least a value k of j and $x_k \rightrightarrows x_l$ for any l, $\min(i, m) \leqslant l \leqslant \max(i, m)$.

We have proved in [6] the following propositions:

Proposition 2. *SD-projectivity* entails *SD-quasi-projectivity*. But the converse is not true.

Proposition 3. *SD-quasi-projectivity* generalises quasi-projectivity.

Proposition 4. Let f be a string not satisfying Tesnière's property. If f is S_1-projective, then f is *SD-projective*. The converse is not true.

Proposition 5. Let f be a string not satisfying Tesnière's property. If f is S_2-projective, then f is *SD-projective*, but the converse is not true.

We have also proved [6] that there is no relation between *SD-projectivity* and S_3-projectivity.

S_1, S_2, S_3 projectivities were introduced by Šreider [7] and they exhibit different types of syntactic regularities.

3. SD-SIMPLE STRING

Among the strings endowed with Tesnière's property there are the simple strings which present the greatest syntactic regularity. We have also observed a similar situation among the strings not satisfying Tesnière's property. For this reason we shall introduce the notion of *SD-simple string*, as a model of such strings.

Let f be a structured string $f = x_1 x_2 \ldots x_n$.

f is said to be an *SD-simple string* if the following three conditions are fulfilled: (a) f does not possess Tesnière's property, (b) there exist two indexes i and j, $1 \leq i, j \leq n$ and only two, such that x_i and x_j depend upon no element of the string f, (c) for any integer k such that $i \neq k \neq j$, x_k is subordinate both to x_i and x_j.

Let us name x_i and x_j the centers of the string f.

An example of *SD-simple string* is given by

(4) 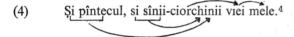Și pîntecul, si sînii-ciorchinii viei mele.[4]

It is obvious that the syntagms *și pîntecul* (and the abdomen) and *și sînii* (and the bosoms) are the centers of this string.

Let us observe there is no logical relation between simple string and *SD-simple string*, since the two definitions contradict each other.

Now we shall try to characterize the graph associated with *SD-simple string*, that is the graph whose vertices coincide with the elements of *SD-simple string* and whose arcs represent the dependence relations between the elements of such a string.

THEOREM 1. Let f be a structured string. If f is an *SD-simple string* with centers x_i and x_j, then its associated graph G_f has the following properties: (1) every vertex of G_f other than x_i and x_j is a terminal vertex to at least one arc of G_f; (2) no arc has its terminal vertex in x_i or in x_j; (3) G_f is connected without circuits; (4) x_i is a center for the subgraph $G_f - \{x_j\}$, while x_j is a center for the subgraph $G_f - \{x_i\}$.

The next theorem proves that the four conditions above are sufficient to come to the conclusion that given such a graph its associated string is an *SD-simple string*.

THEOREM 2. Let G_f be a graph of order greater than 2. G_f is associated with an *SD-simple string* f of centers x_i and x_j if the four conditions appeared in *Theorem 1* are fulfilled. Now it is easy to observe that the

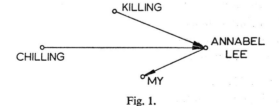

Fig. 1.

reunion between *Theorem 1* and *Theorem 2* gives a necessary and sufficient condition that a string be an *SD-simple string*.

The proofs of theorems above may be found in [6].

Figure 1 represents the graph associated with the string (1) (from Section 2). The order of the words in the plotting above is from the left to the right, that is, their natural order in a string.

It is obvious that this graph satisfies all the properties of Theorem 1.

We shall represent in the plane the graph G_f associated with an *SD-simple string*, in the same way as Marcus constructed the graph associated to a simple string [2]. We obtain the Figure 2.

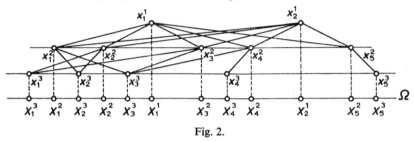

Fig. 2.

On the first line there are the centers of the string f represented by the points x_1^1, x_2^1. On the second line there are those elements of f which depend upon x_1^1 and x_2^1. These elements are disposed from the left to the right in their linear order in f and such that: (a) for every word $a < x_1^1$ ($x_1^1 < a$) the corresponding vertex is situated at the left (at the right, respectively) with respect to the projection line $x_1^1 X_1^1$; (b) for every word $a < x_2^1$ ($x_2^1 < a$) the corresponding vertex is situated at the left (at the right, respectively) with respect to the projection line $x_2^1 X_2^1$; (c) for every word a, $x_1^1 < a < x_2^1$, the corresponding vertex is situated between the projection lines $x_1^1 X_1^1$ and $x_2^1 X_2^1$. On the third line there are represented those elements on f which depend upon at least one element represented on the second line. The disposal of these elements is made in the manner described above. This procedure to draw the graph G_f may be continued until the elements of f are exhausted.

This plotting is more complicated than the former, but we shall use it to prove the next theorem (see [6]).

THEOREM 3. *If f is an SD-simple string with centers x_i and x_j, then f is SD-projective if and only if for any integer s, $i \neq s \neq j$, such that there can* exist many values of k, where $x_k \to x_s$, $1 \leqslant k \leqslant n$, $i \neq k \neq j$, there exists at least one value h of k, $1 \leqslant h \leqslant n$, such that there is no term between x_h and x_s.

An example of *SD-simple string* which is *SD-projective* is given by (4), as we observe in Figure 3.

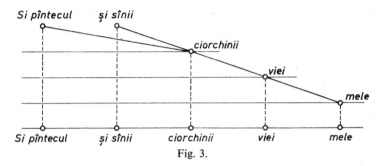

Fig. 3.

It would be interesting to find an example of *SD-simple* string which be not *SD-projective*.

Center for Dialectical and Phonetical Research,
Bucharest

NOTES

[1] Edgar Allan Poe, 'Annabel Lee'.
[2] J. Prevert, 'Fille d'acier' (cf. [5]).
[3] Ch. Baudelaire, 'Paysage' (translated into Romanian by Al. Phillipide; cf [1]).
[4] Ch. Baudelaire, 'Les Bijoux' (translated into Romanian by Al. Hodoş; cf [1]).

BIBLIOGRAPHY

[1] Ch. Baudelaire, *Florile răului* (ed. by Geo Dumitrescu), EPLU, Bucureşti, 1967.
[2] S. Marcus, *Algebraic Linguistics*, Academic Press, New York-London, 1967.
[3] S. Marcus, *Poetica matematica*, Editura Academiei, Bucureşti, 1970.
[4] S. Marcus, 'Trois types d'écarts syntaxiques et trois types de figures dans le language poétique', *Cahiers de linguistique théorique et appliquée* **6** (1970).
[5] J. Prevert, *Paroles*, NRF, Paris, 1949.
[6] L. Schwartz, 'A New Type of Syntactic Projectivity' (forthcoming).
[7] J. A. Šreider, 'Svoistva projectivnostni jazyka', *Naučno techničeskaya informacija* **8** (1964).
[8] L. Tesnière, *Eléments de syntaxe structurale*, C. Klincksieck, Paris, 1959.
[9] *Gramatica limbii române*, Editura Academiei, Bucureşti, 1964 (2nd edition).

JÜN-TIN WANG

ON THE REPRESENTATION OF GENERATIVE GRAMMARS AS FIRST-ORDER THEORIES

As is well known, a generative grammar is considered by Chomsky as a set of rules that, in particular, generates the sentences of a language. In this paper I shall try to make it plausible that for each such generative rule system a corresponding first-order theory can be effectively constructed such that, roughly speaking, what can be generated by the grammar, can be logically derived within this theory. A grammar represented as a first-order theory will provide thus, above all, a deductive nomological explanation of the sentences of a language in the sense of theory of knowledge (Wang, 1971b). We shall treat at first the problem of representation of the well-defined context-free grammar and then that of the less formalized transformational grammar.

1. First-order theory

By first-order theories we mean theories which are formalized within the first-order predicate logic (Shoenfield, 1967). In other words, a first-order theory is a formal system such that
  (1) the language of this formal system is a first-order language;
  (2) the axioms of this formal system are the logical axioms of the first-order predicate logic and certain further axioms, called the nonlogical axioms, which, in our cases, correspond to the given grammatical rules;
  (3) the rules of this formal system are the logical rules of the first-order predicate logic exclusively.
  In order to specify a theory, we have only to specify its nonlogical symbols: function symbols and predicate symbols and its nonlogical axioms; everything else is given by the definition of a first-order theory.

2. Representation of context-free grammars as first-order theories

A context-free grammar is usually defined as a quadruple

$$G = \langle V_T, V_N, S, P \rangle,$$

where V_T is the finite set of terminal symbols; V_N is the finite set of nonterminal symbols and V_N is disjoint from V_T; S is a designated symbol in V_N; P is a finite set of rewriting rules all of the form

$$A \to w$$

with A in V_N and w in $(V_T \cup V_N)^*$.

The set $(V_T \cup V_N)^*$ contains all finite strings including the null string formed from the alphabet $V_T \cup V_N$.

A sequence of strings

$$w_1, \ldots, w_n \quad (n \geq 1)$$

is called a w-derivation of w' if and only if

(1) $w = w_1$ and $w' = w_n$ and
(2) for each $i < n$ there are strings t_1, t_2, B, t such that
   (a) $B \to t$ (i.e. a rewriting rule)
   (b) $w_i = t_1 B t_2$ and
   (c) $w_{i+1} = t_1 t t_2$.

Note that each rewriting rule in a context-free grammar G can be also expressed as follows:

$$A \to A_1 a_1 A_2 a_2 \ldots A_n a_n,$$

where $A, A_1, \ldots, A_n$ stand for nonterminal symbols and $a_1, \ldots, a_n$ for terminal symbols and some of A_i and a_j may be null.

Given now a context-free grammar G, we assign to each rewriting rule of G a nonlogical axiom as follows:

$$\bigwedge x_1 \bigwedge x_2 \ldots \bigwedge x_n (A_1(x_1) \wedge \ldots \wedge A_n(x_n)$$
$$\to A(x_1 a_1 x_2 a_2 \ldots x_n a_n)),$$

where we take $A, A_1, \ldots, A_n$ as unary predicates and $a_1, \ldots, a_n$ as individual constants; the symbols '$\wedge$' and '$\to$' are the sentential connectives: the conjunction sign and the implication sign.

Example. Let $V_T = \{\text{the, man, hit, ball}\}$ and $V_N = \{S, NP, VP, T, N, V\}$. A context-free grammar may consist of the following set of rewriting rules:

$$S \to NP\ VP$$
$$NP \to T\ N$$
$$VP \to V\ NP$$
$$T \to \text{the}$$
$$N \to \text{man}$$
$$V \to \text{hit}$$
$$N \to \text{ball}$$

From this set of context-free rules we obtain then the following nonlogical axioms:

$$\bigwedge x \bigwedge y(NP(x) \wedge VP(y) \to S(xy))$$
$$\bigwedge x \bigwedge y(T(x) \wedge N(y) \to NP(xy))$$
$$\bigwedge x \bigwedge y(V(x) \wedge NP(y) \to VP(xy))$$
$$T(\text{the})$$
$$N(\text{man})$$
$$V(\text{hit})$$
$$N(\text{ball}).$$

Given a context-free grammar G, let T_G denote the first-order theory thus obtained. We have proved then the following theorem (Wang, 1971a):

THEOREM 1. *For each K-derivation of a terminal string w in a given context-free grammar G, where K stands for any nonterminal symbol (category symbol) in G, there is a formal sentence $K(w)$, which is logically derivable in the corresponding first-order theory T_G.*

3. Representation of Transformational Grammars as First-Order Theories

The treatment of the representation of transformational grammars as first-order theories can not be so direct as in the case of context-free grammars. Since, as is well known, the domain of each grammatical transformation has been specified by Chomsky with the notion of structural description and a grammatical transformation is considered by him as a mapping of phrase-markers into phrase-markers, we have to consider first the notion of structural description or phrase-marker in which according to Chomsky the structural description of a generated sentence has been represented.

Epistemologically the phrase-marker of a given terminal string generated by a context-free grammar (or a context-free component of a transformational grammar) can be characterized by a set of formal sentences or formulae. In order to see this fact, it may be helpful to consider here an auxiliary formal system, called primitive formal system. In this primitive formal sytem we can get on the one hand another representation of context-free grammar in which a notion can be defined which will turn out to be equivalent to the notion of phrase-marker. On the other hand, this primitive formal system has a very clear and close relation to first-order theories.

3.1. *Primitive formal system*

A primitive formal system S is given as follows:

(1) The alphabet consists of
  (a) individual constants,
  (b) individual variables,
  (c) predicates,
  (d) the improper symbols: parenthesis and comma.
(2) Definition of *S-term* and *S-formula*
  (a) Definition of *S-term*
    (i) every individual constant and every individual variable is a *S-term*;
    (ii) if t_1 and t_2 are *S-terms*, then $t_1 t_2$ is a *S-term*;
    (iii) the *S-terms* consist exactly of those expressions generated by (i) and (ii).
  (b) Definition of *S-formula*
    (i) if P is a *n*-ary predicate and $t_1, ..., t_n$ are *S-terms*, then $P(t_1, ..., t_n)$ is a *S-formula*. (A *S-formula*, containing no individual variable, is called a *S-sentence*.)
    (ii) The *S-formulae* consist exactly of those expressions generated by (i).
(3) A finite list of axioms (i.e. certain *S-sentences*).
(4) Each rule is of the form

$$F_1 ..., F_m \to F_0,$$

where $F_i (i = 0, ..., m)$ stands for *S-formula*.
If every variable in a rule is substituted by a string of individual con-

stants (for the same variables the same strings of individual constants), we obtain then a rule instance. For a rule or rule instance the $F_1, \ldots, F_m$ are called the premises, F_0 the conclusion.

We can now define the notion of constructibility inductively:

(1) every axiom is constructible;

(2) if all the premises of a rule instance are constructible, the conclusion is also constructible;

(3) a *S-sentence* is constructible if and only if it is constructible by (1) and (2).

A construction (proof) in a primitive formal system is a finite sequence of one or more *S-sentences* such that each *S-sentence* in the sequence is either an axiom or constructible from the preceding *S-sentences* in the sequence.

3.2. *Representation of Context-Free Grammars in Primitive Formal Systems*

In this section we give for each context-free grammar a representation in a primitive formal system.

Given a context-free grammar G, we assign to each rewriting rule of G a rule in a primitive formal system as follows:

$$A_1(x_1), A_2(x_2), \ldots, A_n(x_n) \to A(x_1 a_1 x_2 a_2 \ldots x_n a_n),$$

where we take $A, A_1, \ldots, A_n$ as unary predicates and $a_1, \ldots, a_n$ as individual constants and some of A_i and a_j may be null.

The rules corresponding to the context-free rules have, furthermore, one particular restrictive property; in each rule every variable occurring once in the conclusion occurs exactly one time in one of the premises and, vice versa, each variable occurring once in one of the premises occurs exactly one time in the conclusion of the same rule. A system of rules each of which fulfills this condition is called a construction grammar of type C.

Example. For the context-free grammar given by the example in (2), the corresponding construction grammar of type C can be given as follows:

Rules: $NP(x), VP(y) \to S(xy)$
$T(x), N(y) \to NP(xy)$
$V(x), NP(y) \to VP(xy)$

ON THE REPRESENTATION OF GENERATIVE GRAMMARS

Axioms: T(the)
N(man)
V(hit)
N(ball)

With respect to this construction grammar we can get, for example, the following construction:

T(the)
N(man)
T(the)
N(ball)
V(hit)
NP(the ball)
NP(the man)
VP(hit the ball)
S(the man hit the ball).

In order to show that the phrase-marker of a sentence generated by a context-free grammar is essentially equivalent to a kind of structural description expressed by a system of statements we make a specialization in the notion of construction.

We call a construction dendritic (with respect to a given construction grammar of type C), if with the exception of the last *S-sentence*, every *S-sentence* in the sequence is used exactly once as premise for getting a latter *S-sentence*. The construction given in the example above is dendritic.

From a dendritic construction we can obtain a tree representation too. In this kind of tree we place the formal sentence which is constructed beneath those formal sentences which have been used as the premises of some rule in getting it. For the example considered above we have the following construction tree:

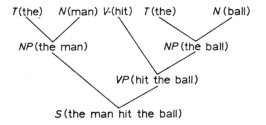

This tree may be compared with the phrase-marker of the sentence 'the man hit the ball':

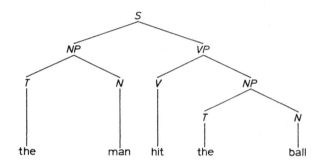

The analogy between the construction tree and the phrase-marker of a sentence is obvious. In fact, we shall prove in the Note to be presented at the end of this paper these interesting relationships, which may be given here as follows:

(1) For each K-derivation of a terminal string t in a given context-free grammar G, where K stands for any nonterminal symbol (category symbol) in G, there is a formal sentence $K(t)$ which is constructible in the corresponding construction grammar of type C.

(2) All the statements concerning the assignment of a grammatical category to a phrase, which are contained in the phrase-marker of a sentence t in a context-free grammar G, constitutes a dendritic construction for the formal sentence $S(t)$ in the corresponding construction grammar of type C.

Due to these relationships, the dendritic construction can be thus considered as a kind of structural description, which is equivalent to the phrase-marker of a sentence generated by a context-free grammar.

3.3. *Construction Grammar and First-Order Theory*

At the next step we want now to establish the relationship between construction grammars introduced above and first-order theories.

According to the definition of the primitive formal system it is obvious that the set of formulae of a primitive formal system is just a proper subset of the set of formulae of a first-order language. Furthermore, from the correspondence induced by a given context-free grammar between the rules and axioms in the construction grammar of type C and the nonlogical axioms of the first-order theory, it is easily to see, that each formal sentence, which is constructible in a construction grammar of type C, can be logically derivable in the corresponding first-order theory too. This can be sketched as follows.

Given a context-free grammar G, let C_G denote the corresponding construction grammar of type C and T_G the corresponding first-order theory. It holds then:

(1) If the formal sentence H is an axiom in C_G, then H is also a nonlogical axiom in T_G; H is then logically derivable in T_G.

(2) If the formal sentence H is in C_G constructible by the using of a rule instance, say,

$$F_1^0, \ldots, F_m^0 \to F_0^0,$$

then $H = F_0^0$. We can use the same substitution in the corresponding nonlogical axiom in T_G and obtain

$$F_1^0 \wedge \cdots \wedge F_m^0 \to F_0^0.$$

F_1^0 is obviously a logical consequence of the set of the formal sentences $\{F_1^0, \ldots, F_m^0, F_1^0 \wedge \ldots \wedge F_m^0 \to F_0^0\}$. H is thus in T_G logically derivable.

In particular, every formal sentence contained in a dendritic construction can be logically derived in the corresponding first-order theory. The conjunction of these very formal sentences contained in a dendritic construction is therefore in the same first-order theory logically derivable too. This very conjunction of the formal sentences contained in a dendritic construction can be thus taken as the representation of the phrase-marker of a sentence generated by a context-free grammar. This fact is important, since grammatical transformations in the sense of Chomsky begin with those phrase-markers which are generated by some context-free component.

3.4. Transformational Grammar as First-Order Theory

With the results thus far obtained, we can now formulate transformational grammars as first-order theories.

As is well known, the notion of grammatical transformation is formulated by Chomsky with the notion of analyzability (Chomsky, 1961). Given a phrase-marker Q of a terminal string t, t is analyzable as $(t_1, \ldots, t_n;$

$A_1, \ldots, A_n$) with respect to Q if and only if t can be subdivided into successive segments $t_1, \ldots, t_n$ in such a way that each t_i is traceable, in Q, to a node labelled A_i. This condition can be expressed, however, by the formal sentence

$$A(t_1 t_2 \ldots t_n) \wedge A_1(t_1) \wedge A_2(t_2) \wedge \ldots \wedge A_n(t_n),$$

where A stands, in general, for the symbol S (sentence). In this way we can use a single statement or formula to specify the domain of a grammatical transformation to the extent that Chomsky has achieved with the notion of analyzability.

It shall be noted that in the characterization of the notion of analyzability as proposed by Chomsky it has been sufficient to use only unary predicates such as $A, A_1, \ldots, A_n$ in forming statements. This is certainly a very special case, since we can generally use a set of statements or formulae built up from predicates or function symbols of any finite degree to characterize a condition. The using of multiplace predicates is at any rate one of the simple and natural ways to treat relations exhibited in a natural language (Wang, 1971c). This possible generalization of the notion of structural description can be naturally expressed in the first-order language.

With these remarks we can state now the principle to formulate each grammatical transformation, which according to Chomsky is a mapping of phrase-markers into phrase-markers, as a nonlogical axiom in the following way: We describe both the applied phrase-markers and the resulting phrase-markers, or parts of them, with formulae which occur

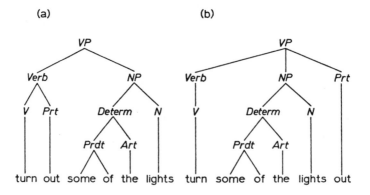

then respectively as premises and conclusion of an implication formula. This very implication formula can be then taken as the representation of the grammatical transformation in question.

As an example, we may consider the permutation transformation. According to Chomsky, it converts the phrase-marker (a) to the phrase-marker (b) (as indicated in the figure above).

For this transformation, which is specified by Chomsky as follows:

structural description: V, Prt, NP
structural change: $x_1 - x_2 - x_3 \rightarrow x_1 - x_3 - x_2$,

we might state the following nonlogical axiom:

$$\bigwedge x_1 \bigwedge x_2 \bigwedge x_3 (VP(x_1 x_2 x_3) \wedge V(x_1) \wedge Prt(x_2) \wedge NP(x_3)$$
$$\rightarrow \cdot VP(x_1 x_3 x_2) \wedge V(x_1) \wedge Prt(x_2) \wedge NP(x_3)$$
$$\wedge \bigwedge y(V(y) \rightarrow Verb(y)).$$

Here both the characterization of the phrase-marker to be applied and that of the phrase-marker to be derived are given by a set of formulae. This nonlogical axiom can be further simplified by cancelling certain redundant formulae and we obtain then

$$\bigwedge x_1 \bigwedge x_2 \bigwedge x_3 (VP(x_1 x_2 x_3) \wedge V(x_1) \wedge Prt(x_2) \wedge NP(x_3)$$
$$\rightarrow \cdot VP(x_1 x_3 x_2) \wedge \bigwedge y(V(y) \rightarrow Verb(y)).$$

In the description of the phrase-marker derived, we have used the formula $\bigwedge y(V(y) \rightarrow Verb(y))$, which in cases may be substituted by another more appropriate one, just in order to remind us that there is a new kind of relation existing between the node labelled '$Verb$' and the node labelled 'V'. This new relation must be justified both linguistically and epistemologically.

It seems evident that the same general principle stated above can be applied to other grammatical transformations as well. We obtain thus a set of nonlogical axioms which represents the transformational grammar in question and from which both the sentences of a language and their structural descriptions can be logically derived. The representation of a generative grammar as a first-order theory provide us thus with a deductive nomological explanation of the sentences of a language. Furthermore, it establishes a relation to the approach as proposed by Montague in his study of the structure of language (Montague, 1970), since it is obvious that for each set-theoretic formulation of syntax presented by Montague we can give an equivalent formulation in the language of predicate logic (Wang, 1971b).

4. NOTE

To show the relationships mentioned in (3.2) between a context-free grammar and its corresponding construction grammar we use the fact established by Chomsky (1963) that with each context-free grammar G we can associate a bracketed context-free grammar G' with the vocabulary of G and the new terminals $]$ and $[_A$, where A is any nonterminal symbol of G. Where G has the rule $A \to A_1 a_1 \ldots A_n a_n$, G' has the rule $A \to [_A A_1 a_1 \ldots A_n a_n]$. G' will generate a string t_1 containing brackets that indicate the constituent structure of the corresponding string t_2, formed from t_1 by deleting brackets, that would be generated by G.

We now proceed to define the notions to be used more formally. Let G be a given context-free grammar. We denote its corresponding bracketed context-free grammar by G'. Let V_T, V_N denote, as above, the set of terminal symbols and the set of nonterminal symbols (category symbols) of G respectively and let V_B denote the set of symbols $]$ and $[_A$, where A is any nonterminal symbol of G. Let w with or without indices denote elements in $(V_T \cup V_N \cup V_B)$ and t, u, v, with or without indices denote elements in $(V_T \cup V_N \cup V_B)^*$.

To characterize the strings generated by a bracketed context-free grammar G' we introduce the notion of well-formed bracketed expressions and their degrees as follows:

DEFINITION 1:

(i) $a_{i_1} a_{i_2} \ldots a_{i_n}$ is a well-formed bracketed expression of degree 0, where $a_{i_j} \in V_T$.

(ii) If $E_1, E_2, \ldots, E_n$ are well-formed bracketed expressions of degree $< s$, at least one of them being of degree $s - 1$, then $[_A E_1 E_2 \ldots E_n]$ is a well-formed bracketed expression of degree s, where $A \in V_N$.

It is easily to see that the strings generated by a bracketed context-free grammar G' are all well-formed bracketed expressions of degree $g \geqslant 1$. Each terminal string t generated by a bracketed context-free grammar G' can be then characterized as follows:

(1) $t \in (V_T \cup V_B)^*$;
(2) t is a well-formed bracketed expression of degree $g \geq 1$.

We now turn to define the notion of debracketing function on strings.

DEFINITION 2: Let d_0 be a mapping from $V_T \cup V_N \cup V_B$ to $V_T \cup V_N$ such that $d_0(w_i) = \begin{cases} w_i & \text{for } w_i \in V_N \cup V_T \\ e & \text{for } w_i \in V_B, \end{cases}$
where e denotes the empty word, i.e., $et = te = t$. The debracketing function d on strings is a mapping (homomorphism) from $(V_T \cup V_N \cup V_B)^*$ to $(V_T \cup V_N)^*$ such that

$$d(t) = d_0(w_1)\, d_0(w_2) \ldots d_0(w_n),$$

where $t = w_1 w_2 \ldots w_n$.

According to this definition it holds obviously that

$$d(t_1 t_2 \ldots t_m) = d(t_1)\, d(t_2) \ldots d(t_m).$$

With the notions thus far defined, we can now define a notion, which, roughly speaking, means that a terminal string in G, denoted by $d(t)$ with the indication that it may be obtained by the debracketization of the corresponding terminal string t in G', is of a certain grammatical category.

DEFINITION 3: The statement $A(d(t))$ holds in a context-free grammar G, if and only if

(1) $t_1 [_A t] t_2$ can be generated in G', where t_1 and t_2 may be null,
(2) $t_1 [_A t] t_2 \in (V_T \cup V_B)^*$,
(3) $[_A t]$ is a well-formed bracketed expression of degree $g \geq 1$.

With the help of this notion we can state the following theorem:

THEOREM 2: If $A(d(t))$ holds in a context-free grammar G, then $A(d(t))$ is in the corresponding construction grammar C_G constructible.

Proof. By induction on the degree g of well-formed bracketed expression $[_A t]$.

(1) $g = 1$. $[_A t]$ is therefore of the form $[_A a_{i_1} \ldots a_{i_n}]$, where $a_{i_j} \in V_T$ for $1 \leq j \leq n$. For this form of well-formed bracketed expression the associated rule in G' is $A \rightarrow [_A a_{i_1} \ldots a_{i_n}]$. To such a rule, which corre-

sponds to the rule $A \to a_{i_1} \ldots a_{i_n}$ in G, we have assigned by the construction an axiom of the form $A(a_{i_1} \ldots a_{i_n})$ in C_G. It is therefore constructible.

(2) $g > 1$. Suppose that this theorem holds for all well-formed bracketed expression of degree $s - 1$, where $s > 1$. Let $[_A t] = [_A E_1 E_2 \ldots E_n]$ be a well-formed bracketed expression of degree s. For $1 \leqslant h \leqslant n$, it is possible that

$$E_h = \begin{cases} [_{A_h} t_h] \\ a_{h_0}, \text{ where } a_{h_0} \in V_T. \end{cases}$$

Since $[_A t]$ is, however, a well-formed bracketed expression of degree s, there is at least one k ($1 \leqslant k \leqslant n$) such that $E_k = [_{A_k} t_k]$ and E_k is a well-formed bracketed expression of degree $s - 1$. There may exist $k_1, \ldots, k_m$ ($1 \leqslant k_1 < \cdots < k_m \leqslant n$) such that $E_{k_j} = [_{A_{k_j}} t_{k_j}]$ ($1 \leqslant j \leqslant m$) of degree g with $1 \leqslant g \leqslant s - 1$. We can therefore write down

$$[_A t] = [_A E_{k_1} a_{k_1} E_{k_2} a_{k_2} \ldots E_{k_m} a_{k_m}]$$
$$= [_A [_{A_{k_1}} t_{k_1}] a_{k_1} [_{A_{k_2}} t_{k_2}] a_{k_2} \ldots [_{A_{k_m}} t_{k_m}] a_{k_m}],$$

where $a_{k_j} \in V_T$ and may be null and at least one of E_{k_j} is not null.

$$d(t) = d(E_{k_1}) d(a_{k_1}) d(E_{k_2}) d(a_{k_2}) \ldots d(E_{k_m}) d(a_{k_m})$$
$$= d(t_{k_1}) a_{k_1} d(t_{k_2}) a_{k_2} \ldots d(t_{k_m}) a_{k_m}.$$

If $A(d(t))$ holds in a context-free grammar G, then $A_{k_j}(d(t_{k_j}))$ with ($1 \leqslant j \leqslant m$) holds in this grammar G too; because the conditions (1) and (2) in the Definition 3 are certainly fulfilled and by our construction in the proof, that $1 \leqslant g \leqslant s - 1$ holds, the condition (3) is equally fulfilled. For the well-formed bracketed expression $[_A [_{A_{k_1}} t_{k_1}] a_{k_1} [_{A_{k_2}} t_{k_2}] a_{k_2} \ldots [_{A_{k_m}} t_{k_m}] a_{k_m}]$, however, the associated rule in G' is $A \to [_A A_{k_1} a_{k_1} \ldots A_{k_m} a_{k_m}]$, which corresponds to the rule $A \to A_{k_1} a_{k_1} \ldots A_{k_m} a_{k_m}$ in G. To such a rule we have assigned by construction a rule of the form $A_{k_1}(x_1), \ldots, A_{k_m}(x_m) \to A(x_1 a_{k_1} \ldots x_m a_{k_m})$ in C_G. By the inductive hypothesis the formal sentences $A_{k_j}(d(t_{k_j}))$ with $1 \leqslant j \leqslant m$ are constructible, because the E_{k_j} are at most of the degree $s - 1$. The corresponding rule in C_G can be thus applied. We obtain $A(d(t_{k_1}) a_{k_1} \ldots d(t_{k_m}) a_{k_m})$, namely $A(d(t))$. Q.e.d.

To characterize the relation between the phrase-marker of a sentence generated by a context-free grammar and the dendritic construction we state, furthermore, the following theorem:

THEOREM 3: All the statements concerning the assignment of a grammatical category to a phrase contained in a phrase-marker of a sentence t in a context-free grammar G, namely $A_i(d(t_i))$ with $u_i[_{A_i}t_i]v_i = [_S t]$ and $[_S t] \in (V_T \cup V_B)^*$, constitutes a dendritic construction for the formal sentence $S(d(t))$ in the corresponding construction grammar C_G.

Proof. According to the Theorem 2, all the statements $A_i(d(t_i))$ in a context-free grammar G can be constructed in the corresponding construction grammar C_G. We need therefore just to show:

(3i) with the exception of $S(d(t))$, every formal sentence $A_i(d(t_i))$ in the sequence is used, in the construction of $S(d(t))$, exactly once as premise for the construction of a latter formal sentence $A(d(t'))$, where $u[_A t']v = [_S t]$.

(3ii) in the construction of $S(d(t))$, only $A_i(d(t_i))$ with $u_i[_{A_i}t_i]v_i = [_S t]$ occurs as premises.

To (3i). For each i with $A_i(d(t_i))$ fulfilling the condition that $u_i[_{A_i} t_i]v_i = [_S t]$, the expression $[_{A_i}t_i]$ occurs obviously in the expression $[_S t]$ at least one time. It can occur more times in $[_S t]$. If this is the case, then there exist $u_{i_1}, u_{i_2}, v_{i_1}$ and v_{i_2} such that (1) $u_{i_1}[_{A_i}t_i]v_{i_1} = [_S t]$, (2) $u_{i_2}[_{A_i}t_i]v_{i_2} = [_S t]$, (3) $u_{i_1} \neq u_{i_2}$ and $v_{i_1} \neq v_{i_2}$. Each occurrence of the expression $[_{A_i}t_i]$ in the expression $[_S t]$ can be therefore characterized by its u_i and v_i uniquely. Let g_i be the degree of $[_{A_i}t_i]$ and g be the degree of $[_S t]$, then from $u_i[_{A_i}t_i]v_i = [_S t]$, it follows that $g \geqslant g_i$.

Case 1: $g = g_i$. From $u_i[_{A_i}t_i]v_i = [_S t]$, it follows that $u_i = 0 = v_i$ and $[_{A_i}t_i] = [_S t]$. Therefore, $S = A_i$ and $t = t_i$. We need therefore not to consider this case.

Case 2: $g > g_i$. There exists then a well-formed bracketed expression $[_A t']$ of the degree $g_i + 1$ such that

$$[_S t] = u_i[_{A_i}t_i]v_i = u[_A t']v$$
$$= u[_A[_{A_{k_1}}t_{k_1}]a_{k_1}\cdots[_{A_{k_p}}t_{k_p}]a_{k_p}\cdots[_{A_{k_q}}t_{k_q}]a_{k_q}]v,$$

where u and v may be null and $[_{A_{k_p}}t_{k_p}] = [_{A_i}t_i]$ for some k_p; it holds then further that

(1) $u_i = u[_A[_{A_{k_1}}t_{k_1}]a_{k_1}\cdots[_{A_{k_{p-1}}}t_{k_{p-1}}]a_{k_{p-1}};$
(2) $v_i = a_{k_p}[_{A_{k_{p+1}}}t_{k_{p+1}}]a_{k_{p+1}}\cdots[_{A_{k_q}}t_{k_q}]a_{k_q}]v.$

The statement $A(d(t'))$ holds then in the context-free grammar G.

According to Theorem 2, it can be constructed in the corresponding construction grammar C_G by the application of the rule

$$A_{k_1}(x_1), ..., A_{k_p}(x_p), ..., A_{k_q}(x_q) \to A(x_1 a_{k_1} ... x_p a_{k_p} ... x_q a_{k_q}),$$

where $A_i(d(t_i))$ is used as one of the premises in the construction of $A(d(t'))$.

Since by $u_i[_{A_i}t_i]\,v_i$ the associated expression $[_A t']$ of the degree $g_i + 1$ is uniquely determined, the statement $A_i(d(t_i))$ characterized by the $u_i[_{A_i}t_i]\,v_i$ with $u_i \neq 0$ and $v_i \neq 0$ has been used, therefore, in the construction of $S(d(t))$, only one time as a premise in getting a latter formal sentence.

To (3ii). In (3i) we have seen that in the construction of $A(d(t'))$, where $u[_A t']\,v = [_S t]$ and u and v may be null, only those $A_{k_i}(d(t_{k_i}))$ with u_{k_i} and v_{k_i} such that

$$u_{k_i}[_{A_{k_i}}t_{k_i}]\,v_{k_i} = u[_A t']\,v = [_S t]$$

occurs as premises. (3ii) holds, therefore, in particular. Q.e.d.

Institut für Kommunikationsforschung,
Bonn

BIBLIOGRAPHY

Chomsky, N., 'On the Notion "Rule of Grammar"' in *Structure of Language and its Mathematical Aspects* (ed. by R. Jakobson), American Mathematical Society, Providence, 1961.

Chomsky, N., 'Formal Properties of Grammars' in *Handbook of Mathematical Psychology* (ed. by R. D. Luce, R. R. Bush and E. G. Galanter), Wiley, New York and London, 1963.

Montague, R., 'English as a Formal Language', in *Linguaggi nella societa e nella tecnica* (ed. by Bruno Visentini et al.), Milan 1970.

Shoenfield, J. R., *Mathematical Logic*, Addison-Wesley 1967.

Wang, J. T., 'On the Representation of Context-Free Grammars as First-Order Theories', *Proceedings of the 4th Hawaii International Conference on System Sciences*, 1971a.

Wang, J. T., 'Wissenschaftliche Erklärung und generative Grammatik', *6. Linguistisches Kolloquium, August 11–14, 1971 Kopenhagen* (ed. by K. Hyldgaard-Jensen), (forthcoming, 1971b).

Wang, J. T., 'Zum Begriff der grammatischen Regeln und strukturellen Beschreibung', in *Probleme und Fortschritte der Transformationsgrammatik* (ed. by D. Wunderlich), Hueber, 1971c.

INDEX OF NAMES

Ackermann, W. 3, 8, 273, 279, 284
Adjukiewicz, K. 172, 210, 213, 246, 272, 276, 278–279, 283
Anderson, A. R. VI, *3–28*
Apostel, L. 177–178
Arbib, M. A. 185, 187–188
Aristotle 46, 62

Bach, E. 151, 179
Bacon, Fr. 81
Bar-Hillel, Y. 106, 210, 222
Barth, E. M. 278, 283
Baudelaire, Ch. 301
Beliş, M. *65–77*
Bellert, I. 177–178
Belnap, N. 7–8, 14, 18–20, 22–24, 39
Bennett, J. 11, 16–25
Benveniste, E. 294
Berge, C. 226, 236
Berkeley, G. 20
Bernays, P. 273
Bertels, K. 177–178
Bierwisch, M. 225
Black, M. 177–178, 210
Bloom, S. L. 54
Blum, M. 125, 128
Bohnert, H. 108–109, 111, 113
Bolzano, B. 20
Boole, G. 83
Borel, E. 66, 77
Borko, H. 179
Bourbaki, N. 254, 278, 283
Brainerd, B. 175, 178
Braithwaite, R. B. 142
Bressan, A. *29–40*
Brouwer, L. T. 31
Buber, M. 119
Bunge, M. 115–116
Bush, R. R. 178, 316

Cajori, F. 82

Cantor, G. 20
Carnap, R. 30–32, 34, 40, 80, 115, 141–142, 165, 178, 185, 187, 209, 273
Cassirer, E. 199–200
Cauchy, A. L. 20
Chabrol, C. 178
Chafe, W. L. 158–159, 178
Chaitin, G. J. 126–128
Chao, Y. R. 177–178, 197, 200
Charney, E. 291
Chebyshev, P. I. 96, 98
Chomsky, N. 172, 177–178, 189, 196, 200, 214, 221, 223–225, 236, 291, 302, 304, 309–312, 316
Church, A. 4–6, 10, 55–62, 274, 278, 283
Clairaut, A. C. 78
Coffa, J. A. 23
Cohen, L. J. *78–82*
Cornman, J. W. 108–109, 113
Craig, W. 109
Cudia, D. F. 236
Curry, H. B. 4, 213, 221–222

Dahl, Ö. 150, 153, 162, 178
Davidson, D. 222, 284
Dedekind, R. 20
Demuth, O. 94, 98
Dennet, R. E. 199–200
Dijk, T. A. van VI, *145–180*
Döhman, K. 243–244
Dressler, W. 177, 179
Drubig, B. 153, 179
Dunn, J. M. 8, 9, 18, 23
Durkheim, E. 197

Edmundson, H. P. 174, 178–179
Egorov, D. F. 84, 94, 98
Einstein, A. 78

Feigl, H. 107
Feys, R. 4, 31, 213, 221

Fillmore, Ch. 158–159, 179
Finetti, B. de 142
Fischer, G. 243
Fischer, W. L. *181–188*
Fitch, F. B. 8, 222, 257, 284
Föllesdal, D. 237, 239, 244
Fort, M. K. 187
Fraassen, B. C. van 206
Frege, G. 5, 6, 27, 32, 47, 345, 260–262, 265, 276–279, 284
Frobenius, L. 199–200

Galanter, E. 178, 180, 316
Gallin, D. 30, 39
Gamkrelidze, Th. von 197, 200
Geach, P. 173, 179, 216, 221–222, 271, 278, 283–284
Gelb, I. J. 196, 200
Gentzen, G. 10, 25
Gernet, J. 199–200
Ginsburg, S. 226, 236
Glivenko, V. I. 83, 98
Golopenţia-Eretescu, S. VI
Goodman, N. 3
Goodstein, R. L. 85, 98
Gram, M. 107
Granet 199
Greimas, A. J. 158–159, 179
Gross, M. 178–179

Halle, M. 196, 200
Halliday, M. A. K. 158–159, 179
Harman, G. 284
Harms, R. T. 179
Harris, Z. S. 146, 170, 179
Harsanyi, J. C. 207
Hartmanis, J. 124, 126, 128
Heger, K. 158–159, 179
Heidolph, K. E. 225
Hempel, C. G. 107–109, 113, 118
Hendricks, W. 177
Hermes, H. 29, 40
Herschel, J. 81
Hesse, M. 177, 179
Heyting, A. 19
Hilbert, D. 20, 273, 279, 284
Hilpinen, R. VII, 107
Hintikka, K. J. VI, 107
Hiż, H. 291

Hume, D. 117, 131
Hocart, A. M. 198, 200
Hooker, C. A. 108–109, 111, 113
Husserl, E. 208
Hutchins, W. 225
Hylgaard-Jensen, K. 316

Ihm 297–298
Ihwe, J. 160, 177–179
Ioanid, M. VI
Isard, S. *189–195*
Isenberg, H. 177, 179
Ivanov, I. *196–200*

Jakobovits, L. A. 179
Jakobson, R. 196, 200, 221, 316
Jaynes, E. T. 74–75, 77
Jeffrey, R. C. 106

Karttunen, L. VII, 152, 177, 179
Kasher, A. 177, *201–107*
Katz, J. J. 177, 237, 244
Kay, M. 225
Kempe, A. B. 198, 200
Keynes, J. M. 67, 77
Khintchine, A. I. 66, 77
Kiefer, F. 225
Kleene, S. C. 56–57, 60, 62
Klemke, E. 107
Klibansky, R. 206
Kolmogorov, A. N. 83, 98, 126, 128
Kossovsky, N. K. *83–99*
Kotarbinska, J. VI, *208–212*
Kreisel, G. 55, 58, 62
Kripke, S. A. 31, 40
Kummer, W. 172, 177–179
Kuno, S. 151
Kuroda, S. Y. 177, 179
Kuryowicz, J. 200
Kyburg, H. E. 106, 112–113, 140, 142

Lakatos, I. 82
Lakoff, G. 153–154, 157, 161, 163, 177, 179, 207
Lamb, S. 225
Lang, E. 177, 179
Langacker, R. W. 151, 179
Laplace, P. S. 71, 74, 77
Lecerf 297–298

INDEX OF NAMES

Lehmann, R. S. 142
Lehrer, K. VI, *100–107*, 109, 113
Leibniz, G. W. 47
Lemmon, E. J. 31, 40
Lentin, A. 178–179
Lešniewski, S. 172
Levi, I. 107
Lewis, C. I. 16–18, 21–22, 25–26
Lewis, D. 207, 218–222
Lewis, H. A. *213–222*
Longuet-Higgins, Ch. *189–195*
Luce, R. D. 178, 207, 316
Lukasiewicz, J. 5
Lyons, J. 179

Mach, E. 29, 40, 118
Marcus, S. VII, 178, 180, 226, 236, 296, 300–301
Markov, A. A. 83, 99
Marshall, B. S. 283
Martin-Löf, P. 83, 99, 127–128
Mauss 197
Maxwell, G. 107
McCawley, J. D. 177
Mel'čuk, I. A. *223–225*
Meredith, C. A. 37, 40
Meyer, R. K. 8–9, 18, 23, 28
Mill, J. S. 81
Miller, G. A. 177, 180
Mises, R. von 126
Montague, R. 39, 312, 316
Musgrave, A. 82
Mönnich, U. 177, 180

Nagel, E. 178
Nauta, D. 177–178, 180
Needham, R. 200
Nelson, R. 62
Neumann, J. von 124–128
Newton, I. 20, 78, 80, 82
Niiniluoto, I. *108–114*
Nöbeling, G. 188

Onicescu, O. 66, 77
Ore, O. 236
Orman, G. *226–236*

Painlevé, P. 29, 30, 35, 40
Palek, B. 168–170, 172, 177, 180, *237–244*

Parsons, T. 221–222
Peirce, Ch. S. 115–116, 141
Peters, S. 151
Petöfi, J. S. 165, 177–178, 180
Pietarinen, J. 107
Poe, E. A. 301
Poincaré, H. 66, 70, 77
Popper, K. R. 79, 112, 114–117, 120
Postal, P. M. 117
Potts, T. VI, 221, *245–284*
Pouillon, J. 198, 200
Prevert, J. 301
Pribram, K. H. 180
Prior, A. N. 37, 40
Przelecki, M. *285–290*

Quine, W. V. 3, 27, 213, 216, 221–222, 237, 244, 268, 273, 284

Raiffa, H. 207
Ramsey, F. P. 109, 142
Reibel, D. A. 177, 179–180
Reichenbach, H. 164–165, 180, 192–193, 195, 293
Resher, N. 163, 180, 205–207
Revzin, I. I. *291–295*
Rieser, H. 177–178, 180
Rogers, H. 125, 128
Rohrer, Ch. 157, 178, 180
Ross, J. 177
Rosser, J. B. 29, 40
Routley, R. 23
Routley, V. 23
Rudner, R. 221
Russell, B. 42, 44, 47, 268, 284, 293

Sanders, G. A. 147–148, 180
Sanin, N. A. 84–87, 89, 91–92, 98–99
Scheffer, I. 108, 114, 221
Schelling, T. I. 207
Schnelle, H. 178, 180
Schoenfield, J. R. 302, 316
Schwarz, L. *295–301*
Scott, D. 31, 39–40
Sellars, W. 25
Settle, T. *115–120*
Sgall, P. 225
Shane, S. A. 177, 179–180
Shimony, A. 142

Short, D. 243
Simon, H. A. 127–128
Skinner, B. F. 291
Sleeman, D. H. 283
Slisenko, A. O. 94, 99
Smokler, H. 106, 142
Sommers, F. 48
Specker, E. 88, 99
Sreider, F. 298, 301
Stalnacker, R. 178, 180
Stearns, R. E. 124, 126, 128
Stechow, A. von 179
Stegmüller, W. 108–114
Stein, H. 62
Steinberg, D. A. 179
Stoianovici, P. 41–48
Suppes, P. 107, 120, 178
Suszko, R. 49–54
Swain, M. 106
Swinburne, R. G. 82
Szaniawski, K. *121–123*

Tarski, A. 5, 178, 189, 216
Tesnière, L. 158–159, 180, 296, 301
Thomas, D. 295

Thomas, W. T. *55–62*
Tratteur, G. *124–128*
Turing, A. 58, 62
Turner, V. W. 198, 200

Vickers, J. M. *129–142*
Vieru, S. VI
Visentini, B. 316
Vygotsky, L. S. 293

Wallace, J. 177, 180, 216, 222
Wang, H. 56, 62
Wang, J. T. *302–316*
Watkins, J. W. 197, 200
Weierstrass, K. 20
Weyl, H. 187
White, R. M. 283
Winograd, T. 195
Wittgenstein, L. 27, 47, 261, 284
Wunderlich, D. 178–180, 316

Zeeman, E. C. 185, 187
Zelinka, B. 185, 188
Zetlin, M. L. 198, 200
Zolkovsky, A. K. 224

INDEX OF SUBJECTS

analyzability (Chomsky) 309–310
anomaly, in science 78–82

Bedeutung 5–6, 12, 27
belief, change in 131, 141

categorial grammar
  Adjukiewicz-type 213, 218
  Fregean 245, 274–277, 280
category 245–252
  and functor 245–246
  and quantifiers 278–279
causal structure, of random processes 65–77
Church's thesis 55–62
coherence, of betting functions 130–132
combinatory logic 4, 213–221, 253
communication system 67–68
complexity, of computation 124–128
computability
  of functions 55–58
  and machines 58–61, 124–128
conceptual change 104–106
confirmation 79–82, 110–113
construction grammar 306–308
  and first-order theory 308–309
  dendritic construction 307
context 191
context-free grammar 226, 302–303
  and first-order theory 302–304
  and primitive formal system 306–308
  bracketed c.-f. g. 312–316

definite description 43–45
definitivization 152–153
determinism 119–120
disjunctive syllogism 8–11, 14–15, 23–26

entailment
  and relevant implication 6–8
  the Lewis paradox 16–26

  systems of e., *see* intensional logic
equivalence
  e. relation 182
  e. structure 182
evidence
  and utility 102
  rule of e. 101
extension
  of language 288–289
  of model 285–286

generative grammar 157, 173, 226, 291, 302–316
  see context-free grammar
grammar
  see categorial, context-free, generative, sentential, text-, transformational g.
graph 223, 226, 292, 299–300
  connex g. 292
  multigraph 227

identity, statement of 41–47
  and definite description 43–45
  and proper name 42–43
implication, see entailment
indeterminism 119–120
induction
  and anomaly 78–82
  and conceptual change 104–106
  and confirmation 79–82, 110–113
  and systematization 108–113
instauration 241
  types of i. 242–243
intension 27, 32, 218–219
  see Sinn
intensional logic 3, 6–27, 29–39
  and axiomatization of physics 29–30, 35
  and truth-values 12–15
  system E (Anderson) 6–8

system E† (Anderson) 6–8
system R (Anderson) 6–8
system R† (Anderson) 6–8
system MC^v (Bressan) 31–34, 38
system MC^v (Bressan) 37–38
interpretation 219, 286–287

law
 dynamical l. 66–68
 natural l. 117–120
 l. of large numbers 95–98
learning theory 291
lexeme 124, 175
logic
 see combinatory, intensional, and predicate-functor logic, sentential calculus, truth-value
lottery paradox 140–141

marker, see phrase-marker
meaning 223
 and text 224
 deictic m. 293
model theory 285–290

occasionality, of expressions 208–212

phrase-marker 154–155, 173, 304–311, 315
pragmeme 202
predicate-functor logic 213–221
presupposition 162–163, 191, 205
probability
 constructive p. 83–99
 objective p. 66–67, 115–120
 subjective 66, 74–75, 100, 104–106, 130–131
 and belief change 141
 and conceptual change 104–106
 and confirmation 79, 110
 and necessity 129, 141
 and propensity 115–120
 and random processes 66, 72–77
 and transparency 129, 131–132
projectivity, syntactic 295–301
pronominalization 150–151
propensity 115–120
proper name 42–46, 245

questions 121–123

random
 r. processes 67–77
 r. strings 126–127
reference 237
 and denotata 238
 and naming 237
 cross-reference 169–170
 referential identity 151

sentential calculus
 non-Fregean (system SCI) 49–54
sentential grammar 150, 168
 and text-grammar 148–149, 157
 s.g. for natural language 168–169
sign
 deictic 291, 293
 egocentric 293
Sinn 5–6, 12, 27
speech act 201–206
stability, of causal processes 69–70
structure
 causal s. 65–77
 deep and surface s. 155, 157, 171, 214, 224, 291–293
 equivalence s. 182
 phonographic s. 274
 semantic s. 153–154, 274
 tolerance s. 185
synonymy 183–185, 187, 240
 definition of s. 187
 models of s. 184
syntagma 224, 292, 299
systematization, of observational statements
 deductive 109, 11–114
 inductive 108–113

tense 192
testability 111–112
text 223–224
 see meaning
text-grammar 145–180
 tasks for t.-g. 150
 for natural language 168–169
 linguistic models of t.-g. 156–160
 logical models of t.-g. 160–167
 mathematical models of t.-g. 167–175

and S-grammar, see sentential grammar
theoretical term 108–109
theory
   first-order t. and grammar 302–316
   and systematization 108–113
   empirical triviality of t. 110–113
tolerance
   t. relation 185
   t. space 185
transformational grammar
   and first-order theory 304, 309–312
transparency, of probability measure 129, 131

truth
   definitions 189, 216–217
   as appropriateness 189–190, 195
truth-value
   intensional interpretation of t.-v. 12–15
   t.-v. calculus 5–6
   truth-valuations 50–53

utility
   u. of evidence 102
   of questions 122–123

vertex 227, 231, 299–300

SYNTHESE LIBRARY

Monographs on Epistemology, Logic, Methodology,
Philosophy of Science, Sociology of Science and of Knowledge, and on the
Mathematical Methods of Social and Behavioral Sciences

Editors:

DONALD DAVIDSON (The Rockefeller University and Princeton University)

JAAKKO HINTIKKA (Academy of Finland and Stanford University)

GABRIËL NUCHELMANS (University of Leyden)

WESLEY C. SALMON (Indiana University)

ROBERT S. COHEN and MARX W. WARTOFSKY (eds.), *Boston Studies in the Philosophy of Science.* Volume IX: *A. A. Zinov'ev: Foundations of the Logical Theory of Scientific Knowledge (Complex Logic).* Revised and Enlarged English Edition with an Appendix by G. A. Smirnov, E. A. Sidorenka, A. M. Fedina, and L. A. Bobrova. 1973, XXII + 301 pp. Also available as a paperback.

K. J. J. HINTIKKA, J. M. E. MORAVCSIK, and P. SUPPES (eds.), *Approaches to Natural Language. Proceedings of the 1970 Stanford Workshop on Grammar and Semantics.* 1973, VIII + 526 pp. Also available as a paperback.

WILLARD C. HUMPHREYS, JR. (ed.), *Norwood Russell Hanson: Constellations and Conjectures.* 1973, X + 282 pp.

MARIO BUNGE, *Method, Model and Matter.* 1973, VII + 196 pp.

MARIO BUNGE, *Philosophy of Physics.* 1973, IX + 248 pp.

LADISLAV TONDL, *Boston Studies in the Philosophy of Science.* Volume X: *Scientific Procedures.* 1973, XIII + 268 pp. Also available as a paperback.

SÖREN STENLUND, *Combinators, λ-Terms and Proof Theory.* 1972, 184 pp.

DONALD DAVIDSON and GILBERT HARMAN (eds.), *Semantics of Natural Language.* 1972, X + 769 pp. Also available as a paperback.

MARTIN STRAUSS, *Modern Physics and Its Philosophy. Selected Papers in the Logic, History, and Philosophy of Science.* 1972, X + 297 pp.

‡STEPHEN TOULMIN and HARRY WOOLF (eds.), *Norwood Russell Hanson: What I Do Not Believe, and Other Essays,* 1971, XII + 390 pp.

‡ROBERT S. COHEN and MARX W. WARTOFSKY (eds.), *Boston Studies in the Philosophy of Science.* Volume VIII: *PSA 1970. In Memory of Rudolf Carnap* (ed. by Roger C. Buck and Robert S. Cohen). 1971, LXVI + 615 pp. Also available as a paperback.

‡YEHOSUA BAR-HILLEL (ed.), *Pragmatics of Natural Languages*. 1971, VII + 231 pp.

‡ROBERT S. COHEN and MARX W. WARTOFSKY (eds.), *Boston Studies in the Philosophy of Science*. Volume VII: *Milič Čapek: Bergson and Modern Physics*. 1971, XV + 414 pp.

‡CARL R. KORDIG, *The Justification of Scientific Change*. 1971, XIV + 119 pp.

‡JOSEPH D. SNEED, *The Logical Structure of Mathematical Physics*. 1971, XV + 311 pp.

‡JEAN-LOUIS KRIVINE, *Introduction to Axiomatic Set Theory*. 1971, VII + 98 pp.

‡RISTO HILPINEN (ed.), *Deontic Logic: Introductory and Systematic Readings*. 1971, VII + 182 pp.

‡EVERT W. BETH, *Aspects of Modern Logic*. 1970, XI + 176 pp.

‡PAUL WEINGARTNER and GERHARD ZECHA, (eds.), *Induction, Physics, and Ethics, Proceedings and Discussions of the 1968 Salzburg Colloquium in the Philosophy of Science*. 1970, X + 382 pp.

‡ROLF A. EBERLE, *Nominalistic Systems*. 1970, IX + 217 pp.

‡JAAKKO HINTIKKA and PATRICK SUPPES, *Information and Inference*. 1970, X + 336 pp.

‡KAREL LAMBERT, *Philosophical Problems in Logic. Some Recent Developments*. 1970, VII + 176 pp.

‡P. V. TAVANEC (ed.), *Problems of the Logic of Scientific Knowledge*. 1969, XII + 429 pp.

‡ROBERT S. COHEN and RAYMOND J. SEEGER (eds.), *Boston Studies in the Philosophy of Science*. Volume VI: *Ernst Mach: Physicist and Philosopher*. 1970, VIII + 295 pp.

‡MARSHALL SWAIN (ed.), *Induction, Acceptance, and Rational Belief*. 1970, VII + 232 pp.

‡NICHOLAS RESCHER et al. (eds.), *Essays in Honor of Carl G. Hempel. A Tribute on the Occasion of his Sixty-Fifth Birthday*. 1969, VII + 272 pp.

‡PATRICK SUPPES, *Studies in the Methodology and Foundations of Science. Selected Papers from 1911 to 1969*. 1969, XII + 473 pp.

‡JAAKKO HINTIKKA, *Models for Modalities. Selected Essays*. 1969, IX + 220 pp.

‡D. DAVIDSON and J. HINTIKKA (eds.), *Words and Objections: Essays on the Work of W. V. Quine*. 1969, VIII + 366 pp.

‡J. W. DAVIS, D. J. HOCKNEY and W. K. WILSON (eds.), *Philosophical Logic*. 1969, VIII + 277 pp.

‡ROBERT S. COHEN and MARX W. WARTOFSKY (eds.), *Boston Studies in the Philosophy of Science*, Volume V: *Proceedings of the Boston Colloquium for the Philosophy of Science 1966/1968*, VIII + 482 pp.

‡ROBERT S. COHEN and MARX W. WARTOFSKY (eds.), *Boston Studies in the Philosophy of Science*. Volume IV: *Proceedings of the Boston Colloquium for the Philosophy of Science 1966/1968*. 1969, VIII + 537 pp.

‡NICHOLAS RESCHER, *Topics in Philosophical Logic*. 1968, XIV + 347 pp.

‡GÜNTHER PATZIG, *Aristotle's Theory of the Syllogism. A Logical-Philological Study of Book A of the Prior Analytics*. 1968, XVII + 215 pp.

‡C. D. BROAD, *Induction, Probability, and Causation. Selected Papers*. 1968, XI + 296 pp.

‡ROBERT S. COHEN and MARX W. WARTOFSKY (eds.), *Boston Studies in the Philosophy*

of Science. Volume III: *Proceedings of the Boston Colloquium for the Philosophy of Science 1964/1966*. 1967, XLIX + 489 pp.

‡GUIDO KÜNG, *Ontology and the Logistic Analysis of Language. An Enquiry into the Contemporary Views on Universals*. 1967, XI + 210 pp.

*EVERT W. BETH and JEAN PIAGET, *Mathematical Epistemology and Psychology*. 1966, XXII + 326 pp.

*EVERT W. BETH, *Mathematical Thought. An Introduction to the Philosophy of Mathematics*. 1965, XII + 208 pp.

‡PAUL LORENZEN, *Formal Logic*. 1965, VIII + 123 pp.

‡GEORGES GURVITCH, *The Spectrum of Social Time*. 1964, XXVI + 152 pp.

‡A. A. ZINOV'EV, *Philosophical Problems of Many-Valued Logic*. 1963, XIV + 155 pp.

‡MARX W. WARTOFSKY (ed.), *Boston Studies in the Philosophy of Science*. Volume I: *Proceedings of the Boston Colloquium for the Philosophy of Science, 1961–1962*. 1963, VIII + 212 pp.

‡B. H. KAZEMIER and D. VUYSJE (eds.), *Logic and Language. Studies dedicated to Professor Rudolf Carnap on the Occasion of his Seventieth Birthday*. 1962, VI + 256 pp.

*EVERT W. BETH, *Formal Methods. An Introduction to Symbolic Logic and to the Study of Effective Operations in Arithmetic and Logic*. 1962, XIV + 170 pp.

*HANS FREUDENTHAL (ed.), *The Concept and the Role of the Model in Mathematics and Natural and Social Sciences. Proceedings of a Colloquium held at Utrecht, The Netherlands, January 1960*. 1961, VI + 194 pp.

‡P. L. GUIRAUD, *Problèmes et méthodes de la statistique linguistique*. 1960, VI + 146 pp.

*J. M. BOCHEŃSKI, *A Precis of Mathematical Logic*. 1959, X + 100 pp.

SYNTHESE HISTORICAL LIBRARY

Texts and Studies
in the History of Logic and Philosophy

Editors:

N. KRETZMANN (Cornell University)
G. NUCHELMANS (University of Leyden)
L. M. DE RIJK (University of Leyden)

LEWIS WHITE BECK (ed.), *Proceedings of the Third International Kant Congress*. 1972, XI + 718 pp.

‡KARL WOLF and PAUL WEINGARTNER (eds.), *Ernst Mally: Logische Schriften*. 1971, X + 340 pp.

‡LEROY E. LOEMKER (ed.), *Gottfried Wilhelm Leibnitz: Philosophical Papers and Letters.* A Selection Translated and Edited, with an Introduction. 1969, XII + 736 pp.

‡M. T. BEONIO-BROCCHIERI FUMAGALLI, *The Logic of Abelard*. Translated from the Italian. 1969, IX + 101 pp.

Sole Distributors in the U.S.A. and Canada:
*GORDON & BREACH, INC., 440 Park Avenue South, New York, N.Y. 10016
‡HUMANITIES PRESS, INC., 303 Park Avenue South, New York, N.Y. 10010